BIOLOGICAL
OCEANOGRAPHY

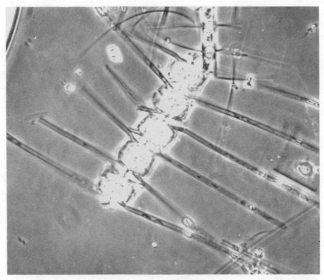

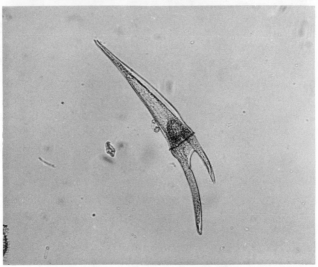

Phytoplankton, namely, the diatom *Chaetoceros rostratum* (above, magnified 400×) and the dinoflagellate *Ceratium schroeteri* (below, magnified 500×), from the Mediterranean Sea. Characterized by Victor Hensen as "blood of the sea," green photosynthetic cells like these were the subject of early work on biological oceanography in Germany, Monaco, Great Britain, and the United States. Photographs courtesy of Guy Léger, Université de Nice, France.

Biological Oceanography

An Early History, 1870–1960

ERIC L. MILLS

Department of Oceanography
Dalhousie University
Halifax, Nova Scotia, Canada

Cornell University Press

Ithaca and London

First published 1989 by Cornell University Press.

Printed in the United States of America

Library of Congress Cataloging-in-Publication Data
Mills, Eric L., 1936–
    Biological oceanography : an early history, 1870–1960 / Eric L. Mills.
        p.   cm.
        Bibliography: p.
        Includes index.
        ISBN 0-8014-2340-6
        1. Marine biology—History.   I. Title.
QH91.25.M55   1989
574.92'09—dc20   89-33048

The paper in this book is acid-free and meets the guidelines for permanence and durability of the Committee on Production Guidelines for Book Longevity of the Council on Library Resources.

Oceanography is not so much a science as a state of mind.

ELIZABETH NOBLE SHOR 1978

In recognition of the pioneers of quantitative biological oceanography

Victor Hensen
Karl Brandt
Hans Lohmann
Alexander Nathansohn
H. H. Gran
Trygve Braarud
W. R. G. Atkins
H. W. Harvey
L. H. N. Cooper
Gordon A. Riley

CONTENTS

ILLUSTRATIONS

TABLES

ACKNOWLEDGMENTS

The metaphor of a voyage is particularly apt for a study of the history of oceanography. This book is the result of many happy, busy days at the end of real voyages to Britain, Europe, and the United States, where, when I began the research, a few of the early explorers of plankton dynamics were still alive. I am particularly grateful to Trygve Braarud (Oslo), L. H. N. Cooper (Plymouth), and Sir Frederick Russell (Goring-on-Thames) for their illuminating personal accounts of their work. My friend and teacher Gordon A. Riley of Halifax told me many of the details of his work, which were later incorporated into his unpublished "Reminiscences of an Oceanographer," a valuable resource for the recent history of biological oceanography. At Plymouth E. I. Butler took pains to ensure that I understood what the real H. W. Harvey was like and why it mattered to the history of research at the Plymouth Laboratory. Also at Plymouth, Sir Eric Smith told me about the recent history of the Marine Biological Association and helped me to feel at home in its laboratory.

The hospitality and enthusiasm of David Edge of the Science Studies Unit, University of Edinburgh, were vital elements aiding my first tentative steps into the history of the Kiel school. When the first roadblock appeared, David Edge surmounted it for me, taking it as a personal challenge to promote my research. Equally, I could not have learned much about the background of the Kiel school without the cheerful help and friendship of Gabriele Kredel, then of the Institut für Meereskunde, Universität Kiel, who unearthed a great deal of unpublished information about Karl Brandt, the Kiel school, and in particular Emil Raben. I will not forget a cold, gray afternoon in March 1982 when, at Frau Kredel's suggestion, we planted a rhododendron on the grave of Victor Hensen in Kiel. Also in Kiel, Bernt Zeitzschel invited me to lecture at the Institut für Meereskunde in April–May 1984. It was there and then, before a German audience, that I began to put together the full story of the Kiel school's contributions to biological oceanography.

Sir Maurice Yonge's magnificent library in Edinburgh contributed many serendipitous finds to my research. I am especially grateful to Sir Maurice and Lady Yonge for their help and hospitality. Moyra Forrest of the Science Studies Unit library in Edinburgh and I. E. Mommsen of the library of the Universität Kiel helped me find obscure or rare material. And the meat and potatoes of research, good hard work in the stacks, was a pleasure at the Cambridge University Library, the Dalhousie University Library, the Library of the Marine Biological Association of the United Kingdom (Plymouth), the Edinburgh University Library, the library of the Musée océanographique de Monaco, the library of the Marine Biological Laboratory (Woods Hole, Massachusetts), and the Archives of Woods Hole Oceanographic Institution.

For help of many different kinds, including advice, encouragement, hospitality, and manuscripts, I thank the following: J. A. Allen (University Marine Biological Station, Millport, Scotland), Margaret Barnes (Dunstaffnage Marine Laboratory, Oban, Scotland), Hans Brattström (Biological Station, Universitetet i Bergen), Jeffrey Brosco (University of Pennsylvania), Adolph Büchmann (Hamburg), Jacqueline Carpine-Lancre and Christian Carpine (Musée océanographique de Monaco), Dean Bumpus (Brownfield, Maine), Frederick C. (Fritz) Fuglister (Woods Hole Oceanographic Institution), Sebastian Gerlach (Institut für Meereskunde, Universität Kiel), Christiane Groeben (Stazione Zoologica di Napoli), G. E. Hutchinson (Yale University), Ralph Jewell (Department of Philosophy, Universitetet i Bergen), Malcolm Nicolson (Science Studies Unit, University of Edinburgh), Susan Schlee (Quissett, Massachusetts), A. J. Southward (The Plymouth Laboratory), and Hjalmar Thiel (Universität Hamburg).

The thankless task of converting my drafts into a real manuscript was accomplished gracefully by Geraldine Hammer of the Department of Oceanography, Dalhousie University. Kathleen Sawler helped at a crucial time. Eleanor Wangersky of Halifax prepared many of the figures. In Monaco, Jacqueline Carpine-Lancre and Cécile Thiéry provided essential help in correcting the final manuscript, and Guy Léger allowed me to use his beautiful photographs for the frontispiece. Financial support for the whole enterprise was provided by two research grants from the Social Sciences and Humanities Research Council of Canada and by a fellowship from the Nuffield Foundation.

Finally, but not least, my gratitude and affection to Anne Mills, who smoothed my path through the research both at home and abroad and who shared the adventures, and to Chris and Karen Mills, who no longer believe that all fathers work every weekend writing books.

Eric L. Mills

Head of St. Margaret's Bay, Nova Scotia

BIOLOGICAL
OCEANOGRAPHY

Introduction:
The Development of
Biological Oceanography

Phytoplankton cells in seas and lakes are the fundamental source of energy on which life in aquatic environments is based. Using the sun's energy, they convert carbon dioxide, water, and dissolved salts into organic compounds. Modern textbooks of oceanography describe how, when light increases in the spring in seas at temperate and high latitudes, phytoplankton cells begin to grow by absorbing the nutrient salts from solution in seawater. When their growth begins to exhaust the dissolved nutrients and when grazing animals increase in numbers, the spring phytoplankton bloom decreases. Production fluctuates during the summer, until, in autumn, when the water is mixed because of falling temperatures and increasing winds, nutrients are renewed in the surface waters. A second somewhat smaller bloom then occurs before the marine production system becomes quiescent during the winter. Even in the seasonless tropics occasional bursts of increased plant production occur when the stable surface waters are enriched with nutrients from below. Events such as these in the annual cycle of plankton are based on the interaction of light, nutrients, physical circulation, and the physiology of the phytoplankton cells, and on the consumption of the plant cells by grazing animals.

In 1911, before this image of the plankton cycle was fully established, the Kiel physiologist Victor Hensen, recognizing the importance of phytoplankton cells to marine life, described them as "this blood of the sea." Hensen's seemingly romantic analogy actually represented his view that the oceans could be studied in a fully quantitative way, just as the systems of an organism could be studied by the physiologist. His viewpoint, rather than being romantically metaphorical, was analytical and reductionistic. It took into account the newly recognized distinction between the amount of living material in the sea (what we now call *standing crop*) and the rate at which that material grew and was replaced, that is, the rate of *production* of marine organisms.

Only a year before Hensen's phrase "this blood of the sea" was printed, the Norwegian research vessel *Michael Sars* had put to sea bearing the eminent student of the deep sea John Murray and the energetic, aggressive Norwegian fisheries biologist Johan Hjort. Their concerted exploration of the North Atlantic Ocean during the summer of 1910 matched Hensen's ideals. It was a significant change from late-nineteenth-century marine biology, which usually had been aimed at describing new species, establishing their places in phylogenetic schemes, and determining their ranges. Murray, Hjort, and their collaborators, whose views appeared in the influential textbook *The Depths of the Ocean* in 1912, and Hensen, who had begun the quantitative study of plankton thirty years before, took the first conscious or unconscious steps toward a new science, biological oceanography, which was born out of late-nineteenth-century natural history, chemistry, microbiology, and physical geography. Phylogenetic speculation, descriptive biology, and biogeography were replaced in practice and in their thinking by attempts to link quantitatively the production of organisms to their chemical and physical surroundings.

Some years ago, while I was writing a review of the history of deep-sea biology, I began to consider how deep-sea exploration had been replaced as a major theme in marine biology by the study of the production cycle in the sea during the first two decades of the twentieth century. I concluded then, considering work done in England after Hensen's day and the cruise of *Michael Sars*, that

> in the 30 years after 1910, when their laboratory had been released from its concentration on applied biology, the Plymouth scientists (along with a number of European scientists doing background work for ICES) established the framework of modern biological oceanography, a discipline that has, since those years, always been strongly oriented toward plankton dynamics. Deep-sea biology, which flourished before the turn of the century, became a side issue (if it were carried on at all), a specialized, difficult, and even scientifically uninteresting residuum of nineteenth century thought.[1]

But what had happened before 1910? Why had plankton dynamics, the study of the production cycle in the sea, become the core study of

1. Mills 1983, p. 63.

biological oceanography, a position it still holds today? How, in particular, did plankton dynamics come to be considered important? Why did it, and biological oceanography in general, take on character as a new field of study, separate from marine biology and general ecology? The answers seemed to lie with Hensen, his students, and his colleagues in Kiel, at least in the beginning. Who was involved, and how did a research group working on highly unorthodox plankton studies develop in Kiel? Furthermore, why did the Kiel group decline? How and why did the study of plankton dynamics then become an English concern, later an American one? How and under what circumstances were biological oceanographic ideas developed or passed from individual to individual, or from group to group, first in Germany then in other nations? How were they related to physical and chemical oceanography, which were evolving at the same time? What, in summary, were the origins of biological oceanography?

This book attempts to answer these questions by examining the contributions of German, Scandinavian, English, and North American scientists to biological oceanography between about 1870 and 1960. Characteristic of scientific change, there was no discernible continuous flow of "progress" from Victor Hensen's early quantitative studies during the 1880s until mathematical models of plankton production were accepted fully into biological oceanography during the 1950s. Instead, there were a few periods of creative new science interrupted by periods of stasis. I admit to being just as fascinated by the latter as the former; in fact, in many ways the dynamics of scientific change have been more evident in my studies of why the research groups at Kiel and Plymouth failed or disbanded than in my studies of why they succeeded. The chapters that follow should make these fascinating transitions clearer. More broadly, though, this book is a contribution to the history of ecology, an attempt to show how a branch of that science grew, albeit haltingly, from origins often quite different from twentieth-century plant, animal, community, and population ecology.

Previous historians of ecology have paid little attention to oceanography as a historically based environmental science. Part of the reason for this is the awkward fit of oceanography into modern ecology. Although modern textbooks in the field are given such titles as *Marine Ecology* or *The Ecology of the Seas*, the historical as well as the current relationship of the marine sciences to the rest of ecology is frustratingly difficult to make concrete. Viewing the situation historically, the American ecologist R. P. McIntosh in his book *The Background of Ecology* has presented the puzzle that in its youth biological oceanography, so evidently an *ecological* subject, seemed to have so little connection with the rest of the field. McIntosh then presents a solution to the problem, suggesting that "the link with commercial fisheries and the attendant institutionalization of marine studies in government-supported organizations separated them from the nascent science of ecology."[2]

There is a grain of truth in this generalization, but the emphasis is wrong. We must look instead—or at least equally—to the kinds of men and disciplines that were brought to bear on the quantitative biology of the seas between the late 1870s and the 1950s. Institutions are founded in response to human interests; the same is true of scientific specialties. What interests and influences led to the development of biological oceanography as a parallel but decidedly independent branch of late-nineteenth–early-twentieth-century ecological science?

In Germany beginning during the 1870s, one answer must surely be physiology. The ideals of physiology—reduction of vital processes to physical mechanisms, flux and process—were at the root of the Kiel physiologist Victor Hensen's approach to life in the sea. Hensen was the first great theoretician of the biology of the seas, establishing a tradition that was developed by his colleagues and students, among them preeminently Karl Brandt, professor of zoology at Kiel. At a time when German animal physiology, cytology, microscopical anatomy, and embryology were in full flower, Brandt followed a course that must have seemed distinctly idiosyncratic if not completely misdirected to many of his colleagues outside the Kiel school of plankton research.

Brandt's distinctive contribution, indeed his genius, was to reach outside his own expertise in protozoan anatomy and physiology into the problems of marine production. Hensen provided the inspiration and the general approach—physiology applied to the organisms of the sea as a group—but Brandt fused the four great strands that gave the Kiel school its success, namely the practical needs of German fisheries, the natural history tradition of classical descriptive biology, the chemical approach revolutionizing European agriculture during the mid and late decades of the nineteenth century, and the dramatic advances in microbiology during the 1880s which elucidated the transformations occurring in the terrestrial nitrogen cycle. Agricultural chemistry, the microbiology of the nitrogen cycle, and good solid, slogging organismal biology were the hallmarks of the Kiel school's work between the 1880s and the 1920s. They were underlain by the theoretical approach and the inspiration provided by Victor Hensen, who early in his career had expressed the belief, soon realized in action, that North German fisheries could be improved by bringing the quantitative approach of the physiologist to bear on the mysteries of fish abundance in North German seas. On this basis

2. McIntosh 1985, p. 115.

the Kiel school, under Brandt's strong direction, provided the first model of how plankton abundance was controlled, invoking plant growth, the regeneration of nutrients (especially nitrogen), and (less emphatically) the effect of herbivorous planktonic animals. Now nearly forgotten, the Kiel school established the foundations of biological oceanography well separated from the realm of late-nineteenth-century ecology.

Agricultural analogies as well as the concrete realities of marine chemistry, physics, and microbiology were familiar to the British scientists at the Plymouth Laboratory who converted the Kiel school's model of plankton dynamics into recognizably modern form during the 1930s. To E. J. Allen, W. R. G. Atkins, H. W. Harvey, L. H. N. Cooper, and others, the English Channel was a "blue pasture," chemically and physically controlled, in which lived planktonic plants and animals that were best studied with methods from the physical sciences. At a time when terrestrial ecology was grappling messily with community concepts, when plant ecology was reaching the summit of Clementsian organicism, E. J. Allen, a zoologist and the director of the Plymouth Laboratory, made the considered judgment that the problems of marine fisheries production were most likely to be solved with physics and chemistry. He appointed men trained in the physical sciences to undertake the expanded work of the laboratory on the production of the English Channel. Thus Atkins (trained in physics and chemistry), Harvey (trained in chemistry), and Cooper (also a chemist) brought their formidable talents to bear on the problem of why the plankton waxed and waned, carrying the rest of the biological production system in the English Channel with it. Zoologists and botanists came later. The institution, and the discipline, separated from mainstream ecology because of Allen's belief, decidedly unusual for the time, that biological problems of large scale in the sea would best be solved by the application of chemistry and physics to biology.

The Plymouth group's work during the 1930s and 1940s elaborated, polished, and greatly extended the qualitative model of plankton dynamics that Brandt and his colleagues at Kiel had developed earlier in the century. In turn, the Plymouth group's elegant application of physical and chemical approaches to phytoplankton photosynthesis, nutrient uptake, and growth, and their quantitative demonstration of the power of grazing by herbivores to control the abundance of phytoplankton in the sea, were the direct inspirations for Gordon A. Riley's mathematical approach to biological oceanography between the mid-1930s and 1960 in the United States. American biological oceanography *did* have roots in ecology too—but in mathematical ecology, which was deeply mistrusted by mainstream American ecologists, who were concerned mainly with community description or with physiological ecology. Riley absorbed from his mentor, G. Evelyn Hutchinson, the use of mathematics as a symbolic language. Moreover, mathematics was an essential element of their mechanism for generating and testing hypotheses, involving techniques and a philosophical approach quite foreign to the majority of American ecologists. Riley and Hutchinson believed themselves embattled prophets of a new, purer approach to the complexities of organisms within their environments. American mainstream ecology did pay some attention to the marine environment—Victor Shelford and W. C. Allee were only two who used marine organisms or assemblages to advance their individual approaches—but American biological oceanography drew its inspiration from Plymouth, from quantitative traditions in English science, including the physicist Harold Jeffreys' unorthodox use of deductive reasoning in science and the statistical approach inherited from British biometricians, and from limnology, not from any American tradition of ecology. The result, in Riley's work, was the first quantitative formulation of the processes that govern plankton abundance in the sea and the framework of plankton dynamics, the core of biological oceanography, which still exists today.

Thus the story that I tell is more than a narrowly scientific one, held within the restricted compass of "the ecology of the seas." It shows how a new scientific discipline, often regarded as a branch of ecology, grew under numerous, disparate influences between the early 1880s and 1960. At various times Prussian imperialism, agricultural chemistry, microbiology, the problems of German universities, failures of commercial fisheries, the development of analytical chemistry, the establishment of international scientific organizations, and sheer scientific curiosity played their roles. The work and ideas of the story's main actors—Victor Hensen, Karl Brandt, Hans Lohmann, Alexander Nathansohn, H. H. Gran, Trygve Braarud, W. R. G. Atkins, H. W. Harvey, L. H. N. Cooper, and G. A. Riley, among others—are inseparable from many of those seemingly unrelated factors. The blood of the sea, like its physiological counterpart, is a complex entity, impossible to isolate from its surroundings without loss of vitality. So too, biological oceanography and its practitioners, from Hensen to Riley, have proved to be worthy of study within their own context, the development of a quantitative science of the sea which unites biology with physics and chemistry to explain, again in Hensen's words, the fertility of the waters.

PART I

The Origin of Biological Oceanography in Germany and Scandinavia during the Late Nineteenth and Early Twentieth Centuries

Meine ursprüngliche Absicht war es, ein Urtheil über die Planktonproduktion des Meeres zu gewinnen. [My original aim was to arrive at an estimate of the plankton production of the sea.]

VICTOR HENSEN 1887

1 "This Blood of the Sea": The Origin of Quantitative Plankton Biology in Germany, 1870–1911

> You all know the occurrence of which it is said: the water is blooming! Then the surface of quiet, brackish bays appears to be covered with a green layer which consists of minute spheres or threads. . . . At such times the water is filled with these primitive kinds of plants.
>
> VICTOR HENSEN 1906

That the sea bloomed in spring at high northern latitudes became obvious to the many biologists who studied marine algae, especially the panorama of planktonic species, throughout the late nineteenth century.[1] But what made the sea bloom? This question was first asked seriously during the 1870s, after many decades of work on the systematics of marine organisms. Answers were slow to emerge. By 1912, when H. H. Gran reviewed contemporary knowledge of plant life for *The Depths of the Ocean*,[2] the first complete textbook of oceanography, there were two "interesting theories," admittedly unverified ones, about the control of plankton production. Forty years before, there had been none.

Between 1870 and 1911 the quantitative study of plankton biology was born almost entirely outside the Darwinian framework that dominated, by aim or default, much of late-nineteenth-century biology. Its impetus came from practical concerns such as the welfare of coastal communities and the economics of fisheries, and from scientific attitudes brought to marine biology from physiology and chemistry. Systematics

---

1. For brief accounts of early knowledge of planktonic organisms, especially phytoplankton, see Gran 1912a and Taylor 1980.
2. Murray and Hjort 1912.

and biogeography were left behind or shouldered aside when plankton biology developed, much as nineteenth-century natural history was made anachronistic by Darwinian genetics and population biology during the early twentieth century.

The introduction of quantitative techniques into plankton biology was a significant and well-defined cultural phenomenon. It resulted in a clearly drawn model of the forces that bring about seasonal change in the sea. Just as with Darwinism, the origins of quantitative plankton biology may be dated, analyzed, and attributed to scientific attitudes, social forces, and the institutions that supported them. The early development of quantitative plankton biology derives from the German state, German universities, and especially, the application of reductionist quantitative techniques by a few powerful Prussian professors and their young colleagues.

Victor Hensen (1835–1924) of Kiel is the central figure with whom this story begins, for he provided practical tools to measure the abundance of plankton. Moreover, and equally important, he set out a theoretical view of the ways in which the distributions, amounts, and rates of production of plankton organisms could be determined. Hensen and his co-workers, especially Karl Brandt and Hans Lohmann, created a new branch of marine biology between 1870 and 1911. Grounding their work in the peculiar excellences of the German university system they achieved a coherent hypothesis about the factors controlling the annual plankton cycle in temperate seas. But, nearly inevitably, the restrictions inherent in the German professoriate and its most distinctive nineteenth-century creation, the university research institute, did not allow German scientists to remain leaders in quantitative marine biology. After a remarkable period of development in Germany, other nations and differently organized institutions carried on from the foundations established in the unified German states.

The development of quantitative techniques in nineteenth-century plankton biology is closely linked with Victor Hensen's early life in the Danish-governed but largely German duchy of Schleswig.[3] Hensen (Figure 1.1) was born in 1835 in the city of Schleswig, where his father, a lawyer who came from a succession of agriculturalists and farm managers, directed an institution for the deaf and dumb. His high school career was interrupted by a brief exile in Friedrichstadt and Berlin in 1848 when young Schleswig-Holsteiners were inducted into the Danish army during the Danish attempt to fully annex the disputed duchies.

At the completion of his high school years in the Gymnasia of Schles-

---

3. The most useful biographical accounts are Porep 1970 and Brandt 1925d. See also Brandt 1925c; Porep 1968, pp. 202–206; Porep 1976; Reibisch 1926; and Rothschuh 1972.

FIGURE 1.1. Victor Hensen in 1903. From Murray and Hjort 1912, published by Macmillan Publishers Ltd.

wig and Glückstadt, Hensen spent five semesters during 1854–1856 with the eminent medical faculty of Würzburg in Bavaria, where he studied under the organic chemist Johan Josef Scherer (1814–1869), the physiologist Albert Koelliker (1817–1905), and the great pathological anatomist Rudolf Virchow (1826–1902), among others. Working first in Scherer's laboratory and then in Berlin with Virchow in 1856, Hensen isolated glycogen from liver preparations. His publications documenting the discovery appeared only a few weeks after Claude Bernard's announcement of the isolation of glycogen early in 1857.[4] He also visited the shore at Kiel with his friend Carl Semper in 1856 to collect plankton, using, in all likelihood, the modified butterfly nets introduced to marine biology by Johannes Müller (1801–1858) in the 1840s. Upon the completion of a semester in Berlin as a clinical student, where he must have been influenced by Müller, Hensen returned to the modestly sized university at Kiel,[5] to complete his medical studies. After his final examination

4. Hensen 1857. See Young 1937, p. 52, and Porep 1970, pp. 78–81, for accounts of Bernard's and Hensen's discovery of glycogen in the liver.

5. According to Porep 1970, pp. 22–23, the medical faculty at Kiel had only five Ordinarien (full professors) in 1857. It is not clear why Hensen returned to Kiel to complete his degree. Possibly it was to be near his widowed mother.

The great influence of Johannes Müller's teaching on German biological thought has been only partially explored. See especially Koller 1958, Lohff 1977, and Steudel 1974.

in 1858 and the completion of his thesis on the urine of epileptics (prepared during voluntary work in an insane asylum directed by his grandfather), in 1859 Hensen was appointed Prosektor in anatomy; very shortly thereafter he was granted the Habilitation (see Chapter 6 near n. 8) based on his thesis and a formal disputation, then appointed Privatdozent in anatomy.

During the early 1860s, despite his work as a very busy Privatdozent and assistant to the professor of anatomy at Kiel, Hensen began research on a variety of morphological and physiological subjects. However, after only four years in his first university position, political events in Schleswig-Holstein conspired to increase Hensen's power and his ability to follow new paths, both scientific and political. In 1863 the professor of physiology at Kiel, the Dane P. L. Panum (1820–1885), perhaps sensing that confrontation between Denmark and Prussia was imminent, sought and was awarded the chair of physiology in Copenhagen. Hensen was appointed to replace Panum in 1864, as Professor extraordinarius and interim director of Panum's Physiological Institute. This coincided with the accession of Christian IX as King of Denmark, whose claim as duke of the duchies of Schleswig and Holstein was disputed by the German states (including Austria). After the defeat of Denmark during the Danish-Prussian war of 1864, Schleswig was administered by Prussia, Holstein by its ally Austria. Bismarck engineered a war against Austria in 1866, resulting in the total annexation of Schleswig-Holstein by Prussia under the Treaty of Prague. Thus Kiel, Schleswig, and other northern German cities were fully incorporated into the newly emerging German nation.

Hensen's sentiments were nationalistic. Coming from a family that had roots in the land, involved in the largely agricultural economy of Schleswig, he stood for election as a Liberal representative during the elections of 1867 and became a representative for the city of Schleswig in the Prussian Landtag during 1867–1868. But his overt political ambitions were short-lived, for in 1868 Hensen was appointed Professor ordinarius of physiology at Kiel and full-time director of the Physiological Institute. He gave up his political career with the intention of expanding the institute and beginning new lines of research that might help the North German economy, in decline toward the depression of the 1870s.

Hensen rapidly rose to eminence in the scientific community and conducted nonmarine research for many years. Only an accident of publication while he was a student had robbed him of being credited for the discovery of glycogen. His work on the structure of the inner ear,[6] commemorated by the name *Hensen's duct*, was widely noted. Although

6. Hensen 1863.

he published relatively little on embryology, his observations of early development, particularly of mammals, were significant[7] and resulted in the term *Hensen's node.*[8] Hensen's observations on the structure of striped muscle resulted in the designation of the H-band (probably so called by Koelliker) and other morphological features.[9] His later work extended to physiological acoustics and the physiology of speech.[10]

Hensen's 119 publications, beginning in 1857 and ending in 1921, span much of late-nineteenth-century German zoology and physiology. His was a powerful and wide-ranging intellect, relatively little noted in the English-speaking world because his publications were not translated into English. It extended to marine biology as well as physiology and morphology; a third of Hensen's published works were on marine biology, most of them between 1890 and 1901. But in the early years, before 1871, the foundations of both Hensen's physiology and his marine biology were laid. Economic interests, the unitary self-assurance of the nascent German state (in which Prussia was a dominant member), and Hensen's ambitions as Professor ordinarius at Kiel were combined in 1870, when a commission for the study of the German seas was formed. Its activities between 1870 and 1887 allowed Hensen to carry on work on fisheries research, then plankton biology, in addition to his physiological and morphological studies, and gave a distinctive form to German marine biology up to and after the turn of the century. At first it was the economic difficulties of his homeland that turned Hensen's attention to the sea.

Although overall the German economy was ebullient in 1870, the fisheries and economy of Schleswig-Holstein were depressed.[11] Schleswig and Holstein, largely agricultural, had been shuttlecocks between Denmark and Prussia for most of the century until the Prussian Anschluss of 1866. Even though the Prussian economy expanded throughout the nineteenth century, the relative contributions of agriculture and fisheries were decreasing. In absolute terms the numbers employed in those sectors fell rapidly just before 1870. The German states, not yet a colonial power in the mold of Great Britain, could not break the trade monopoly of the colonial nations; German ports, among them certainly Kiel, Schleswig, and Hamburg, were economically underdeveloped.

---

7. Hensen 1876.
8. For details see Porep 1970, pp. 83–84.
9. Hensen 1869. See Porep 1970, p. 76, for discussion.
10. Porep 1970, pp. 85–94.
11. For a general discussion of economic history in Germany during the nineteenth century see Borchardt 1973. Some details of the fisheries of Schleswig-Holstein (but only the North Sea portion) are given by Illing 1923a, b. Dittmer (1902) gives a broader account.

The complete unification of Schleswig-Holstein with Prussia in 1867 appears to have encouraged local politicians and societies involved in the economy to look for state help to improve fisheries and agriculture. Among these was the Deutsche Fischerei-Verein, a nonscientific organization devoted to increasing fisheries catches, which petitioned the Prussian government to establish a scientific commission that would collect information on the physical and biological conditions governing fisheries in the western Baltic and coastal North Sea areas. With the help of the Prussian agriculture ministry a scientific commission, the Kommission zur wissenschaftlichen Untersuchung der deutschen Meere in Kiel (hereafter called the Kiel Commission), was established in 1870, with a mandate to collect information on the depths, tidal levels, salinity, and chemical content of the Prussian seas, to determine the animals and plants present in the fishing areas, and, in particular, to concentrate on the biology of commercial fishes, notably their distribution, abundance, food, reproduction, and migrations.[12] The first chairman of the Kiel Commission, H. A. Meyer (1822–1889), Hamburg merchant and physicist, was joined by Karl Möbius (1825–1908), professor of zoology at Kiel (later at Berlin); Gustav Karsten (1820–1900), professor of physics and mineralogy at Kiel; and Victor Hensen.

The commission began with two projects, to establish a chain of permanent observation stations along the German Baltic and North Sea coasts, and to conduct a scientific cruise to examine Baltic fisheries. Both were delayed by the outbreak of the Franco-Prussian War in 1870; but in 1871 stations, at which routine physical and biological observations were made, were established at Sonderburg, Friedrichsort (at the entrance to Kiel Fjord), Fehmarnsund, and Lohme auf Rügen. By 1887 another eight had been added, creating a chain from Denmark to the Russian border. On the North Sea coast three stations were established in 1872 at Sylt, Helgoland, and Borkum and another was added in 1875 at the Weser-Aussen-Leuchtschiff (Weser Lightship).[13] Observations were published regularly;[14] later the stations established by the Kiel Commission, at which observations were made monthly on weather, currents, and properties of the water and plankton, were easily adapted to the requirements of an international program for the exploration of the seas begun in 1901–1902.[15]

12. Meyer et al. 1873, also Reincke 1890. On the Deutsche Fischerei-Verein see Henking 1910. Heincke (1889) extols the virtues of Hensen's early work to the practically minded members of the Deutsche Fischerei-Verein.
13. Karsten 1887, pp. 135–157.
14. Anonymous 1874–1895.
15. The International Council for the Exploration of the Sea, made up after 1903 of representatives from Great Britain, Belgium, Denmark, Germany, Finland, the Netherlands,

The Kiel Commission's first cruise took place in June–August 1871 using the naval dispatch steamer *Pommerania*, which followed a route within the southwest Baltic area from Kiel to the Grosser Belt, Kattegat, Skagerrack, and Öresund to Bornholm, Åland, and Stockholm, then south and westward along the east Prussian coast via Danziger Bucht back to Kiel. The scientific staff were H. A. Meyer, whose book on the physical oceanography of the Baltic[16] provided a basis for the work on *Pommerania*, Karl Möbius, P. W. Jessen of Kiel, and the botanist V. Magnus, but not Hensen. Their observations included the specific gravity, salinity, and chlorinity of the water; wind, cloud cover, and tide height; sulfate, carbonate, and magnesium salt content of seawater; the petrology, mineralogy, and chemical content of sediments; lists of benthic algae and higher plants; the distribution of the invertebrates; and a list, compiled by Hensen,[17] of the fish collected at each location using a wide variety of nets and lines.

During 1872 *Pommerania* was again at sea. The Nordsee Expedition from July 21 to September 9 extended the previous year's work to the North Sea along a 600 nautical mile track from Kiel through the Kattegat and Skagerrack, along the Norwegian coast to Bergen, then to Peterhead, Edinburgh, the Dogger Bank and Great Yarmouth, returning via Texel and the north European coast.[18] Once again the scientific staff, this time including Hensen, concentrated on the physics and chemistry of seawater and the distribution of commercially useful fishes.

The two expeditions of 1871–1872 rapidly established the Kiel Commission as a scientific force, capable in only two summers of greatly enlarging knowledge of fish distributions in the Baltic and of beginning a study of the physical oceanography of the North Sea to match that of the longer studied Baltic. But despite its early successes, the commission did not mount another major cruise, the Holsatia Expedition, until 1885 (see below), probably because of tight economic conditions during the great European depression that began in 1873. Instead the commission extended its network of observation stations and concentrated on routine data gathering. Hensen, for example, the resident biologist of the commission, organized and distributed a series of fisheries questionnaires in coastal communities. These he regarded as complementary to the results

---

Norway, Russia, and Sweden, resulted from international conferences begun in 1899 by Otto Pettersson and King Oscar II of Sweden (see Chapter 3 near n. 29). Kofoid (1910) describes national contributions to the ICES program, including the role of the Kiel Commission by 1908–1909. In Chapter 3, I discuss the German contribution in more detail.

16. Meyer 1871.
17. Hensen 1873, pp. 155–159.
18. Meyer (1875) includes a description of the cruise.

from observation stations,[19] since the total fish catch from the area was a necessary component of any attempt to evaluate what he called the general metabolism of the sea.

Hensen's first 700 questionnaires, administered by local officials, recorded the number of fishermen, number of boats, number of voyages, area of the fishing ground, distance from shore, catches of various species and other details[20] in several major fishing areas along the western Baltic coast. The results appeared first as statistical tables in 1875 and 1878; in the latter, Hensen expressed the catch of herring, salmon, and flatfish, and total catches, catches per day, and catches per boat-day (the first use of catch-per-unit effort in fisheries biology), and recommended that German fishing effort be transferred from the Baltic to the North Sea, which was likely to prove richer.[21] Despite the fact that he was busy with the planning of a new Physiological Institute at Kiel (opened in 1879), Hensen also spent time trying to raise herring eggs and to determine the conditions under which ordinarily pelagic fish eggs would float when exposed to Baltic seawater, which went through wide fluctuations of salinity.[22]

Provided fish eggs floated, and provided their distribution was not unduly patchy, Hensen felt sure that the abundance of fish eggs could be determined using nets. Early in 1883 he began a series of one-day cruises using a variety of nets, especially a vertically hauled plankton net, to collect fish eggs. On one cruise he released a number of weighted glass spheres (the bulbs of specific gravity hydrometers), which rapidly separated at a rate of about 3 meters in 10 minutes. Later he used 10 weighted glass balls, which also dispersed rapidly, and, as he thought, evenly.[23] Based on this observation and his theoretical arguments that the surface waters of the ocean were well mixed due to wave action, currents, and the rotary action of waves, Hensen concluded that it should be possible to make at least a minimum estimate of floating eggs and larvae. Because the fecundity of average-sized flatfish (turbot) and cod was known, it might be possible to relate the abundance of the eggs to the abundance of the adult stocks, a possibility that he had first attempted in 1873: "With this, one gets the possibility of obtaining in the course of the year, from the abundance of fish eggs and embryos, partly also by direct observation of spawning animals, comparative information on the abundance, size, increase and decrease of fishes."[24]

19. Hensen 1875, p. 346.
20. Hensen 1875, p. 349.
21. Hensen 1875, pp. 353–370; Hensen 1878.
22. Hensen 1884, esp. pp. 302–305.
23. Hensen 1884, p. 311; Hensen 1887b, p. 2.
24. Hensen 1873, p. 159. All translations mine.

By 1883, when Hensen began his three-year series of 34 one-day cruises to the Kiel Bight, he had developed a clear theoretical view of the way that the fisheries could be investigated scientifically. In 1875, using a physiological analogy, he described how "the animal should be regarded as a measure of the stream in which the organic material of the seas circulates, and of which it forms a part."[25] But even though physiology might be used to measure the metabolism, nutrition, and excretion of a single organism, Hensen did not believe it adequate to deal with the larger problem of the *Gesamtstoffwechsel* (the general metabolic cycle) of the sea. Instead, he stated that the only feasible approach was to determine the "Anfangsmaterialen und Endprodukte" (initial materials and end products) until sufficient details could be accumulated to give a full picture of the structure and metabolism of marine communities. Such a study could be conducted best in an enclosed sea such as the Baltic, where external influences (such as the effects of larger seas) were as small as possible.

Concentration on initial materials and end products, combined with classical biological work, was the hallmark of German biological ocean-ography for at least forty years after Hensen's early statements in 1875. Hensen was convinced that marine production could be measured, with a physiologist's precision, using instruments having predetermined charac-teristics. He proceeded, quite deliberately, beginning in 1883, to design, build, and use new nets and laboratory equipment intended to put the study of pelagic fish eggs, larvae, and, not incidentally, the floating animals and plants, on a fully quantitative basis. The trick was to build sampling devices that were reliable, easy to use, and quantitative.

Hensen's experiments with horizontally towed nets in the water col-umn and with dredges on the sea floor convinced him during the early 1880s that these techniques were nearly impossible to make quantitative. By designing a new net and using it vertically rather than horizontally, thus making it possible to calculate the volume of water that had been filtered, he satisfied himself that he had solved the problem of collecting fish eggs quantitatively.[26] Thus, in early April 1883, Hensen's *Eiernetz* (egg net) (Figure 1.2A) was born. It was soon used for general plankton investigations in a series of short cruises to the outer Kiel Fjord between August 1883 and August 1886, and on a major investigation of the North Sea on the steamer *Holsatia* in July–August 1885.[27] Hensen's mono-

25. Hensen 1875, pp. 343–345.
26. Hensen 1884, pp. 307–309.
27. The expedition left Kiel on July 25 and returned August 2, 1885, after making 61 stations during the nine days along a track across the northern North Sea into the Atlantic between the Outer Hebrides and Rockall. A short log of the voyage is given by Hensen 1887b, pp. 30–33.

A

B

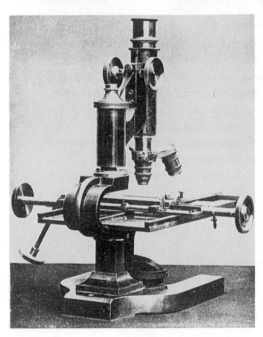

FIGURE 1.2 *A*, Egg nets, bucket, and silk mesh. *B*, Plankton microscope, with large stage. From Hensen 1887b, Tafel I.

graph *Über die Bestimmung des Planktons oder des im Meere treibenden Materials an Pflanzen und Thieren*, published in 1887, gave an account of this, his first quantitative work with plankton, and described his equipment. It also established the term *plankton* for the floating animals and plants of the sea, organisms he later described as "dies Blut des Meeres" (this blood of the sea).[28] The word *plankton*, replacing Johannes Müller's word *Auftrieb*, was coined, at Hensen's request, by his colleague Richard Förster (1843–1922), professor of classical philology and elocution at Kiel. It was intended to signify both living and dead organisms of the water column, for as Hensen said, "I mean thereby everything that drifts in the water, whether shallow or deep, living or dead."[29] But, curiously, Hensen never used the word in this broad sense; instead he used it to represent only the living organisms accessible to his nets.

The essence of the new method[30] was to use large plankton nets of known filtering characteristics which could be towed vertically from the bottom in shallow seas (or from some carefully established intermediate depth in deeper water) to the surface. The floating eggs, plant cells, and animals could be expressed quantitatively as numbers or volume in a water column below a square meter of sea surface.[31] By restricting the opening of the net (usually to 0.1 m²) in the largest nets[32] with a truncated conical canvas collar between the mouth ring and the net proper (see Figure 1.2A) inadvertent losses of plankton caused by the rolling of the ship could be reduced; in addition, the net was less likely to collect mud if it hit bottom.

The new vertically towed nets incorporated an important technical innovation. As Hensen stated, it was important that the mesh size should be equal to or less than the size of the smallest significant organism, that the pores be of equal size, and that they be many. Cotton netting, which had been in use since Johannes Müller's first collections with nets in the 1840s, shrank, clogged easily with particles, was hard to clean, and was so loosely woven that the pore size changed during use. After experimenting with several types and sizes of netting, in 1884 Hensen replaced the

28. Hensen 1911, p. 5.
29. Hensen 1887b, p. 1; Porep 1968, pp. 196, 200. Hensen (1887b, p. 6) claimed that the word was derived from the wanderings of Odysseus as described by Homer.
30. My account is based largely on Hensen 1887b and the virtually verbatim summaries in Jenkins 1901 and Dakin 1908a. Jenkins' account, in particular, has a number of small numerical errors. Hensen's methods evolved over the years. The definitive account is Hensen 1895.
31. The concept of expressing abundance below an area of sea surface first appears in Hensen 1884, p. 308. He later credited John Murray with the first use of such a measure (Hensen 1887b, p. 2).
32. Hensen's largest and most frequently used vertical net had a mouth area of 0.1 m² (diameter 38 cm). Occasionally he used a shorter net, which paradoxically had a larger mouth, 0.5 m² (diameter 80 cm).

cotton mesh with silk, fine flour-miller's cloth (hence *Beutelgaze*; more frequently called *Müllergaze*). Throughout his work thereafter he mainly used Müllergaze no. 20, which had a pore size of 50 μm and an interlocking weave that resisted stretching or shrinkage. The wisdom of his choice is evident in the fact that silk nets were used until the mid-twentieth century, when monofilament nylon netting became available.

Because the water passing through the entrance of a plankton net is impeded by the netting, the volume filtered cannot be determined merely as the product of the mouth area and the length of the tow. Hensen approached the problem of estimating the volume of water filtered by using a combination of direct measurement and theoretical computation[33] which can be attributed directly to empirical methods widely used in German physiology. The basis of his method was to determine the pressure resisting the entrance of water to the net, and to calculate the actual (unmeasurable) velocity of water into it, assuming that inflow to the net must equal outflow through the mesh. Using a laboratory device that measured the flow through silk mesh at various pressures (much like devices in use to measure blood flow), Hensen tabulated pressure and flow rate for a variety of meshes used in plankton nets. His ingenious approximations,[34] made using laboratory equipment and widely known physical principles, (since, at the time, there were no current meters small enough to put in nets) and with various corrections for net shape, were calculated for each net being used.[35]

Once in the laboratory the plankton samples, which were usually preserved in alcohol on shipboard, were treated according to a rigid protocol. First the total volume of the sample was determined by allowing it to settle in a graduated cylinder for 24 hours. In later developments of the technique[36] Hensen or his colleagues used a displacement method for more accurate determinations, or, if the organisms were very small and sparse, sometimes centrifuged them in narrow-tipped graduated tubes.

---

33. Hensen 1887b, pp. 4–5, 10–13, Tafel II; Hensen 1895, pp. 71–101 and Tafel V. The only account in English is by Dakin 1909.

34. See sample calculations given by Dakin 1909, pp. 229–238.

35. The calculation was a tedious one, and as Dakin (1909) emphasizes was far from accurate, being dependent on approximations related to the shape of the net. But more important than this, the capture of plankton on the mesh would change its effective filtering area and thus the volume filtered. For this reason Dakin, and probably eventually Hensen himself, favored an empirical method of determining filtration coefficients, for example by comparing a net catch with the amount of plankton pumped from a known volume of water and captured on fine filter paper.

36. Schütt 1892a, Jenkins 1901, Dakin 1908a. Franz Schütt accompanied Hensen on the *Holsatia* expedition in 1885 and was responsible for the Kiel Commission's regular plankton collections, 1885–1888. He was Privatdozent in botany at Kiel, 1887–1895, then became Professor ordinarius of botany at Greifswald (see Chapter 6 near n. 4).

Once volumes had been determined, the sample was shaken to disperse the organisms then subsampled using a series of newly developed quantitative piston-pipettes (*Stempelpipetten*). Hensen himself at first, later the young assistants to the Kiel Commission or doctoral students, then counted all the organisms in each subsample, using a specially designed low-power compound microscope with a large gridded glass well-plate on the stage (Figure 1.2B). Using a variant of the Poisson distribution developed by Ernst Abbe (the proprietor of the Zeiss Werke at Jena after the death of Carl Zeiss in 1888),[37] the investigators counted enough subsamples so that the addition of another did not change the mean number in all the subsamples by more than 5%.[38] Once counting and identification were finished, the figures were converted to abundances per square meter, usually throughout a 20-meter water column during Hensen's early work in the western Baltic Sea.[39] Each finished sample represented the investment of much tedious labor, frequently up to a week's work by a trained laboratory assistant.

Although Hensen emphasized the careful identification and enumeration of planktonic organisms, he began quite early to attempt chemical determinations of plankton. These began with wet and dry weights. Detailed chemical analysis was difficult. In *Über die Bestimmung des Planktons* Hensen gave it brief attention[40] so that he could estimate the weight of living material separately from inorganic substance (ash) in bulk samples of plankton. Detailed studies of the chemical composition of plankton began with the work of Hensen's young colleague Karl Brandt ten years later (see Chapter 2).

By the time Hensen's *Über die Bestimmung des Planktons* was published in 1887, Hensen had established a regime of well-defined and tested techniques for plankton work. The monograph (which is prescriptive as well as descriptive) suggests that he intended his techniques to be widely adopted. Moreover, he intended to apply them to the open ocean, where he could both test his assumptions about the pattern of distribution more fully and examine the possibility that newly discernible laws governing plankton distribution applied on a worldwide basis. Late in the 1880s he turned his attention away from the German seas to the distant reaches of the open Atlantic.

37. Hensen 1895, pp. 166–169; Lussenhop 1974, pp. 325–326. Jenkins (1901) and Dakin (1908a) also discuss this procedure. On Abbe see Bulloch 1938, p. 349; Stahlberg 1940; and Günther 1970.

38. Jenkins 1901, p. 301.

39. See as examples the Fangverzeichnisse I–V in Hensen 1887b, which includes specific gravity, temperature, and depth at each station, the volume of settled plankton, and volume after the water had been pressed out, then counts group by group (e.g., ostracods, chaetognaths) or species by species (the most abundant dinoflagellates and diatoms).

40. Hensen 1887b, pp. 34–38.

The German Plankton Expedition of 1889, occurring only thirteen years after the return of HMS *Challenger* from its 1872–1876 voyage,[41] is recognizably modern in conception and execution. Ostensibly, Hensen's purpose was to use a modern steel steamship with electric lights, refrigeration, and steel wire cables[42] to answer scientific problems concerning the nature of plankton distributions in the open ocean and the levels of production far from land. But his plea for support for the expedition suggests a strong nationalistic motivation as well.

There is no question that Hensen had compelling scientific reasons for investigating the open ocean.[43] He was convinced that plankton was distributed regularly in the sea, but the Baltic might not be representative of the open ocean far from land, where physical uniformity was far greater. "Wie ist die Vertheilung?"—what is the distribution?—was the dominating question that Hensen believed could be answered by a major expedition. Previous German expeditions, even the one on *Holsatia* in 1885, had never been more than 100 miles from land, so that an expedition to the open Atlantic would enable near-shore conditions, including the richness of the fish fauna, to be compared with those in the open sea. Further, an expedition traveling over the great abysses might suggest answers to the difficult question of how food reached the deep-sea floor.[44] And ultimately, in the scientific sense, a quantitatively conducted expedition to the open Atlantic could shed light (in ways that Hensen never clearly specified) on production in the seas, the cycle of material which he regarded as the driving force of marine life.

But Hensen did not emphasize his scientific concerns in his request for support. Writing to Kaiser Friedrich III in 1888, Hensen, Brandt, and the botanist Franz Schütt (see Chapter 4, n. 9) pointed out, in order, the importance of plankton as the base on which life in the sea depended, the hypothesis of even distribution, and the need to assess the production of the sea. These scientific concerns took second place, however, to the sociopolitical importance of such a voyage to the German state. Using nationalism as a lever, Hensen described how Britain, the United States, Sweden, Italy, and France had already organized great deep-sea expeditions, whereas by comparison "Germany is behind other nations in its contribution to the domain of open-sea investigations."[45] The Akademie

41. The British *Challenger* Expedition (1872–1876) around the world, under Charles Wyville Thomson, was widely admired and envied; it led to a number of state-supported voyages by European nations later in the century. For an assessment of early oceanographic exploration, notably the precursors of *Challenger* and that voyage itself, see Mills 1983.

42. See Mills 1983 for an assessment of technological change in research vessels during the nineteenth century.

43. Hensen 1892b, pp. 1–17.

44. See Mills 1983 and especially Mills 1980, pp. 360–372.

45. Hensen 1892b, p. 12.

FIGURE 1.3 SMS *National* at Kiel, 1889, before the Plankton Expedition. From Krümmel 1892, p. 48.

der Wissenschaften in Berlin supported Hensen in this. Its officials A. Auwers and Emil Du Bois-Reymond (1818–1896) (physiologist and friend of Hensen) wrote to the kaiser that "scientific investigation of the sea is due above all to the English, then the French, Italians and other nations."[46] Germany by contrast lay behind because its navy had developed late and because the state had provided very little money for marine investigations. It was now in Germany's interest to support Hensen's proposal—indeed it would be contrary to its interest to lose such an opportunity, according to their rhetoric.

By negotiating privately a donation of 24,600 marks from the Humboldt-Stiftung für Naturforschung und Reisen, a newly founded independent branch of the Akademie der Wissenschaften, Hensen made the project seem viable. The kaiser (now Wilhelm II, for Friedrich III had died in the summer of 1888) responded with 70,000 marks to be used for a ship and its personnel. To this the Deutsche Fischerei-Verein added a further 10,000 marks, and a wealthy landowner donated 1,000 marks to pay the salary of a marine artist.

After considering a small German naval vessel, Hensen arranged to charter the new steamer *National* from Paulsen and Ivers of Kiel (Figure 1.3). This vessel, 58 meters long, 835 tons gross weight, was outfitted with steam winches, a Kelvin-Sigsbee-type sounding machine, electric lights, and a refrigerator, the last two introduced to research vessels by the U.S. Fish Commission steamer *Albatross* in 1882. Hensen's large plankton nets were the main collecting gear around which the scientific studies were arranged, but in addition the ship had an *armamentum* of

46. Hensen 1892b, p. 15.

open plankton nets, Chun-Petersen closing nets, a pelagic otter trawl (modeled on one used by the Prince of Monaco, Albert I), a Blake trawl, and a variety of fish nets, some designed by Hensen.[47] Hensen's description of the early arrangements, the ship itself, and its outfitting is one of the most exacting technical accounts of an oceanographic expedition.[48]

National departed Kiel on July 15, 1889, during the Kiel University holidays, bearing Hensen, his assistant Karl Brandt (who had just replaced Möbius as professor of zoology and director of the Zoological Institute at Kiel), the zoologist Karl Friedrich Dahl (1856–1929), Franz Schütt, the geographer and oceanographer Otto Krümmel (1854–1912), the bacteriologist Bernhard Fischer (1852–1915), and the artist Richard Eschke, whose drawings and paintings illustrated the narrative.[49] Their departure, which according to one account was "Tagesgespräch in Kiel,"[50] was noted by Kaiser Wilhelm, who in a telegram wished the expedition "the greatest of science, glory and honour to the Fatherland," a phrase that evoked German cultural and political aspirations throughout the preceding seventy years.[51]

The financing of the cruise was very tight from the start. At one time Hensen had considered selling coal in Brazil to buy rubber, which might be sold at a profit in Europe. Fortunately,[52] he rejected that scheme and plotted a conservative route across and around the North Atlantic Ocean which would take advantage of wind and currents to reduce coal consumption. Between July 15 and November 7, when the Kiel scientists had to return to their teaching, *National* steamed 15,649 nautical miles along a figure-eight track from Kiel to Greenland, southwest to the Bermudas (avoiding the Gulf Stream), southeast to the Cape Verdes and Ascension Island, along the axis of the South Equatorial Current to Pará (now Belém), Brazil, then home nearly directly via the Azores and the English Channel (Figure 1.4). In all, Hensen and his colleagues occupied 278 stations at 50- to 200-mile intervals on the open ocean including 126

47. For a discussion of equipment on voyages during this era see Mills 1983, also McConnell 1982.

48. Hensen 1895 may be compared with accounts describing the equipment on HMS *Challenger* (Tizard et al. 1885) and on the U.S. Fish Commission steamer *Albatross* (Tanner 1897).

49. Krümmel 1892, pp. 47–69, 80–104, 113–134, 150–167, 185–203, 210–231, 315–330.

50. Porep 1970, p. 109.

51. The interplay of international affairs, state, and universities in nineteenth-century Germany may be followed through the arguments of Hoppe 1966, Ben-David 1971, Turner 1971, and McClelland 1980. See also Chapter 6.

52. When the ship ran aground in Brazil, damaging its propeller, it was refloated by jettisoning coal. After this, it was not certain that the ship could steam to Europe without coaling in the Azores.

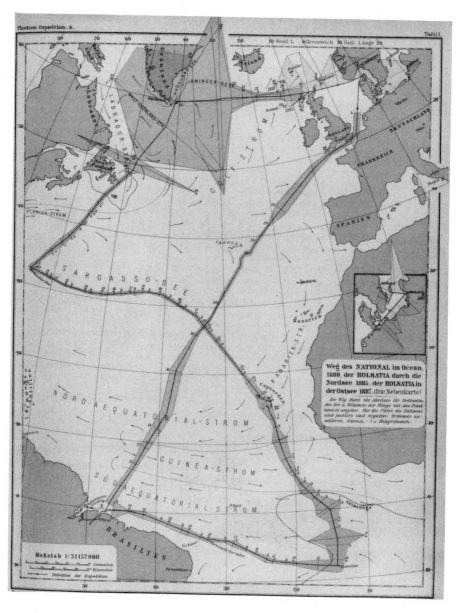

FIGURE 1.4 Cruise track of *National* on the Plankton Expedition, July 15–November 7, 1889 (15,649 nautical miles and 278 stations). Shaded areas show the plankton volumes determined for Hensen by Franz Schütt. Note that the volumes are large in the northern and northwestern Atlantic, very small in the tropics, and somewhat larger in the Equatorial Current region. From Hensen 1892a, Tafel I.

FIGURE 1.5 A pair of Hensen's vertically hauled nets being washed down just after return to the surface. Most hauls were made with single nets; double hauls like this one were used by Hensen and Lohmann to assess the uniformity of plankton distribution. From Krümmel 1892, p. 54.

quantitative vertical plankton hauls, most of them from 200 meters to the surface[53] (Figure 1.5).

The Plankton Expedition on *National* appears to have been a great adventure for the young academics on board (their average age, excluding Hensen, was 34), most of whom had never been farther afield than the Mediterranean. The cruise itself was not uneventful; *National* ran aground twice in the Amazon, damaging its propeller, and this, along with problems with fresh water supply and the steam system, reduced both the speed and the working ability of the vessel on its return journey. In particular, deep hauls with the closing nets, Brandt's responsibility, had to be cut short between South America and Europe because there was no steam for the winches.[54] Nonetheless, by the standards applied to modern expeditions the Plankton Expedition was highly successful, especially considering that Hensen had vetoed the possibility of a short trial cruise because it would be too expensive. In terms of its scientific results, Hensen believed that his hypothesis had been vindicated, although both the cruise and its results soon involved him in controversy as well as a twenty-year preoccupation with the scientific problems raised by his fifteen weeks at sea during 1889.

The plankton samples taken during the expedition were sorted in

53. Hensen 1892a, pp. 18–46 (see Tafel I); Hensen 1895, pp. 5–11 and Tafel I. First brief notice of the results of the expedition was Hensen 1890, pp. 243–253, read to the Akademie der Wissenschaften by Emil Du Bois-Reymond on March 13, 1890.

54. As a result Hensen could not contribute significantly to the debate between Carl Chun and Alexander Agassiz about the existence of a midwater ("intermediate") fauna in the open ocean. See Mills 1980.

Brandt's Zoological Institute at Kiel by his doctoral students and post-doctoral associates, especially Carl Apstein (see Chapter 6 near n. 5), and the assistants appointed to the Kiel Commission, who along with their scientific work managed the library and business affairs of the commission. At first Hensen and Apstein directed the sorting, then Apstein, and in later years a series of young zoologists working their way up the difficult ladder of German academic life, including Johannes Reibisch (1868–1948) and Hans Lohmann (see Chapter 4 near n. 41).[55] Hensen and Brandt combined the resources of their professorial positions, their research institutes, and access to doctoral students, with space and salaried staff provided by the German government through the Kiel Commission. Even with these combined forces it took three years to sort 106 of the quantitative vertical plankton hauls on which verification of Hensen's hypotheses might depend,[56] and the taxonomic work of description took far longer. Hensen, busy with his institute, physiological work, administration in the university, and governmental affairs, did not fully summarize the results until 1911, when his great monograph "Das Leben im Ozean nach Zählung seiner Bewohner" appeared, distilling many decades of thought and work on plankton biology. But his basic ideas were expressed much earlier, in 1892, in a brief, rather disorganized work, "Einige Ergebnisse der Expedition."

The results of the expedition set out in "Einige Ergebnisse" may be summarized as follows:

1. The distribution of plankton in the open ocean was indeed uniform. As Hensen said, "Great evenness is being irrefutably established by the counts along the transects."[57] But even a casual examination of the early results made it clear that on the geographical scale plankton was *not* distributed evenly—for example, the collections made off Greenland were outstandingly large, those in the Sargasso Sea uniformly small (Figure 1.4). This paradox could be explained, according to Hensen, if the unusual volumes were regarded as exceptions, caused by local circumstances not representative of the majority of oceanic locations. Even moderate variations of catches within the zones of uniform distribution (such as the Sargasso Sea) could be attributed to sampling error, caused, for example by the ship's drift.[58] By excluding all instances contradicting his theory of evenness (uniformity of distribution) Hensen greatly in-

55. Brandt 1925d, pp. 77–78.
56. Brandt 1925d, p. 75.
57. Hensen 1892a, p. 19.
58. Hensen 1892a, pp. 32–40.

creased the power of his nearly tautological explanatory system, as I will discuss later.

2. The oceans were, in general, very poor in plankton. This was especially true of the tropics, exemplified by the Sargasso Sea. "The poverty of volume is a noteworthy, newly emerging, certain fact,"[59] supported by the quantitative vertical plankton hauls. This unexpected finding totally contradicted the widespread belief that the tropical seas were the marine equivalent of the tropical rainforest, verdant and luxuriant. The weight of emphasis in Hensen's report, reflecting his view of the general state of the oceans, was on the poverty of the warmer seas, rather than the richness of the northern ones.[60]

3. Although the possibility of calculating oceanic production had been one of the dominant scientific reasons for the expedition, Hensen's sanguine expectations were soon dashed. The sheer volume of material, the long lists of species, and the difficulty of making calculations from collections representing two seasons militated against studies of production. Hensen retreated into platitudes about the incremental advance of knowledge; estimating production would have to await the work of others.[61]

Despite its failures and apparently equivocal results, the early results of the Plankton Expedition were highly influential. Hensen's young colleague Franz Schütt, still a Privatdozent in botany at Kiel, built his reputation on his monographs "Das Pflanzenleben der Hochsee," which has been described as the "first general work on the ecology of marine phytoplankton,"[62] and *Analytische Plankton-Studien*, which described quantitative plankton methods. He also developed a mathematical technique for assessing errors around the means of the collections.[63] Hans Lohmann, who studied the appendicularians collected on the expedition, noted that they were able to feed on exceedingly small phytoplankton cells in the range 1 to 20 μm. These he came to call "nannoplankton."[64] Karl Brandt, who attacked the problem of why plankton, apparently paradoxically, should be more abundant in high latitudes than in the tropics, redirected the Kiel school to the study of nutrient cycling, leaving plankton as an epiphenomenon of material cycles in the sea. However, the results, and particularly Hensen's evaluation of the performance of

59. Hensen 1890, p. 250.
60. Hensen 1892a, p. 35.
61. Hensen 1892a, p. 43.
62. Taylor 1980, p. 509.
63. For a discussion of Schütt see Chapter 6.
64. Lohmann 1896. Lohmann 1909a, p. 233, first uses the word *nannoplankton*. Etymologically the correct spelling is *nanoplankton*.

his nets, remained a controversial and difficult subject that directed his attention away from the general problems of production to those of technique during three decades following the expedition.

Both Hensen and his methods of collecting plankton were criticized from without and from within beginning immediately after the Plankton Expedition returned. From without, Ernst Haeckel (1834–1919), professor of zoology at Jena, attacked the foundations of Hensen's beliefs as well as the adequacy of his techniques. From within, Hans Lohmann, who had received his Ph.D. at Kiel during the year of the expedition, carried out a series of investigations that threw doubt on the ability of vertical net tows to collect the most abundant organisms in the surface waters. Haeckel's attack was frontal, Lohmann's of necessity (since he relied on Hensen and Brandt for support) was more subtle, but, in the long run, more effective and long-lasting.

Haeckel, who had accompanied Johannes Müller to Helgoland in 1854 to study plankton, and who was now at work on the Radiolaria collected by the *Challenger* Expedition, unleashed a broad attack in response to the early reports of the Plankton Expedition's results.[65] Through an acolyte, the journalist Carolus Sterne (Ernst Krause), he accused Hensen and his associates of the misuse of funds, a charge refuted by Du Bois-Reymond and Brandt.[66] Earlier, in a widely read monograph, *Plankton-Studien*, which was soon translated into English, Haeckel had systematically attacked the basis of Hensen's ideas as well as his collecting techniques.[67]

In *Plankton-Studien*, Haeckel combined a sharp criticism of Hensen, his methods, and his ideas with a review of the organisms found in the open sea. The essence of Haeckel's approach was to submerge Hensen's newly reported findings in a welter of his own observations, made over many years in the Atlantic, the Mediterranean, and the Indian Ocean. In truth, however, the difference lay deeper than disagreement about observations, for as Haeckel said, "*The fundamental fault of Hensen's plankton theory* in my opinion lies in the fact that he regards a highly complicated problem of biology as a relatively simple one, that he regards its many oscillating parts as proportionally constant bulks, and that he believes that a knowledge of these can be reached by the exact method of mathematical counting and computation."[68]

65. Du Bois-Reymond 1890; Hensen 1890.
66. Porep 1970, pp. 110–114; Porep 1972; Brandt 1891.
67. Haeckel 1890a, b; Haeckel 1893. Taylor (1980) comments on the fact that Haeckel's monograph was translated into English, assuring maximum impact outside Germany, while Hensen's monographs were never translated.
68. Haeckel 1893, p. 637. Haeckel's antipathy to "statistics," that is, the quantitative approach, has been discussed by Stauffer 1957 and by Porep 1972, pp. 79–80.

On the basis of his observations, not on quantitative collections, Haeckel was convinced that the plankton of the open sea was distributed in patches and that the tropics could not be poorer than the cold seas, for it was manifest that the tropics were species rich and in all likelihood had as many individuals as the high latitudes.

> The pelagic fauna and flora of the tropical zone is richer in different forms of life than that of the temperate zone, and this again is richer than that of the cold zone of the ocean. . . . The wealth of individuals can in none of these regions be called absolutely greater than in the others, since the quantitative development is very dependent upon local and temporal conditions and, according to time and place, is on the whole extremely irregular. Estimation of individuals can in this relation prove nothing.[69]

Given irregularity of distribution, it was quite unrealistic for Hensen to expect to collect plankton quantitatively using vertical net hauls. If the plankton were distributed evenly in space, vertical migrations combined with the uncertainties of using nets from drifting ships could only result in an illusory certainty, "for one can never certainly know what considerable changes in the plankton of this column of water one or more undercurrents have caused during the drawing up of the vertical net."[70]

Even if the organisms in the sea could be counted accurately, Haeckel doubted the utility of counting organisms in determining the overall production of the seas: "The only thorough method of determining the yield, in planktology as in economy, is the determination of the useful substance according to mass and weight and subsequent chemical analysis."[71] Of what possible utility were the laborious, time-consuming counts that Hensen reported when the basic currency of production was substance, not individuals, and when rates of change and not standing stocks of individual species were the basic data to be sought?[72]

In Haeckel's view, Hensen's attack on the problem of planktonic production should be turned in reverse order, for the plankton was far less important in the seas than the food supplied by the rivers and by benthic algae. Equally, the large predators had a far greater effect on the economy of nature than small plankton cells, "for a single large fish which daily devours hundreds of pteropods or thousands of copepods exerts a far greater influence on the economy of the sea than the hundreds of small animals which belong to the plankton."[73] Hensen's "oceanic

69. Haeckel 1893, pp. 620–621.
70. Haeckel 1893, p. 625.
71. Haeckel 1893, p. 634.
72. Haeckel 1893, p. 635.
73. Haeckel 1893, p. 580.

population statistics," based on the use of nets and counts of individual species, was vitiated by nature's characteristics, especially by the irregularity of plankton distribution and the absence of mathematically certain relations among the elements of the natural world. "Mathematical treatment of these does more harm than good, because it gives a deceptive semblance of accuracy, which in fact is not attainable."[74] He accused Hensen, implicitly, of lack of thought, recommending to him "that method which Johannes Müller, the discoverer of this field, always employed in a manner worthy of imitation: simultaneous 'observation and reflection.' "[75] What Haeckel had set out to show, "that the general results obtained . . . are not only false, but also throw a very incorrect light on the most important problems of pelagic biology,"[76] was manifest as long as one viewed Hensen's results as products of unfounded theory and defective or unsuitable collecting techniques aimed at an aspect of nature that, unlike physiology, was intrinsically nonquantitative.

All Hensen's work on plankton after 1890 was, more or less, a response to Haeckel's criticism. Beginning in 1891 with a polemical reply-in-kind, *Die Plankton-Expedition und Haeckel's Darwinismus,*[77] Hensen and his colleagues repeatedly claimed to show that the regularity of plankton in the sea was not merely the theoretical basis of their endeavors (as Haeckel, quite correctly, had stated) but a demonstrable fact, and that their new collecting equipment, especially large vertical nets, could adequately, that is quantitatively, characterize it. After the first results in "Einige Ergebnisse" (1892), Hensen returned in his monographic work on the methods (1895) and especially, years later, in his masterwork "Das Leben im Ozean" (1911), to the problems that Haeckel had emphasized. Franz Schütt and Karl Brandt also became involved in the enterprise for a time. Schütt's monograph, *Analytische Plankton-Studien* (1892) seems to have been written at great speed, under Hensen's demands. It is a simple account of Hensen's techniques, clearly intended to be used as a laboratory manual and to spread the word about quantitative studies to the skeptical, but it also contains a sophisticated mathematical assessment of the errors inherent in sampling.[78]

The crux of Hensen's approach was his assumption of regularity in the sea. This, as Haeckel clearly saw, underlay all else. It was based on Hensen's belief that because the physical properties of the sea—tempera-

74. Haeckel 1893, p. 637; Stauffer 1957; Porep 1972.
75. Haeckel 1893, p. 639.
76. Haeckel 1893, p. 572.
77. Hensen 1891. The dispute was noted for its bitterness.
78. The mathematical basis of Hensen's sampling and its context are discussed by Lussenhop 1974.

ture, salinity, and dissolved gases—indicated that it was thoroughly mixed, the small organisms must also be thoroughly mixed, at least in the surface waters, subject to wind and wave.[79] *Mixing yields regularity* was the axiom from which all else followed. The results of the Plankton Expedition modified this view only slightly: they showed that plankton organisms were distributed in zones, roughly corresponding with latitude, within which regularity, evenness, of distribution and abundance was maintained. Only the disturbing influence of coastlines or currents (the Gulf Stream and the Equatorial Current system) could upset the basic regularity.

Hensen's putative tests of evenness using samples had started early. In 1887 he used the difference between the geometric mean and the median values of a series of paired samples to assess samples from the Baltic,[80] a technique that was carried further by Schütt with samples from the Mediterranean collected explicitly for that purpose along with those from the Plankton Expedition.[81] From this work it appeared that there was a relatively constant collecting error of about 20% which had to be accounted for by the performance of the nets, ship's drift, and other variables not attributable to nature. Errors were also associated with subsampling for laboratory counts; here Hensen used Abbe's derivation of the Poisson distribution to determine the errors contributed by laboratory analysis and the number of subsamples needed to keep errors within reasonable bounds.[82]

In the main statement of his approach and theory, "Das Leben im Ozean" (and a later paper), Hensen showed how regularly distributed organisms (assumed to be at the center of closely packed hexagonal columns) could appear to be irregularly distributed as nets of various sizes passed at angles through the columns (Figure 1.6). From this he calculated tables of frequency distributions arising from the imperfect sampling of regularly distributed organisms and showed that vertical hauls using his nets could be expected to give results differing by up to 25% solely on the basis of sampling error.[83] W. A. Herdman's results in the Irish Sea, which showed extreme patchiness, could be accounted for in this way, especially if the unknown influence of different nets (Herdman was using a Nansen closing net) and the confused hydrography of

    79. This view is mentioned in Hensen 1884, p. 311; Hensen 1887b, p. 2; and Hensen 1911, p. 2.
    80. Hensen 1887b, pp. 22–23. See Lussenhop 1974 for detailed analysis of these statistical techniques.
    81. Schütt 1892a, pp. 56–60, 76–81, tables 6–9.
    82. Hensen 1895, pp. 166–169; Lussenhop 1974.
    83. Hensen 1911, pp. 22–29; Hensen 1912.

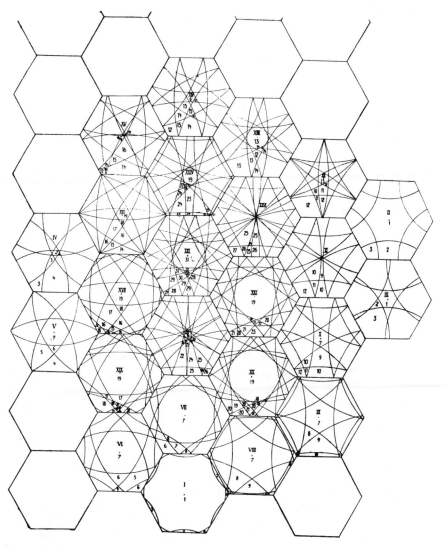

FIGURE 1.6 Hensen's hexagon theory of plankton sampling. The dot at the center of each hexagon represents a single plankton organism, each being uniformly spaced with relation to the others and maximally packed into the space under these conditions. The curved lines represent the passage of plankton nets of various sizes at various angles through the population. Using this model of distribution and sampling, Hensen calculated the frequency distribution of catches to be expected using nets of various sizes. Hensen 1911, figure 4.

the Irish Sea were invoked.[84] On the basis of this sampling theory Hensen criticized Herdman's claims to have found patchy plankton: "It is understandable that unsuitable nets and various errors should make regularity irregular, but the reverse would be a remarkable discovery in the light of the foregoing powerful demonstration" (that the hexagon theory could account for sampling error).[85]

Hensen's approach to Herdman's results was characteristic of his use of data. Because regularity was the norm, irregularity must be due to defective technique when not due to hydrographic variations. He frequently disregarded or discarded observations that contradicted the theory of regularity. In the absence of theoretical statistical models (see Lussenhop 1974) Hensen provided his own means of validating samples, comparison with the ideal of regularity. Only during and after 1911, when the hexagon theory was published, did this theoretical model have a mathematical basis. The perilous circularity of Hensen's approach is evident in his discussion of Hans Lohmann's claim[86] that adjacent plankton hauls in the Baltic (see Figure 1.5) often differed by large amounts. Hensen stated, in effect, that the difference in two hauls *was a measure of the variability of the physical environment.*[87] Use of the theory of regularity of distribution allowed him to assess the uniformity or variability of the physical environment, that is hydrographic variability of the kind often seen in the Baltic Sea. Hensen never rigorously tested the theory of regularity; in fact, his assumption of the uniformity of distribution in seemingly uniform water masses never could be tested using this logic, since the proportions and abundance of species in samples could always be suited to the axiom *mixing yields regularity* or ignored. Nor did Hensen ever apply the appropriate criteria for a test. The similarity of hauls was always evidence for uniformity; dissimilarity was never evidence against it. The concept of regularity, giving mathematical certainty, was a powerful, indeed overwhelming, organizing force in all Hensen's plankton work; he never transcended its limitations.

To Hensen, the necessity of counts was as basic as the concept of regularity. He believed that the parts would affect the overall system, that only individual estimates, involving single species, could be combined to give production estimates and to illuminate the interactions of the component species.[88] The interrelation of abundances could be used as an

84. Herdman, professor of zoology, later professor of oceanography, at Liverpool, began an extensive study of plankton in the Irish Sea in 1907. See esp. Herdman and Scott 1908, also Herdman 1922.

85. Hensen 1911, p. 17.

86. Lohmann 1908b, p. 214.

87. Hensen 1911, p. 19.

88. Hensen 1895, pp. 154–155.

index of the state of the whole system.[89] It was unjust of Haeckel to criticize Hensen for ignoring chemical methods, for he had introduced them to plankton biology;[90] equally, it must have been galling that after Haeckel's remarks about counts and Hensen's arguments for their utility, a number of North European nations accepted the technique of estimating plankton abundance rather than Hensen's numerical techniques. After Apstein's careful comparison of counts with relative estimates of plankton abundance,[91] Hensen wrote scathingly that "80% of the estimates were wrong, 4% nearly correct and 16% at least not altogether false."[92] In general, counts, being numerical, must be preferable to inexact methods. Hensen's colleague Schütt integrated this view into a holistic theory of plant ecology, based perhaps on August Grisebach's work (1872), in which associations of species were the products rather than the sum of individual interactions.[93] The resulting superorganism (Gesammtorganismus) was made up of components linked like the parts of a clock ("die Räder einer Uhr"), but in which the components affected the metabolism as well as the structure of the whole. It is unlikely that Schütt's analogy appealed to the intellectually austere Hensen, but its mechanical certainty was of a piece with his view of necessary relations in nature.

It is ironic that an attempt by one of Hensen's colleagues to increase the numerical accuracy of the Kiel group's collection methods produced the most serious threat to Hensen's cherished quantitative technique. By so doing, Hans Lohmann opened a new field of study but probably undermined his chances of a lengthy career at Kiel (see Chapter 6 near n. 6). He had observed tiny cells (later he called them "nannoplankton") in the guts and filtering apparatus of appendicularians while studying the Plankton Expedition collections, and had speculated then (1896) that they might be an important source of food in the poorer seas. At the same time, Charles Atwood Kofoid (1865–1947), working at the Illinois Biological Station, compared plankton net catches with those gathered by pumping water through a fine filter that would retain the very small cells. He showed that the majority of cells passed through the plankton net, and that it was more effective to capture the abundant small cells by centrifuging and then filtering water samples. To check these conclusions about the tiny cells that Kofoid held to "constitute a fundamental link in

89. Hensen 1911, p. 12ff.
90. Hensen 1887b, pp. 33–38.
91. Apstein 1905. The technique adopted by ICES was not directly attributable to Haeckel; it was proposed by the Swedish planktologist P. T. Cleve.
92. Hensen 1911, p. 13. The early ICES investigations (except in Germany) used estimates of abundance rather than counts, which were perceived as too time-consuming.
93. Schütt 1892a, pp. 8–10.

the cycle of aquatic life,"[94] Lohmann set out to collect nanoplankton from Kiel Fjord.

Hensen had claimed in 1887, on the basis of visual observations but not counts, that although losses from his nets certainly occurred, they did not appear to be large and would certainly be insignificant in relation to captures of large organisms.[95] Lohmann's initial results[96] contradicted this in startling fashion, for he claimed that although large organisms, such as the dinoflagellate *Ceratium tripos* and many zooplankton, along with fish eggs, were retained by the net mesh, more than 90% of the small cells, by number, passed through. His tables showed that 60% of the absolute plankton volume was lost. Moreover, because of the changing porosity of the nets as organisms accumulated in them as well as the nature of the organisms, it did not appear to be possible to establish a coefficient of loss by which the net catches could be corrected to absolute values.

In later years Lohmann elaborated his work on the losses of small cells from Hensen nets.[97] The results were consistent: his studies in the Mediterranean, the open Atlantic, and the Baltic showed uniformly that very small cells, nanoplankton (in the range approximately 1–20 μm), were ubiquitous; the sea contained from 5 to 100 times as much plankton as the nets indicated. If a true assessment of the numbers, volume, and chemical composition of the plankton were to be achieved, the nanoplankton would have to be considered. He developed a method of pumping water from a standard water column into a filter, and even ingeniously used the catches made by appendicularians in combination with net hauls and the catches on silk taffeta filters to give the total number and volume of the organisms in Mediterranean water off Syracuse.[98] Centrifugation would certainly be needed to reliably remove the variety of very small cells; it was a technique he applied widely on a cruise from Germany to Buenos Aires during the German Antarctic Expedition of 1911,[99] when it became clear that nanoplankton were indeed widely distributed in the oceans. Lohmann concluded that the sea was much richer than net hauls had indicated and that a combination of methods, not just net hauls, would be needed for an accurate assessment of its plankton populations. Although he diplomatically claimed that his results "neither affect the value of quantitative plankton investigations nor

94. Kofoid 1897, p. 832. See also Reighard 1898 on Hensen's methods and Kofoid's criticisms of them.
95. Hensen 1887b, p. 14.
96. Lohmann 1901, pp. 46–66.
97. Lohmann 1903a; Lohmann 1908a, b; Lohmann 1912.
98. Lohmann 1903a, p. 72, Tabelle XIII.
99. Lohmann 1912, 1920.

the merits of its founder,"[100] it was clear that here was a significant challenge to the hegemony of Hensen's collection methods.

Hensen's reaction to Lohmann's work was ambivalent. Lohmann was one of the Kiel school; his work had been set in motion by Hensen's reaction to Kofoid's criticisms. Nonetheless, his results, as Hensen admitted, were "personally not pleasing,"[101] and Hensen considered their importance to be somewhat overstated. Nonetheless, they could not be ignored, and they set Hensen on a new though brief course of speculation in his last writings on plankton research. Lohmann's results might help to explain the paradoxical poverty of plankton in the tropics, for, as Hensen suggested, it was possible that if small cells were abundant in warm waters tropical production might actually be as high as that in the temperate regions; this would be particularly the case if the small cells were productive through a longer water column than the turbidity of the water and limited sunlight at high latitudes would allow.[102] Lohmann soon disproved Hensen's hypothesis that nanoplankton were more abundant in the tropics; on his cruise on *Deutschland* in 1911, using his centrifugation and filtration techniques, he found that nanoplankton followed the general abundance pattern of the plankton captured in the nets.[103] But Hensen's suggestion that tropical production might be distributed throughout a long water column was an inspired guess that arose independently and was verified thirty years later.

Hensen's reactions to Lohmann's work are characteristic of his approach to scientific method. He was anxious to make quantitative observations, to refine field techniques and the analytical tools used in the laboratory, and to apply mathematics to every aspect of his investigations, an approach unfamiliar if not wholly repugnant to many German biologists during the late nineteenth century. But despite his quantitative ideals, Hensen frequently abandoned the rigor of mathematical analysis, even of logic, if the exact stood in the way of the general. He was willing to use the imprecision of his net hauls for lack of better information, preferring an estimate to ignorance, "since no one can afford to wait for the however indispensable detailed investigation."[104] Throughout Hensen's work he consistently generalized well beyond the logical entailment permitted either by his results or his stated philosophy of knowledge, aiming single-mindedly toward a well-defined theory of production in the oceans.

100. Lohmann 1903a, p. 78.
101. Hensen 1911, p. 17.
102. Hensen 1911, pp. 369–370, 390.
103. Lohmann 1920.
104. Hensen 1897, p. 91. For more about Hensen's views on generalization see Hensen 1875, p. 344; Hensen 1906; and Brandt 1925d, p. 98.

Hensen's early writings contain explicit statements of his purposes in investigating the sea. From the beginning he intended to determine the *production* of organic material, that is the fluxes of materials (initial materials and end products) made manifest as organisms such as fish or phytoplankton.[105] His approach to this topic and to his estimates of abundance show an extreme reductionism which, although not always followed in practical matters, underlies all his work. In 1897, in an essay titled "Ueber die Fruchtbarkeit des Wassers,"[106] Hensen outlined his philosophy of scientific investigation. Opening with the statement that "Wissen aber ist Macht" (knowledge is power), his essay mixes the practical (estimates of fish eggs) with the theoretical (an epistemology of natural knowledge). Rejecting the philosopher Hermann Lotze's (1817–1881) contention that science has moral meaning, uniting the physical world with human aspirations and obligations,[107] Hensen stated that such a conception was of merely "secondary interest," since it had no numerical foundation.

> It is important that the NECESSITY of events in nature be tested, or rather that their necessity be probable, which is only possible with the aid of amount and number. . . . For those . . . who through their studies of nature come upon only formed and formless masses, but no lawgiver, the final purpose must be the mathematical establishment of necessity. . . . The amounts and forces mentioned at the outset are no lawgiver. Scientific investigation of nature does not go beyond these. The physical laws are in truth mathematical formulas and that's fine [*und daher so gut*]!

These ideas might easily have come directly from his friend Du Bois-Reymond's famous, indeed notorious, lecture of 1880, "Die sieben Welt-räthsel," in which that eminent physiologist relegated to the transcendental such difficult problems of existence as the essence of force and matter ("das Wesen von Materie und Kraft"), consigning them thereby to the consciousness of the individual as distinct from scientific investigation.[108] Hensen's statements appear to fit very neatly into the pragmatic chemicophysical reductionism of the mechanistic physiologists, the so-called 1847 group, which included, as well as Du Bois-Reymond, Hermann von Helmholtz (1821–1894), Ernst Brücke (1819–1892), and Carl Ludwig (1816–1895). His views are both explicable and paradoxical in these terms. Paradoxically, the 1847 group (to the extent that it was

---

105. Hensen 1875, p. 343.
106. Hensen 1897, pp. 79–91.
107. Hensen 1897, p. 79. For remarks on Lotze, see Rothschuh 1973b.
108. Du Bois-Reymond 1886 (lecture delivered 1880; for an English translation see Du Bois-Reymond 1882). See also comments by Rothschuh 1971.

a group, rather than a way of thought) had largely abandoned their original intention of relegating all physiological processes to the molecular and the mathematical, largely on the pragmatic grounds of the empirical difficulty of the enterprise. Thus Hensen's prescription was anachronistic by at least twenty years. Even Du Bois-Reymond, despite the uneven dualism of "Die sieben Welträthsel," felt compelled to acknowledge the subjective reality, the cogency, of the individual's sense of free will and his need for ultimate explanations. Indeed, there was a striking difference between the aims stated by the most vociferous theorists of the 1847 group and the less ambitious empirical aims of the most successful practitioners such as Ludwig, Helmholtz, and Brücke.[109]

Hensen's mathematical reductionism is explicable in terms of his individual scientific interests—to simplify the complexities of living marine systems—but also in larger terms. Hensen was in direct competition, intellectually and practically, with two of the 1847 group: Du Bois-Reymond, whose institute at Berlin specialized in neurophysiology and sensory physiology, and Carl Ludwig, whose institute at Leipzig, the most eminent in late-nineteenth-century Europe, was devoted to a wide array of problems centered on the physiology of respiration and circulation. Ludwig's institute opened its doors in 1869, Du Bois-Reymond's in 1877, Hensen's enlarged one at Kiel in 1879. Kiel remained an undistinguished school, despite widespread respect for Hensen's ability in physiological anatomy and sensory physiology, and he often must have felt overshadowed, particularly by the size, eclecticism, and influence of Ludwig's enterprises. In this context, the mathematical mechanist credo expressed Hensen's desire to compete on equal terms, in addition to expressing his inborn proclivity to use mathematics and physics as tools of reason.

Curiously, none of Hensen's works after 1897 directly explains his philosophy of nature in terms of mathematical necessity. Nonetheless, neither his view of the construction of the natural world nor his empirical approach to it changed radically during the next two decades, although he could and did subordinate the mathematical to practical goals. Very early Hensen set out a scheme in which the organism was part of a larger cycle of matter and energy. The expression of this is limited to his

---

109. As Adolf Fick of Carl Ludwig's school said in 1874, "The absolute dominance of the mechanistic-mathematical orientation in physiology has proved to be an Icarus-flight" (Sulloway 1979, p. 66). On the 1847 group see Cranefield 1957, 1966; Mendelsohn 1965; and Rothschuh 1973a. The pervasive influence of the group is suggested in Sulloway 1979. The background to their viewpoint is drawn in detail in Lenoir 1981 and Lenoir 1982, chapter 5. A sophisticated account of the group, indicating their link with German romanticism and their qualified materialism, can be found in Culotta 1975.

monographs in 1875 and 1887 and a lecture published in 1906. Later, in the great concluding monograph on the Plankton Expedition, he restricted himself almost entirely to the sampling problem, not to the philosophy of nature.

Even in 1875, Hensen had stated clearly that the problem of production was that of determining the pathways followed by materials such as carbon and nitrogen in organisms and in their inorganic milieu. "A special way must be followed, which, referring continuously to knowledge gathered other ways, proceeds from the study of initial materials and end products" because the intervening processes, going on in a multitude of organisms, were both too complex to understand readily and at the wrong level to allow a general picture to emerge.[110]

Using a graphic analogy, Hensen described the living system within its physical setting as like a warm stream of water ascending from the sea floor. As it ascends, the warm layer thins, shedding eddies to the cold, inorganic milieu (the physical world) surrounding it, or into the core of the stream (from organism to organism).

> Finally, only a small stream of warm water reaches the surface, the whole moving or living mass moving as the inside of a conical, blunt-ended stream. The cold sea, the inorganic material, surrounds the stream and is drawn into its base; in this are found the lowest organized beings, mainly killed, before they advance. The apex of the stream represents the highest animal forms. The parts in the middle of the cone can pass along various paths, ascending or descending via eddies, or being pushed to the cold inorganic periphery.

Fish are "the truncated apex of our cone of metabolism" and their matter will be recirculated, via predation and decay, directly to the inorganic realm.

> One can regard the fish as the summit of the stream of metabolism in the sea, through which passes on average the most extensive flow of carbon and nitrogen in living organisms before they (C and N) return to the inorganic state from which they have come.

Modifying this analogy, but enlarging its geometrical and numerical character, Hensen explained that he viewed the carbon and nitrogen of organisms as passing through a cylinder representing metabolism.

> The cross-section of this cylinder must then correspond with the quantity of fish existing at the moment being considered. . . . Thus it would be

110. Hensen 1875, pp. 343–345.

sufficient to know this [the cross-section] and the total height of the cylinder to obtain a satisfactory picture of relative metabolism [i.e., of production]. This is the task which will make precise the statistics of metabolism in the sea.

Given this image of the marine ecosystem, especially its living component, it is clear that Hensen set out systematically to determine "the cross-section of the cylinder," whether as standing stock of commercial fish or the amount of phytoplankton, as the first measure, needing only a time component (the height of the cylinder) to give an estimate of organic production. He clearly distinguished stock (which was static) and production (which was dynamic); furthermore his analogy specified that stock must be determined within reasonable limits, before the goal of determining production was achieved. All else, even his stipulation of accuracy, was subordinated to achieving this goal. When the stock of organisms is known, "to estimate production we must either try to learn the speed of destruction of preexisting material or the rate of origin; of course it would be best to know both."[111]

A further consideration giving force to his views was Hensen's belief that there must be a constant ratio between various kinds of organisms.

The quantity of land plants must correspond with the quantity of animals (humans included), in the sense that there can be no more animals than the food available for them. This food originates directly or indirectly, thus ultimately, from plants.... [As] much digestible plant material as the animals need must be produced yearly, because at present the number of animals is in equilibrium—the number of animals neither increases nor decreases. The same is necessary in the sea.[112]

Because, unlike Haeckel, Hensen believed that the input of food to the North Sea from rivers was inadequate to support its fish stocks, he invoked the plankton as primary food material leading up the pyramid of production to fish. "The only question is how often this quantity renews itself."[113] Furthermore, "if we should succeed in determining the consumption or reproduction of food plants, especially certain phytoplankton, then the amount of animal material in the sea may be determined or at least delimited."[114]

Victor Hensen single-mindedly devoted his marine biological work to the determination of fish stocks and to assessing the quantity of plankton

111. Hensen 1887b, p. 94.
112. Hensen 1897, p. 80.
113. Hensen 1897, p. 81.
114. Hensen 1906, p. 367.

in the sea, both way stations toward the goal of calculating marine production. His development of nets, the early expeditions in the Baltic and the North Sea, the great Plankton Expedition of 1889, and his incessant mathematical analysis of his nets and their collecting power were part of this larger framework.[115] Yet clearly, he did not achieve his goal. He chose or was forced to concentrate on methods when the assessment of the seas had been his aim. His goals were taken up by young colleagues, particularly Brandt and Lohmann, who approached the chemistry and production of marine organisms from a viewpoint that owed its origin to Hensen but concentrated not only on the origins and endpoints of materials but on the details of their transformation in material cycles. Hensen himself had directed attention to the problem of inorganic nutrients, when, commenting on the paradoxical results of the Plankton Expedition, he raised the possibility that a lack of inorganic nutrients might be responsible for the poverty of tropical seas.[116] The Kiel group, under Brandt, now brought chemistry to bear on the problems of marine plankton production.

115. Porep (1968, pp. 206–207) makes the plausible claim that Hensen's quantitative plankton biology was the natural outcome of German physiology and microscopic anatomy, which, unlike zoology in the 1880s, were not preoccupied with the description of new species.
116. Hensen 1890, p. 250.

2 The Control of Metabolism in the Sea: Karl Brandt, the Nitrogen Cycle, and the Origin of Brandt's Hypothesis

> The suggested investigations upon the cycle of materials in the sea depend not merely upon one another but are also intimately related to the results of hydrographic investigations and form an indispensable foundation for knowledge of the nutrition of all marine animals including commercial fish.
>
> KARL BRANDT 1902A

Karl Brandt (1854–1931) came to Kiel in 1887 as successor to the great faunal biologist and pioneer of community ecology, Karl Möbius. Although Möbius's name is better known, Brandt's aims and accomplishments were as important and distinguished as those of his predecessor. Brandt took on the problem of why plankton abundance should be so high at high latitudes and so low in the tropics. Using ideas and methods from agricultural science he developed a theory of chemical control of the plankton cycle and of geographical variations in plankton abundance. Brandt's theory reached its peak of development during the 1920s, when, despite his initial reluctance, he combined his ideas with those of physical oceanographers who had developed widely applicable schemes of horizontal and vertical circulation in the oceans.

Brandt single-mindedly developed a school of plankton research, guided by the implications of an idea and assisted by doctoral students, postdoctoral associates, and Privatdozents. He cleverly used the facilities—laboratories and ships—provided by the state for the university, for the Kiel Commission, and for a program of international fisheries investigations in which Germany played a major part, to study the chemical control of plankton abundance. Brandt was truly an entrepreneur of marine science, using its varied institutions, personnel, and ideas to further his research.

When Brandt retired in 1922, his ideas were past their intellectual zenith. In fact, by 1915, when World War I lured most of his students away to the carnage of Flanders and the Somme, Brandt's school of plankton research at Kiel had begun to collapse. But despite the decline of Brandt's influence and ideas throughout the 1920s, the Kiel school's four decades of work on the plankton cycle were the foundation of all later work, chemical, physical and mathematical, on that ubiquitous phenomenon of temperate and polar seas, the spring bloom, and of our understanding of global marine production.

Karl Brandt (Figure 2.1) was born at Schönebeck, near Magdeburg. After his school years he studied zoology at Berlin, then at Halle, where he received the doctorate in 1877 for a thesis on the heliozoan *Actinosphaerium*.[1] Between 1878 and 1882 Brandt was an assistant in the Mikroskopische Abteilung of Emil Du Bois-Reymond's physiological institute in Berlin, where he developed his knowledge of Protozoa by working on their cytochemistry and on the then problematic inclusions in their cytoplasm now called zooxanthellae and zoochlorellae. His work revealed that the symbionts of Radiolaria were yellow-green algae, while those of freshwater hydra, sponges, and flatworms were green algae.[2]

Brandt continued his work on Radiolaria at the Stazione Zoologica in Naples between 1882 and 1885. There he showed that the buoyancy of the cells was controlled by fluid-filled vacuoles. His monograph of 1885, *Die coloniebildenden Radiolaria*, summarized his physiological investigations and a series of other studies on the ecology of the group. It was this important publication that served as Brandt's Habilitationsschrift (see Chapter 6 near n. 8) at Königsberg under the zoologist Carl Chun in 1885.

Brandt spent only two years as a Privatdozent in Königsberg. In 1887 Karl Möbius, who had been Professor ordinarius of zoology at Kiel since 1868, was appointed to direct the new Museum für Naturkunde in Berlin. He left early in 1888, after a few months with Brandt in Kiel, relinquishing both his chair of zoology and the directorship of the Zoo-

1. Brandt's life and academic career deserve more study. His life is described briefly by Reibisch (1931, 1933) and Remane (1968, pp. 167–169), but documents are scarce, despite searches in Kiel and Schleswig. Brandt's books and papers may have been destroyed during World War II when his house at Düppelstrasse 3 in Kiel was bombed, killing the occupants, his two unmarried daughters. His son Günther, who was implicated in Walther Rathenau's assassination in 1922, died in Bayreuth during the 1970s (Gabriele Kredel, Kiel, personal communication).
2. Reibisch 1933, p. ii.

FIGURE 2.1 Karl Brandt, professor of zoology and director of the Zoological Institute at Kiel from 1888 until 1922. From Reibisch 1931, with permission of the Conseil International pour l'Exploration de la Mer.

logical Institute to Brandt.[3] Brandt was convinced, certainly correctly, that his work at Naples, which combined physiology and sound natural history, had been the main reason for his appointment. Further details of Brandt's appointment are obscure, but Hensen is likely to have been the main force in appointing him, for Brandt combined impeccable physiological credentials (from his association with Du Bois-Reymond) and obvious talents for marine natural history. Hensen's and Möbius's collaboration had been a fruitful one; now a young physiologically trained successor to Möbius could carry on Hensen's ideal of quantitative marine investigations.

Brandt's career at Kiel was a full one. In addition to the duties of his chair, he lectured in the Marine Academy (until 1913) and took over Möbius's membership in the Kiel Commission. With Carl Apstein, he began a busy program of plankton sampling in the Baltic based on Hensen's earlier work, and accompanied Hensen, Schütt, Dahl, and the others on the Plankton Expedition of 1889, where his special responsibility was the vertical distribution of the zooplankton. With Hensen, he supervised the sorting of the collections brought back on *National* and

3. On Möbius's career and move to Berlin see König 1981, esp. pp. 12–13.

directed the work of a succession of students who used the plankton from the great expedition or from the Kiel Commission's surveys in their doctoral research.[4] Brandt himself began research on the effects of changing salinity on animals, studied the environmental effects of dumping dredge spoil, and from 1895 to 1907 assessed the effect of the newly built Kaiser Wilhelm Kanal (now the Nord-Ostsee Kanal) on benthic animals, their larvae, and the marine plants of the areas influenced by the canal, which brought Baltic water into the Elbe estuary.[5] In 1900, Brandt was appointed to the newly founded Deutsche wissenschaftliche Kommission für internationale Meeresforschung when Hensen refused to serve; thereby he came to be at the center of German work under the umbrella of the International Commission for the Exploration of the Sea. Germany's contribution, work centered at a new Internationale Meereslaboratorium in Kiel (which opened in 1902), was both hydrographic and biological. Brandt, who directed the biological work, found himself with an international voice in marine affairs and the ability to control and direct most of the marine biological work in Kiel, whether at the university, in the ICES laboratory, or under the aegis of the Kiel Commission.

The account that follows is a summary of how Brandt developed a few important ideas on the control of plankton abundance in the sea between 1888 and the 1920s. Brandt's influence throughout those years was very great. Even after his retirement in 1922, he became chairman of the Kiel Commission when Hensen died in 1924; in addition he continued to expand and develop the implications of work done in Kiel in major monographs,[6] two of which remained unwritten or unfinished when he died in 1931. Like Hensen, Brandt was preoccupied with the metabolism of the sea. Unlike Hensen, the circumstances of his position allowed him to explore unstintingly the chemical and biological environment of plankton organisms. His power and prestige allowed him to develop and publicize a well-articulated theory of plankton dynamics, which in turn stimulated attacks by younger scientists attempting to explain the same phenomena in new terms.

Brandt based his first quantitative work on the plankton of the Baltic

4. Sorting and counting 106 hauls taken by the Plankton Expedition took three years in the hands of students and assistants in the Zoological Institute. Hensen and Apstein directed the work at first, later Apstein alone was in charge, followed by Hans Lohmann and Johannes Reibisch, who became involved in extensive investigations of the Baltic and North seas once the Plankton Expedition samples were sorted and being studied by specialists. See Brandt 1921a, pp. 77–78. Lohmann (1918, pp. 76–77) describes the excitement he and others felt during the early years they worked with Hensen and Brandt.

5. Brandt (1921a) outlines this work and the later work of the Kiel school in great detail. On Stettin Harbor, see Brandt 1896.

6. Brandt 1927, p. 206.

Sea using collections made for him by Carl Apstein on more than 60 steamer cruises in Kiel Fjord and Kiel Bight between 1888 and 1893.[7] They provided vast numbers of plants and animals to document seasonal changes off Kiel, allowing Brandt to extend Hensen's quantitative approach to the economy of the oceans. Apstein's collections, made at all seasons, quickly showed that the plankton had a striking seasonality, later known as the spring and fall blooms.[8] But it was chemistry that interested Brandt during these early years, especially the chance to apply agricultural chemistry to the plankton.

Victor Hensen, in his famous monograph of 1887, *Über die Bestimmung des Planktons*, had made the point that the significance of the plankton in a marine economy could be given broad meaning only if marine production was compared with production on land.[9] To do this, and to explore the best means of expressing the abundance of plankton, he used several measures of plankton abundance, including wet weight, displacement volume, dry weight, ash content (the residue after combustion), and the silica content of the cells. Brandt, influenced directly by Hensen, followed and expanded this line of action in his early work. In two early papers, published in 1897 and 1898, he concentrated especially on the chemical content of organisms, using it as an index to assess the food value of plant and animal plankton relative to forage crops on land. His general approach, as he stated, was to compare sea and land. "The goal of quantitative plankton research according to Hensen's method consists of determining the yield of the water in relation to the yield of the land."[10] This could be done best using the most modern techniques of agricultural chemistry. He turned for advice to his friend Hermann Rodewald, director of the seed-testing station of the Agricultural Institute at Kiel.[11] With the help of Rodewald's associates Orth and Brandes, the latter working in the Zoological Institute specifically on the analysis of plankton, by 1898 Brandt had assembled a large amount of information on the chemical composition of plankton, including not only the simple measures dry weight and ash, but also elemental analyses of

7. Brandt 1897, p. 11; Brandt 1902b, p. 31ff.; Brandt 1925b, p. 68.
8. First documented for the Kiel Fjord by Brandt 1897, pp. 29–31, including figure 3.
9. Hensen 1887b, p. 33ff.
10. Brandt 1898, p. 46.
11. Hermann Rodewald (1856–1938), born in Hannover, received a doctorate in plant physiology in 1879 from Göttingen. In 1891 he became Professor extraordinarius für Landwirtschaft in Kiel, later director of the Agricultural Institute. He was appointed Professor ordinarius in 1920 and retired in 1922 (the same year as Brandt). Brief biographical details are given by Volbehr and Weyl 1956 and Blohm 1968. Blohm claims that Rodewald was among the first to use mathematical methods in agricultural research.

carbon, hydrogen, and nitrogen plus lipids, cellulose, and silica, as well as estimates of protein and carbohydrate.

These results allowed Brandt to compare land with sea. In the sea at Kiel the dominant organisms were diatoms (mainly *Chaetoceros*), dino-flagellates (mainly *Ceratium*), and copepods. They occurred in surprisingly constant ratios, thus, despite the occasional aberrant sample, Brandt believed it would be possible to express the abundances of these organisms on a common basis. From this early work Brandt developed a simple expression relating the total dry weight of a plankton sample to the number of diatoms, dinoflagellates, or copepods likely to be present. Such a formulation originated in Hensen's claim that there must be constant ratios among the abundances of organisms in marine communities. Brandt hoped that he had shown how any group of organisms could be used to assess the abundance of others.[12] Each group of organisms could then be compared with the main forage crops on land, thus demonstrating its value as food for marine animals. For example, based on their chemical composition, fall and winter plankton, which were dominated by dinoflagellates and copepods, lay between lush pasture and medium-quality green lupine fodder. Dinoflagellates alone had no direct counterpart on land because of their low lipid and high carbohydrate (mainly cellulose) content; nonetheless, they could be compared with straw or meadow hay if the carbohydrates were considered, or with good meadow hay or green rye if the protein content alone was at issue. Spring plankton, in contrast, because of the silica content of its diatoms, was difficult to compare to terrestrial forage plants. The diatoms themselves had protein and carbohydrate contents close to those of peas, lupines, vetch, and grass, but far higher lipid content.[13] If only the organic content was considered, diatoms appeared to be excellent food. Thus Brandt maintained that diatoms were likely to be important food items for planktonic animals, despite Hensen's reservations that they were indigestible because of their silica shells.[14] He suggested feeding experiments, similar to those Hensen had done with copepods and *Ceratium*, to determine the true role of diatoms as food for planktonic grazers.

Brandt clearly believed that the plankton was far more important than any other source of food in the sea, contradicting the widely held view that the main source of food for marine animals was attached algae or

12. For Hensen's ideas see Hensen 1897, p. 80. Brandt's development of Hensen's ideas was not carried further.

13. Brandt 1898, pp. 87–90. Comparisons of terrestrial crops based on the common currency of their nitrogen content were begun by J. B. Boussingault between 1836 and 1838 (Aulie 1970, pp. 444–446).

14. Hensen 1887b, p. 99.

plant remains from shallow water or the land. Plankton was abundant everywhere, unlike the coastally located attached algae and seagrasses; in addition, Brandt's knowledge of animal morphology convinced him that the majority of invertebrates were filter feeders and would have to depend on broken-up marine plants or on small organisms such as plankton.[15] It was his view, originating with Hensen, that small particles of varying food value (mainly plankton) were essential for the nutrition of marine animals, which impelled Brandt to look beyond plankton abundance to its chemical composition. In his position at Kiel, aided by the fortunate circumstance of his friendship with Rodewald, he stepped outside the bounds of classical biology into agricultural chemistry to assess the food value of the plankton. Beyond that, once the chemical composition of the plankton was known, that is, when the elements necessary for its mainte-nance could be readily estimated, he could attempt to solve the problem of what chemical factors controlled the abundance of marine planktonic plants and animals.

Brandt's first tentative steps toward solving the problem of plankton abundance were taken in 1897 and 1898. Shortly afterward, agricultural chemistry and, equally important, the concept of Liebigian limiting fac-tors ("the law of minimum") began to play a major role Brandt's thought (see Chapter 3 at n. 92). The rigor of physiology was replaced by the chemical analyses and nutritional concepts of agriculturalists in his ap-proach to marine production.

Early in 1899, Brandt delivered a rectoral address at Kiel University titled "Ueber den Stoffwechsel im Meere" (published privately and by the Kiel Commission that year) in which these ideas were developed for a nonscientific audience. The text of this address, among the most widely quoted of his publications,[16] contains the embryos of the ideas he devel-oped between 1899 and 1927. Thus it is the starting point for the Kiel school's major contributions to plankton dynamics during the early twentieth century.

---

15. Brandt 1898, pp. 45–46. An outspoken proponent of the idea that attached plants provided the main source of food for marine animals was C. G. J. Petersen (1860–1928), director of the Danish Biological Station 1889–1926. See Petersen 1918.

16. Late in 1899 Brandt's address was republished in Kiel's *Wissenschaftliche Meeresunter-suchungen*, then in 1901 in English as "Life in the Ocean" in the annual report of the Smithso-nian Institution, which with the report of the U.S. Fish Commission, published a number of important translations of German biological works around the turn of the century. The English version, based on a French one in 1899, appears to have been particularly influential in England, but to have had relatively little effect in North America. It contains a number of errors in translation and does not include the lengthy footnotes that accompany the German versions. The two German texts were later referred to collectively as the first in Brandt's series of three monographs with the same title.

Brandt's shift of emphasis from natural history to physiology and physiological chemistry was set out early in the lecture. Claiming that natural history would prove inadequate to determine "the general laws which govern the phenomena of marine life," he recommended applying "the fundamental principles and approved methods of animal and vegetable physiology which have been deduced from the terrestrial world."[17] In particular, these fundamental principles, he claimed, should be applied to "the totality of exchanges (Stoffwechsel) of matter in the ocean," these following cyclical pathways from the inorganic environment to animals and plants then back again. In particular "it is . . . of the first importance to follow the cycle of transformation of nitrogen in nature"; nitrogen, he implied, is both indispensable and relatively low in abundance in nature.[18]

Referring to recent work by E. Schulze and E. Kramer on the chemistry and microbiology of the nitrogen cycle in nature, Brandt described the nitrogen cycle on land, involving the oxidation of ammonia to nitrite and nitrate (nitrification), eventual reduction of nitrate to free nitrogen of the atmosphere (denitrification), and the fixation of atmospheric nitrogen by legumes. He then took the large and significant step of applying this cycle to the sea, for, as he said, "as far as we know, the cycle whose essential steps have just been sketched is performed in the sea as on land."[19] Bacteria like those active in the terrestrial nitrogen cycle must be present in the sea,[20] since one would expect the nitrogen (mainly nitrate) from the land to be washed into the oceans, where over millenia the nitrate content of seawater could rise to poisonous levels.[21] But the sea appeared to contain relatively small amounts of combined nitrogen compared with land. How could this be? Brandt suggested that "according to the current state of our knowledge, this inconsistency will only be made comprehensible by the hypothesis that *denitrifying bacteria in the ocean remove the*

17. Brandt 1901, pp. 493–494; Brandt 1899a, pp. 4–5.

18. As evidence, Brandt cited work showing that increasing the nitrogen content of carp ponds resulted in increased yield (Brandt 1899a, pp. 9–11, 27–28 nn. 11, 12), and that lakes in Holstein with the highest nitrate content had the highest abundance of plankton (based on the work of Carl Apstein, summarized in Apstein 1896).

19. Schulze 1888, 1890, 1891; Kramer 1890. The quotation is from Brandt 1901, p. 496; Brandt 1899a, p. 7. E. Schulze (1840–1912) was professor of agricultural chemistry at the Zurich polytechnic.

20. At the time of this lecture Brandt knew only of Fischer's work (1894a, b) showing that marine bacteria were ubiquitous and Vernon's (1898) suggestion that nitrifying and denitrifying bacteria occurred near shore in the Gulf of Naples.

21. Brandt 1899a, pp. 7–8, 26–27 (n. 7). He estimated that the world's rivers would contribute 30 g of combined nitrogen to the sea per cubic meter in a million years, which as ammonia or nitrate would make plant and animal life in the sea impossible.

excess of nitrogen compounds and that it is they that restore the prevailing equilibrium of nature's household."[22]

This hypothesis, that denitrifying bacteria were responsible for maintaining the balance of nitrogen in the sea, became the conceptual and practical core of all Brandt's work on the marine nitrogen cycle. The idea appears to have originated in Schulze's suggestion in 1888 that the increase of combined nitrogen in the sea must be offset by the conversion of organic nitrogen compounds and the ammonia from decomposition into nitrogen gas, which could be released into the atmosphere.[23] Brandt combined this idea with observations that bacteria of the nitrogen cycle were temperature-dependent, that is that their activity was greater at high temperatures than at lower ones. Thus, it was likely that at the near-freezing temperatures of the deep sea year-round and in high-latitude shallow seas in winter bacterial metabolism would be very low. Both Schulze and Kramer (the latter in a contemporary textbook) indicated that nitrifying bacteria in terrestrial soils could function only above 5°C;[24] Brandt assumed that all the bacteria of the marine nitrogen cycle were likely to be similar. Then he made another great hypothetical leap: "If denitrifying bacteria can not perform their function in cold waters, it follows, almost necessarily, that polar seas must be richer in nutritive substances than tropical seas," for the bacteria would continue to strip combined nitrogen from seawater in the tropics year-round, while at high latitudes they would cease activity during the winter and allow nitrogen salts to accumulate in solution.

The outcome was clear. "It thus seems that the cause of richness of cold waters and of the poverty of warm waters should be sought in the difference of development of the bacteria of putrefaction in the largest sense of the term [meaning denitrifying bacteria], and in the influence of these bacteria on the proportion of nitrogenous compounds in the water."[25]

The paradox that tropical plankton was impoverished while that at high latitudes was abundant could be resolved if denitrifying bacteria were more active at high temperatures than at low ones. In Brandt's view, Hensen's quantitative methods, which provided the paradoxical results, could be combined with chemical techniques and microbiology to explain both regional variations in plankton abundance and the nature of

22. Brandt 1899a, p. 8.
23. Schulze 1888, p. 83.
24. Schulze 1890, p. 115; Kramer 1890, p. 25.
25. Brandt 1901, p. 506. The translations, though not exact, capture Brandt's meaning adequately.

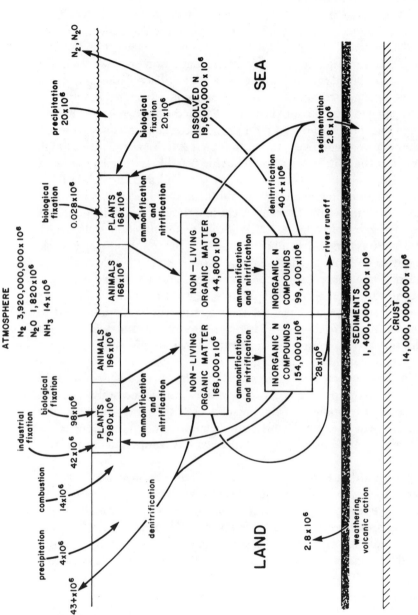

FIGURE 2.2 The global nitrogen cycle, including its major reservoirs and fluxes. Amounts in reservoirs (capital letters) in metric tons of nitrogen. Fluxes (shown along arrows) are metric tons of nitrogen per year. Estimates of the amounts and fluxes in the nitrogen cycle change frequently; this diagram is intended to indicate only orders of magnitude and the major routes nitrogen follows in the biosphere. Based on Delwiche 1970, 1981 and Hattori 1982.

Table 2.1

The main reservoirs of nitrogen, on which the nitrogen cycle is based

Reservoir	Amount of N (metric tons $\times$ 10^9)
Atmosphere	3,920,000
Dissolved N_2 in oceans	19,600
Nonliving terrestrial organic matter	168
Terrestrial inorganic nitrogen	154
Marine inorganic nitrogen	99.4
Marine nonliving organic matter	44.8

Note: Excluded are deep soils and marine sediments (ca. 1400×10^{12} metric tons of nitrogen) and the earth's crust (ca. $14,000 \times 10^{12}$ metric tons), which contribute material to the nitrogen cycle slowly and in small amounts.
Source: Based on Delwiche 1970, 1981 and Hattori 1982.

the cycle of materials, especially nitrogen, in specific locations in the sea. "Brandt's hypothesis" on the active role of denitrification in marine production entered the scene in 1899 destined to play a paradigmatic role in Brandt's thinking and in the way he directed the research of the Kiel school for nearly two decades.

The history of plankton dynamics after 1899 is largely the history of knowledge of the nitrogen cycle. Thus an understanding of modern knowledge of the nitrogen cycle is necessary before I describe Brandt's contribution and explain the difficulties he encountered in making sense of the nitrogen cycle in the sea.

The great global reservoirs of nitrogen (Figure 2.2) are the atmosphere, dissolved nitrogen gas in the oceans, nitrogen-containing nonliving organic matter in the soil and seawater, and inorganic nitrogen compounds (mainly nitrate and ammonium salts) in soil and the oceans (Table 2.1).[26] The importance of the cycle is based not on these static quantities but on fluxes of nitrogen into, through, and out of the elements of the biosphere, namely plants, animals, and the microbiota. Some major fluxes, listed in Table 2.2, are the uptake of ammonia and nitrate by marine organisms; terrestrial and marine fixation of atmospheric nitrogen gas; terrestrial and marine denitrification returning nitrogen gas to the atmosphere or to solution; river run-off to the ocean carrying organic nitrogen, ammonium salts, and nitrates; and precipitation carrying small amounts of ammonia and larger amounts of nitrate (the latter originating mainly in lightning discharges) to land and sea.

Nitrogen enters the sea or reaches the land biota mainly by nitrogen

26. Data in the tables and the discussion are based, with modifications, on work by Delwiche (1970, 1981), Hattori (1982), Capone and Carpenter (1982), and Carpenter and Capone (1983). Estimates of reservoirs and fluxes involved in the nitrogen cycle are constantly changing. The figures are intended to give an idea of magnitudes, not illusory accuracy.

TABLE 2.2
The main fluxes involved in the nitrogen cycle

Process	N flux (metric tons $\times$ 10^6 per year)
Uptake of ammonia in sea by plants	6500
Uptake of nitrate in sea by plants	1500
Terrestrial N_2 fixation	98
Marine N_2 fixation	20
Terrestrial denitrification	43
Marine denitrification	42
River run-off to ocean (inorganic salts only)	28
Precipitation to ocean	20

Note: The figures are intended to show orders of magnitude rather than accurate amounts.
Source: Based on Delwiche 1970, 1981 and Hattori 1982 except for marine nitrogen fixation, which is from Capone and Carpenter 1982.

fixation, that is by biological conversion of the gas to ammonia,[27] and its incorporation into the protein of cells. Smaller amounts of nitrogen arrive via rainfall, either as nitrate produced by lightning discharge and cosmic rays or as dissolved ammonia; this may be absorbed by land and marine plants. The nitrogen in organisms is lost mainly upon their death, when ammonia is produced by bacterial breakdown of proteins (ammonification). In addition, in the sea, biologically significant amounts of ammonia are produced by zooplankton in their excreta; this is rapidly reincorporated by phytoplankton, mainly to synthesize proteins. On land and in marine sediments, as well as in the marine water column, unincorporated ammonia is oxidized, probably usually first to nitrite (NO_2^-) salts, then to nitrate (NO_3^-) salts; this process and its components is called *nitrification*. Nitrate then becomes available as a nitrogen source for plant growth on land and in the sea, supplementing ammonia and a few other much less abundant nitrogen-containing compounds also involved in the nitrogen cycle.

Nitrogen is returned to solution in the oceans and to the atmosphere by reduction of nitrate, often to nitrogen gas; lesser amounts become nitrous oxide. This process, called *denitrification*, occurs mainly where oxygen is low, for example in anoxic terrestrial soils, below the top few centimeters of marine sediments, and in special but very extensive regions of the marine water column, the oxygen minimum zones found at depths of a few hundred meters in tropical oceans. In denitrification, nitrate rather

27. Throughout this discussion of the nitrogen cycle, the term *ammonia* refers to free ammonia, ammonium salts, and the ammonium ion indiscriminately. In the sea, non-ionized ammonia is usually very scarce (Whitfield 1974).

TABLE 2.3

Energy produced (or required) in the reactions of the nitrogen cycle

Process in nitrogen cycle	Energy yield or requirement (kilocalories/mole)
Denitrification of nitrate to nitrous oxide (N_2O) with glucose as carbon source	+545
Denitrification of nitrate to nitrogen gas (N_2) with glucose as carbon source	+570
Ammonification of glycine (CH_2NH_2COOH)	+176
Nitrification	+106
Ammonia to nitrous acid	+83.5
Nitrite to nitrate	+17.5
Nitrogen fixation	−160
Splitting N_2 molecule	−147.2
Reduction of N atoms to NH_3	+12.8
Respiratory oxidation of glucose (major energy sources of heterotrophic organisms)	+686

Note: The oxidation of glucose (respiration) is included to indicate the magnitude of energy yields in the nitrogen cycle. In general, the processes of the nitrogen cycle produce less energy than conventional respiratory oxidation (last row) but enough to be metabolically useful under special conditions such as low levels of oxygen. Note that nitrogen fixation requires energy to proceed, rather than releasing it for metabolic purposes.

Source: Based on Delwiche 1970.

than gaseous oxygen serves as the source of oxygen in the energy-yielding respiration of bacteria.

All the biological processes of the nitrogen cycle except fixation yield usable energy to organisms for metabolic processes (Table 2.3); this, combined with the universal need for nitrogen to synthesize proteins and nucleic acids accounts for the ubiquity of nitrogen-based processes in the biosphere. Even nitrogen fixation, which requires large amounts of metabolic energy to proceed, is selectively advantageous (and has become widespread) if organisms can obtain energy for fixation from plant photosynthesis, as, for example, do the root nodule bacteria of legumes, the marine cyanobacterium *Oscillatoria*, and the symbiotic bacteria of some marine diatoms.

The general cycle of nitrogen (Figure 2.2) shows some special features in the sea. Figure 2.3 illustrates the cycle and recent estimates of fluxes in the oceans. In the sea, ammonia is a far more important plant nutrient than on land. In addition, the restriction of plant photosynthesis to, at most, the top 100 meters of the ocean imposes a spatial component on the nitrogen cycle. In general, ammonia is recycled within the lighted surface waters, being released by zooplankton and absorbed by plants, while

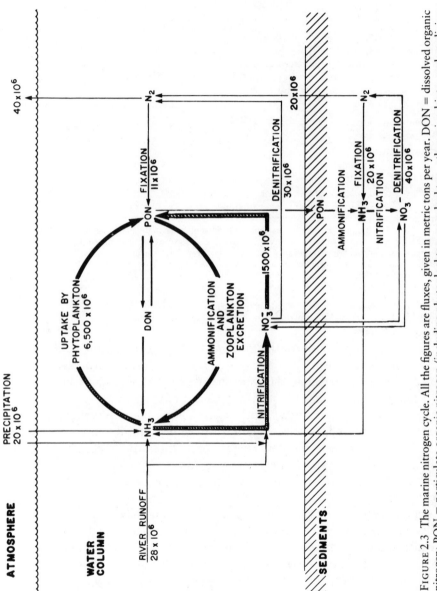

FIGURE 2.3 The marine nitrogen cycle. All the figures are fluxes, given in metric tons per year. DON = dissolved organic nitrogen; PON = particulate organic nitrogen (including phytoplankton, zooplankton, other microbiota, and nonliving organic particles in the water column, benthic organisms, and organic detritus in the sediments). Many of the estimates are uncertain, especially rates of denitrification. Redrawn from Hattori 1982, with permission of the Oceanographical Society of Japan.

nitrate is produced by bacterial nitrification of organic remains in deep water and the sediments and must be transported into the surface waters by vertical water movement before it becomes available. Denitrification occurs mainly in anoxic sediments and in mid-water regions, the oxygen minimum zones, where both molecular nitrogen and nitrous oxide are produced. Although the full extent of marine denitrification is unknown, it is presumed to balance the nitrogen entering the sea from rivers (mainly as nitrate and dissolved organic nitrogen), rain, and by nitrogen fixation, which is carried out by pelagic cyanobacteria (mainly *Oscillatoria*, previously called *Trichodesmium*) and the symbionts of some diatoms.[28]

Because the main reservoir of nitrogen-containing nutrients for marine phytoplankton is the deep water of the oceans, biological oceanographers since Brandt have struggled to explain the vertical distribution of the main nutrient salts. Only recently has a full explanation been provided. Figure 2.4 shows both static profiles of the main nutrient salts and the processes that produce these profiles.

It is known now that nitrifying bacteria are found throughout the water column, but that they are inhibited by light.[29] Thus the conversions $NH_3 \rightarrow NO_2^-$ and $NO_2^- \rightarrow NO_3^-$ occur mainly below the euphotic zone, where light is 1% or less of the surface value. Moreover, the conversion of NO_2^- to NO_3^- by bacteria is more light-sensitive than the oxidation of ammonia to nitrite. As a result a small peak of nitrite occurs at the base of the euphotic zone, to which is added nitrite excreted by plant cells. Nitrate is virtually absent in the euphotic zone but increases rapidly with increasing depth below it, giving a characteristic profile of that nutrient. Both nitrite and nitrate are absorbed rapidly by plant cells in the euphotic zone, resulting in very low levels of both in the surface waters under normal conditions. Ammonia, which is relatively scarce throughout the water column and varies erratically, is produced mainly in the euphotic zone by zooplankton excretion and in deeper water by the bacterial decomposition of plankton. Especially in the surface waters it is taken up very rapidly by phytoplankton.

During the nineteenth century, long before the details of the marine nitrogen cycle were known, the chemistry and microbiology of the terrestrial nitrogen cycle were worked out in qualitative though not fully quantitative terms. Unfortunately, there is no comprehensive account of the development of knowledge about the nitrogen cycle which sets nineteenth-century developments in a complete economic, social, and scientific context, although R. P. Aulie's and F. W. J. McCosh's mono-

28. For more details see esp. Hattori (1982), Carpenter and Capone (1983).
29. See esp. the work of Olson (1981a, b) and Ward et al. (1982), on which this discussion is based.

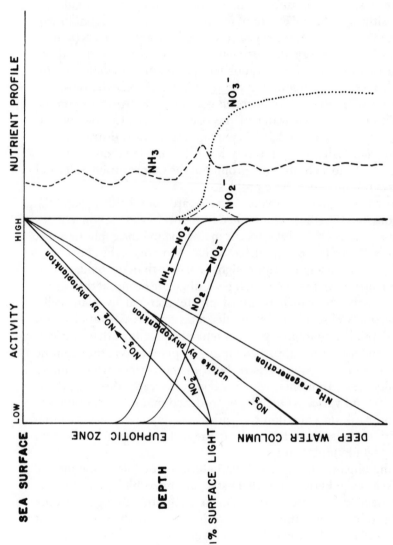

FIGURE 2.4 The marine nitrogen cycle in vertical profile. The abcissa indicates depth. The upper (lighted) waters, where phytoplankton net photosynthesis occurs (the euphotic zone), ends by definition at the 1% light level. On the right are typical profiles of ammonia, nitrite, and nitrate, expressed as relative concentration. The curves on the left indicate processes of the nitrogen cycle, described in more detail in the text, which give rise to the profiles. From Olson 1981a, with permission of R. J. Olson.

graphs on the French agricultural chemist, J. B. Boussingault (1802–1887) are a major step in this direction.[30] Thus what follows is merely a sketch of the context in which Brandt's early work took place. Brandt made a great inductive leap from his knowledge of the nitrogen cycle on land to the presumably similar cycle in the sea. Nitrogen was the key both to Brandt's hypothesis and to the work of the Kiel school until the 1920s. What did Brandt know about the nitrogen cycle in 1899 to justify such a leap and such an obsessive concentration on nitrogen? Why did the nitrogen cycle rather than another cycle of materials capture his attention?

By about 1880 most of the major problems that had occupied agricultural chemists concerned with the chemical transformations of nitrogen had been solved. It was clear by then that land plants depended for nitrogen mainly on salts of nitrate, though ammonia (as ammonium salts) played some role in their nutrition. It was also known that nitrate was produced by the nitrification of organic matter in the soil, and that plants could grow on inorganic nitrate alone, independent of organic matter and the cation accompanying the nitrate ion, as J. B. Boussingault showed in 1855.[31] By the same date J. Liebig's hypothesis that the mineral components of the fertilizers were the active components, not nitrogen, had been decisively refuted by the work of Boussingault and of J. B. Lawes and J. H. Gilbert.[32] Atmospheric ammonia, which Liebig believed was the source of the nitrogen in terrestrial plants, was clearly too scarce to be important, as Boussingault, Lawes and Gilbert, and especially Thomas Way, had shown between 1853 and 1863.[33] Moreover, when Boussingault noted in 1860 that nitrate accumulated in garden soil despite the negligible input of nitrogen in rain, he was forced to conclude that "it is the soil and not the plant that fixes the nitrogen."[34]

Nitrogen fixation, in fact, was *the* great remaining problem of agricultural chemistry during the middle decades of the nineteenth century. Boussingault had begun experiments in 1838 on the nitrogen-fixing powers of legumes, a capacity that had been observed by farmers and chemists for many years.[35] But it was his experiments during the period 1850 to 1856, in opposition to the chemist Georges Ville, that demonstrated

---

30. The history of research on the nitrogen cycle during the nineteenth century may be pieced together from Aulie (1970) along with the works of Browne (1944), Bulloch (1938), Collard (1976), Holmes (1973), Krohn and Schäfer (1976), McCosh (1984), E. J. Russell (1942, 1966), Lechevalier and Solotorovsky (1965), Rossiter (1975), Waksman (1952), and Zo Bell (1946).
31. Aulie 1970, pp. 464–465.
32. Aulie 1970, pp. 454–455.
33. Aulie 1970, pp. 452–458.
34. Aulie 1970, p. 471.
35. Aulie 1970, pp. 446–449.

that legumes growing from seed in artificial soil did not fix nitrogen. It was this, combined with his observation that nitrate accumulated in garden soils that led him to believe that the soil, or some factor in the soil, fixed atmospheric nitrogen, not the plants themselves.[36] Eventually, between 1866 and the 1890s Boussingault's paradoxical observations, based on highly controlled experiments, were resolved when it became clear that both legumes and the soil contained bacteria that could fix atmospheric nitrogen. Boussingault had been the victim of his own careful experiments, for the calcined (sterilized) soils he used did not contain the nitrogen-fixing bacteria necessary to infect the root modules of the legumes.

Boussingault, whose scientific work ended in the 1870s, emphasized the chemical nature of the soil. As a chemical system, the soil provided the main mineral nutrients, especially nitrogen, which he had demonstrated between 1857 and 1860 to have greater growth-promoting effects than phosphorus.[37] He recognized the presence of bacteria in the soil, but had concentrated his work on chemistry, not on microbiology. It was advances in microbiology, beginning in 1877, that set the stage for Brandt's work.

In 1877, J.-J.-T. Schloesing (1824–1919) and Achille Muntz published the results of experiments showing that sewage allowed to percolate slowly through a sand column produced large amounts of nitrate.[38] They were able to stop nitrification by adding chloroform to the column, then to reestablish it by adding a little garden soil once the chloroform passed through, demonstrating that nitrification was caused by "ferments organiques," that is, bacteria. In 1889, S. N. Vinogradsky (1856–1953),[39] working in Zurich, isolated a nitrifying bacterium responsible for the conversion of ammonia to nitrite. In 1891, he reported the isolation of another bacterium, this one completing nitrification by converting nitrite to nitrate in culture.[40] By 1898 there were indications, from the work of H. M. Vernon, a student of E. Ray Lankester (of University College,

36. Aulie 1970, pp. 458–464.

37. Aulie 1970, pp. 466–468.

38. J.- J.- T. Schloesing and A. Muntz 1877; see also Lechevalier and Solotorovsky 1965, pp. 260–263, and concluding comments on Boussingault by Aulie 1970, pp. 474–475. Jean-Jacques-Théophile Schloesing (1824–1919), whose son Alphonse-Théophile (1856–1930) also worked in this field, was professor of agricultural and analytical chemistry in the Ecole d'applications annexées à la manufacture and in the Conservatoire des Arts et Métiers, Paris; Muntz had studied with Boussingault.

39. On Vinogradsky (also spelled Winogradsky) see Waksman 1946, 1953 and Gutina 1976. Apt excerpts from his work are given by Lechevalier and Solotorovsky 1965, pp. 263–274.

40. Vinogradsky 1890, a, b; Vinogradsky 1891. See also Lechevalier and Solotorovsky 1965, pp. 263–274.

London) working at Naples, that both nitrifying and denitrifying bacteria occurred in the sea, although Vernon isolated neither.

Nitrogen fixation too, Boussingault's bête-noire, was shown, between 1860 and 1899, to be caused by bacterial agents. In 1866, bacteria were observed in the root nodules of legumes by M. Woronin, who regarded them as pathogens. More than a decade later (1879), experiments showed that root nodules were caused by nitrogen-fixing bacteria penetrating the root hairs of legumes, and that sterilization of the soil prevented nodule formation—the experimental factor that had vitiated Boussingault's results during the 1850s. But the association between the nodules and nitrogen fixation was not evident at first. J. Hellriegel and H. Wilfarth, who examined the problem in 1885, concluded that plants with nodules could fix nitrogen and those lacking them could not, prima facie evidence for nitrogen fixation by the bacteria. But it was still uncertain whether the fixation was carried out ultimately by the bacteria or the plant itself in association with bacteria.[41] Quantitative uptake of nitrogen gas by legumes was demonstrated by Alphonse-Théophile Schloesing and E. Laurent in 1890 (the death blow for Boussingault's conclusions). Martinus Beijerinck (1851–1931), the great microbiologist of Delft, had isolated root nodule bacteria in 1888, but it was still not clear from experiments what they did.[42] Nonetheless, by implication, root nodule bacteria were clearly involved in the fixation of atmospheric nitrogen by legumes. Free-living soil bacteria too could fix nitrogen, as both Vinogradsky and Beijerinck showed in a series of publications between 1893 and 1901, although they found it difficult to get free-living nitrogen-fixing bacteria to live in pure culture.[43]

In 1899 denitrification was the least known of the major processes in the nitrogen cycle. In his great work, *Elements of Agricultural Chemistry*, Humphry Davy had noted that nitrogen gas could be produced during the decomposition of organic material, but F. Goppelsröder is credited with direct reference to bacterial denitrification much later, in 1862,[44] followed in this conclusion by C. F. Schoenbein in 1868 and E. Meusel in 1875.[45] In 1882, only seventeen years before Brandt's first exploration of the nitrogen cycle, P. P. Dehérain and L. Maquenne showed definitively

41. Waksman 1952, pp. 14, 211. See also Vinogradsky's comments quoted in Lechevalier and Solotorovsky 1965, p. 270. An accessible account of their earlier work is given by Hellriegel and Wilfarth (1889).

42. Aulie 1970, p. 475; Waksman 1952, p. 212. On Beijerinck see Bulloch 1960, p. 352; Lechevalier and Solotorovsky 1965, pp. 275–279; and Van Iterson et al. 1983.

43. Lechevalier and Solotorovsky 1965, pp. 270–277; also Vinogradsky 1893, 1894, 1895.

44. Lloyd 1931, p. 530; Goppelsröder 1862.

45. Waksman 1952, pp. 155, 185.

that bacterial denitrification occurred in agricultural soil. In the same year, U. Gayon and G. Dupetit isolated a denitrifying bacterium, *Bacillus denitrificans*, from soil; later (1886) they suggested the equations representing the reduction of nitrate to nitrite. Marine denitrification was noted, or received a little attention during the 1890s,[46] but little attention was paid to the process, and no marine denitrifying bacteria had been isolated when Brandt began his work.

Perhaps it is not strange that so little was known of marine denitrification, for the process had not received much attention on land, but another factor also played a significant role. As Brandt noted in 1902,[47] there was a widespread belief, apparently based on remarks by Boussingault in 1860, that the sea, a great reservoir of nitrogen, contained mostly ammonia, which was constantly being released to the atmosphere. Both Schloesings, *fils et père*, subscribed to this view in 1875. Alphonse-Théophile Schloesing summarized their view of the nitrogen cycle as follows:

> One must depict in entirety the global circulation of nitric acid and of ammonia thus. Nitric acid produced in the atmosphere arrives sooner or later in the sea; there, after passing through living things, it is converted to ammonia; from then on the combined nitrogen has taken on the most appropriate state for diffusion; it passes into the atmosphere, and, travelling with it, like carbonic acid, becomes available to non-locomotory organisms, to whose nutrition it must contribute. On its way, it is fixed when it encounters plant foliage, or equally, arable land capable of absorption through tillage or by the presence of humus. Thus nitrous production in the air, the nitrous contribution of the air to continents and the sea, return of nitrates from the continents to the sea, transformation of these salts to ammonia in the marine environment, passage of the alkali [ammonia] into the atmosphere and transport to the continents ought to be the circulation of mineralized nitrogen compounds.[48]

His father described how it should be possible to devise mathematical formulations of the diffusion of ammonia from water to the air, so that its partitioning between soil, water, and air could be precisely calculated. From experiments in his laboratory he concluded that ammonia would diffuse from the sea to the atmosphere by the dissociation of amonium carbonate (which was viewed widely as the main form of combined ammonia in the sea), and that because of differences in alkalinity between tropical and high-latitude seas there should be more ammonia in tropical

46. See Beijerinck 1890; H. L. Russell 1893; Fischer 1894a, b; and Vernon 1898.
47. Brandt 1902b, pp. 46–47.
48. A.-T. Schloesing 1875, p. 177.

air than in air over the colder seas. In effect, their view of the nitrogen cycle precluded the need for denitrifying bacteria to return nitrogen from the sea to the atmosphere; as long as their view was accepted, as it was by Konrad Natterer, whose work on the nitrogen content of seawater appeared in the 1890s, denitrification could be ignored. Despite this, Brandt made denitrification the central tenet of his view of the nitrogen cycle, basing his ideas on the proposition that ammonia was scarce in the sea and that dissociable ammonium carbonate was less likely to be responsible for the release of nitrogen than denitrifying bacteria of the kind known on land. In this, Brandt followed the change of fashion from purely chemical views of the nitrogen cycle to microbiological views that prevailed during the 1870s and 1880s, preferring to speculate that biological processes rather than purely chemical ones must be responsible for completing the global cycle of nitrogen.

Brandt's early conception of the nitrogen cycle (Figure 2.5) incorporated many of the ideas summarized above. It was based on six major ideas, derived mainly from contemporary texts and a few review articles in the agricultural literature which are likely to have been available only to Rodewald, as a senior member of the Agricultural Institute of the university.[49] To these Brandt added his general knowledge of the terrestrial nitrogen cycle, which had been given modern form by studies in microbiology of the soil and marine sediments between 1877 and 1898.

1. Inorganic nitrogen, especially ammonia, nitrites, and nitrates, controlled production in the sea. As Brandt stated, "All the life on the globe depends absolutely upon the existence of these compounds."[50] Boussingault had shown in 1857 that nitrogen was of greater importance than phosphorus in controlling the growth of crops. Similarly, in the sea, nitrogen, not phosphorus, was the substance that controlled production because there was less nitrogen than phosphorus in solution, but organisms contained more nitrogen than phosphorus, and their need for it must be greater in relation to abundance.[51] Using Apstein's data on the volume of plankton found in the lakes of Holstein, Brandt found that the lakes with the richest plankton also had the highest nitrate content (though he made no attempt to relate plankton and phosphorus).[52] Carbon dioxide could not be limiting because its relative abundance in the sea was higher than in the atmosphere; land plants showed no evidence of CO_2 limita-

49. Notably Schulze 1888, 1890, 1891; Kramer 1890.
50. Brandt 1901, p. 495; Brandt 1902b, pp. 65–67.
51. Aulie 1970, p. 466.
52. Brandt 1899a, pp. 20, 34; Brandt 1901, p. 504.

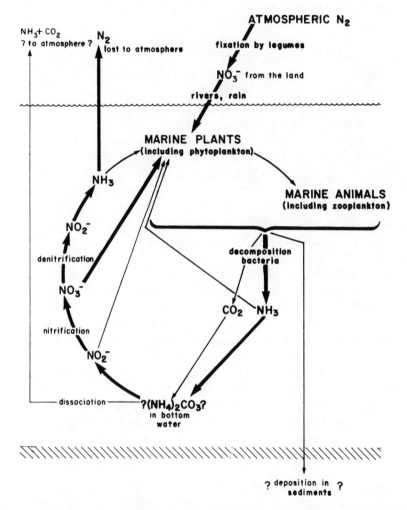

FIGURE 2.5 The marine nitrogen cycle, as envisioned by Brandt in 1899. Broad arrows show what he believed to be the major pathways of nitrogen flow. In 1899 Brandt had not recognized the existence of free-living nitrogen-fixing bacteria on land or in the sea. Question marks indicate Brandt's skepticism that significant amounts of ammonium carbonate formed in the sea and that this dissociated, releasing ammonia and carbon dioxide to the atmosphere. Instead, he believed that the input of nitrogen to the sea was balanced by denitrification, shown on the left. He was also skeptical of Ardoynaud's (1875) belief that significant amounts of nitrogen left the water column and were stored in the sediments. The great importance of ammonia to marine plants was not recognized until the middle of the twentieth century. Based on concepts expressed by Brandt in 1899, 1901, and 1902.

tion but were limited by nitrogen.[53] And even silica, which was obviously needed in large quantities by diatoms, was abundant in the coastal waters at Kiel compared with the amount found in diatom tests.[54]

2. The nitrogen in organisms was lost as excreta or was released when they decomposed, producing ammonia. Bacteria of decomposition (Fäulnissbakterien) were responsible for this step.[55] This process was likely to be completely analogous to decomposition of organic matter on land.

3. The ammonia from excreta and decomposing organisms was nitrified in two steps:

$$NH_3 \rightarrow NO_2^-,$$
$$NO_2^- \rightarrow NO_3^-,$$

as on land. Very little nitrite accumulated because the process ended with the production of nitrate.[56] Nitrate, as on land, along with smaller amounts of nitrite and ammonia, was a major nutrient of phytoplankton and was rapidly absorbed.

4. The salts of ammonia and nitrate not absorbed by plants on land dissolved and were carried by rivers to the sea. Using the flow rate of the Rhine and the nitrogen content of the water as examples of the worldwide flux of nitrogen to the oceans, Brandt calculated that after a million years, the oceans should contain 36 g of ammonia or 135 g of nitrate per cubic meter, amounts that would poison organisms. But far less ammonia and nitrate were present in the oceans, although the information was scanty.[57]

5. Nitrogen, especially ammonia and nitrate, was lost from the sea by bacterial denitrification, following the sequence

$$NO_3^- \rightarrow NO_2^- \rightarrow NH_3 \rightarrow N_2,$$

by which dissolved nitrogen in excess of the needs of plants was eventually released as nitrogen gas to the atmosphere. Denitrification prevented the poisoning of the oceans, but it could not proceed without countervailing processes, for "though the greater part of the organic nitrogen returns to living organisms, a certain portion of it is lost by the action of the denitrifying bacteria. The quantity of living organisms

53. Brandt 1902b, pp. 46–47.
54. Brandt 1902b, pp. 69–72.
55. Brandt 1899a, pp. 5–6; Brandt 1901, p. 495.
56. Brandt's view of nitrification is derived mainly from Schulze 1890, 1891, and Kramer 1890; both were aware of Vinogradsky's work.
57. Brandt 1899a, pp. 7–8, 26–27; Brandt 1901, p. 496.

would, therefore, be gradually diminished were there no other source of combined nitrogen to make up for the loss."[58]

6. The "other source" of combined nitrogen reaching the sea, balancing the losses to denitrification, was nitrogen fixation, brought about mainly by lightning. On land, according to Brandt, nitrogen fixation was carried out only by legumes; he seems to have been unaware in 1899 that free-living bacteria could fix nitrogen, as Beijerinck and Vinogradsky had demonstrated beginning in 1892.[59]

On these grounds, Brandt committed himself and his colleagues to the study of the marine nitrogen cycle, expressing his cautious belief that "according to the current, admittedly incomplete state of our knowledge of life and metabolism in the sea, the hypothesis that nitrogen compounds, as the result of the activities of nitrogen bacteria, take first place in controlling the intensity of production, even though not exclusively, has the greatest probability."[60]

Brandt used a three-part strategy in developing his denitrification hypothesis. First, did the results of the Plankton Expedition stand up? Were there indeed larger plankton volumes (and by inference greater production) in the seas at high latitudes than in the tropics? In his vicious attack Haeckel had accused Hensen of an egregious error.[61] Second, were bacteria of the nitrogen cycle, especially denitrifiers, present in the oceans, and were they abundant enough to be important? Third, was the distribution of nitrogen-containing plant nutrients in the oceans consistent with the pattern of plankton abundance and with the distribution of denitrifying bacteria? Proceeding on these fronts, Brandt threw the resources of his colleagues at Kiel into the study of the nitrogen cycle.

The first question was answered very early. The results of the Plankton Expedition did stand up. Brandt combined information on plankton volumes gathered for him by his younger colleagues with information from collections made by ship captains and surgeons all over the world to document that Hensen's conclusions applied to more than just the Atlantic Ocean at a single time of year. In addition, they showed fascinatingly different patterns of seasonality that required explanation.[62]

The first information came from Franz Schütt, who visited Naples during the winter of 1888–1889. There, between October and March, he

58. Brandt 1899a, pp. 6–7; Brandt 1901, p. 495.
59. Brandt 1899a, pp. 6–7, 25; Brandt 1901, pp. 495–496.
60. Brandt 1902b, p. 72.
61. Haeckel 1890a, b. On Haeckel 1893 see Chapter 1, esp. nn. 67, 68.
62. Brandt 1899a, pp. 19, 34–35; Brandt 1902b, pp. 23–24.

took vertical plankton hauls using a large Hensen net.[63] Outside the Mediterranean, Apstein and Brandt's five years' worth (1888–1893) of plankton samples from the Baltic were available for analysis. And at high latitudes, Ernst Vanhöffen, who accompanied the expedition of the Gesellschaft für Erdkunde to West Greenland in 1892–1893, took fifteen samples most months from August to the following July in the Karajak Fjord (now Karats Fjord) at 70°N using an Apstein middle plankton net.[64] The second Nordsee Expedition (1895) provided information on plankton volumes in the North Sea.[65] This information was increased when Apstein sampled for Brandt in the Gulf of Naples from October 1895 to January 1896; later Lohmann made similar collections at Messina during a visit between April 1896 and February 1897 and in the open Mediterranean off Syracuse during the winter of 1900–1901.[66] Most exotic and distant of all were the collections made by Friedrich Dahl, who had been with Brandt and Hensen on the Plankton Expedition, at Ralum in New Pommerania (now Dalum, New Ireland, in the Bismarck Archipelago), those by Augustin Kramer in Samoa, New Zealand, and New South Wales,[67] and Apstein and Vanhöffen's collections around the world during Carl Chun's German Deep-Sea Expedition on *Valdivia* in 1898–1899.[68] Later, this accumulation of plankton volumes was increased by Lohmann's collections on the German cable layer *Von Podbielski* when it made a round trip from Hamburg to New York in 1902, and by his work on Filchner's expedition to Antarctica on the ship *Deutschland* in 1911.[69] And after 1902 a vast storehouse of plankton samples from the quarterly German cruises for the International Council for the Exploration of the Sea in the Baltic and western North Sea was available to document the plankton volumes and seasonal changes in those seas.

Brandt had no need to modify the opinion he expressed in 1902 that plankton volume increased from the equator toward the poles.[70] The average annual volume of the more or less monthly samples taken at

63. The basic information is in Schütt's *Analytische Plankton Studien* (1892a).

64. Vanhöffen 1897, pp. 254–292. On Vanhöffen, who had a lengthy career at Kiel, see Lohmann 1918; on the expedition itself see Drygalski 1897.

65. Hensen and Apstein 1897.

66. Lohmann 1899, 1909b, for Messina. On Syracuse, see Brandt 1902b, pp. 37–38.

67. Brandt 1902b, pp. 38–40. Dahl's life is outlined by Damkaer and Mrozek-Dahl 1980.

68. On the expedition and Chun's role in plankton biology see Mills 1980 and Mills 1983, pp. 39–44. The results of the *Valdivia* expedition were not available immediately, but by 1906 there was enough information to allow Hensen (1906, p. 371) to comment that the Indian Ocean showed the same pattern as the North Atlantic.

69. Summarized in Lohmann 1903b, 1920.

70. Brandt 1902b, p. 44.

Messina, in the Bismarck Archipelago, at Kiel, and in West Greenland increased in the ratio 1:2:10:20. Messina, because of its unusual oceanic location, had the lowest volume of all, otherwise the results were just as expected from Hensen's survey and the denitrification hypothesis.[71] Brandt's results, then and later, led to his conclusion, confirming Hensen's results, that "the Arctic regions are very rich in summer while the tropical regions are poor in plankton the whole year round."[72]

The second question—were bacteria of the nitrogen cycle present and important—also required new work. In 1899 Brandt had only slight evidence that denitrifying bacteria might occur in the oceans. Beijerinck in 1890 had noted that some of the luminescent bacteria he found near the Dutch coast appeared to be capable of reducing nitrate. Three years later, a broad survey of marine bacteria by H. L. Russell, who worked at Naples and Woods Hole, indicated that denitrifiers were found there too. H. M. Vernon, working at Naples on the conversions of nitrogen occurring in aquaria using simple chemical methods, found several complex changes in ammonia, nitrite, and nitrate in enclosed seawater, including the decrease of nitrite in water samples kept in the dark. Noting this, and observing that nitrite was usually low in seawater, Vernon concluded that denitrification was occurring.[73]

The crucial evidence Brandt needed came in 1900–1901 when Lohmann left Kiel for eight months to study the plankton of the open Mediterranean off Syracuse. Brandt replaced him with Erwin Baur, a bacteriologist, who in 1901 isolated two species of marine denitrifying bacteria using media containing seawater that had been enriched with potassium nitrite. The nitrite in these media, when inoculated with slime and small worms from the Kiel aquarium, decreased in three days at room temperature. The denitrifying bacteria that Baur isolated appeared to be facultative anaerobes, reducing nitrite quite readily at normal oxygen tensions. They were widespread in the mud and water of the Kiel Fjord, needed a protein or carbohydrate source, and had high-temperature preferences (most denitrified at the highest rate at 25°C, even 35°C).[74] Baur's findings were reinforced a few months later when the Norwegian botanist H. H. Gran, who had gone to Beijerinck's laboratory in Delft to

71. See results tabulated by Brandt 1902b, p. 43.
72. Brandt 1901, p. 505.
73. Beijerinck 1890; H. L. Russell 1893; Vernon 1898 (mentioned briefly by Waksman, Hotchkiss, and Carey 1933, p. 148). Vernon determined nitrate and nitrite using diphenylamine plus sulfuric acid and the potassium iodide-starch test, respectively; Brandt and others believed his results were ambiguous.
74. Baur 1902. Baur concluded that denitrifying bacteria were "widespread" at Kiel. In fact, his data are more modest: they show that strong denitrification occurred in only 25% of his 15 seawater samples and in 33% of his 12 mud samples.

study marine bacteria, isolated and characterized three additional species of denitrifying bacteria. Gran's samples were taken from the North Sea off Texel, indicating that denitrifiers occurred in the open coastal sea; moreover they were truly marine for they grew best on media containing seawater. Like Baur, Gran found that his marine denitrifying bacteria had high-temperature optima, but he expressed doubts that there was enough organic matter in seawater distant from the coasts to act as a carbon source for open-sea denitrification. Thus, according to Gran, denitrification might proceed near the coasts, but it was unlikely to be important in the open sea, where energy sources, such as organic detritus, appeared to be low.[75]

When Baur left the Zoological Institute in 1901 he was replaced by Rudolph Feitel, an assistant in the institute, who set out to determine whether or not Baur's results applied to the open sea as well as to isolated embayments such as Kiel Fjord. Using samples that Apstein had taken for him on *Holsatia* in 1901 and on the first three cruises of the new research vessel *Poseidon* in 1902, Feitel began a lengthy program of work on denitrifying bacteria found in the Baltic and in the North Sea.[76] This lengthened into a two-year study of denitrification by Brandt and Feitel beginning in 1903 which dealt with the horizontal and vertical distribution of denitrifiers, their relative abundance, nutrient preferences, temperature relations, and response to oxygen.[77] The results were clear. Denitrifying bacteria were widespread in the coastal and open waters of the Baltic and North seas, both near the surface and in the sediments. And the effect of temperature was just as Brandt had predicted: at 0°C denitrification took months rather than days. Feitel had no doubt of either the general distribution or the importance of denitrifying bacteria in the sea. As he said, "the result of the investigation must be considered a full confirmation of the conception expounded by Brandt."[78]

Opportunities to look for denitrifying bacteria in far more remote oceanic locations also arose during the years. During the preparation of Erich von Drygalski's German South Polar Expedition in 1901, Erwin Baur trained the ship's physician, Hans Gazert, in bacteriological techniques. During the voyage (1901–1903) Gazert took samples throughout the water column and from deep-sea sediments which he examined for bacterial abundance, nitrification, and denitrification. His results (pub-

75. Gran 1902b. Gran and Baur used similar but not identical techniques. Baur's media contained protein (sometimes carbohydrates or alcohols) as energy (carbon) sources, Gran's contained only organic salts.
76. Feitel 1903, pp. 94, 97–98.
77. Brandt 1921a, p. 181.
78. Feitel 1903, p. 101.

lished in 1903, 1909, and 1912) showed only limited evidence of denitrification; in particular there was none in sediment samples and very
little in bottom or muddy waters. Only 11 of 78 samples from the open
Atlantic showed noticeable denitrification. Gazert concluded that true
denitrifying bacteria were scarce in the oceans, whether one looked in the
tropics or at high southern latitudes.

Gazert's results, which displeased Brandt and were apparently dis-
trusted by Gazert himself, made it imperative to try again. Another ship's
physician was pressed into service, this time Dr. Gräf, who accompanied
SMS *Planet* during its circumnavigation in 1906–1907. Recruited by
Brandt, then trained by Feitel, Emil Raben, Bernhard Fischer, and Apstein, Gräf did chemical and bacteriological studies on board; the latter revealed 17 new species of denitrifying bacteria in surface waters
throughout the world, some of which occurred in bottom water deeper
than 4000 meters.[79] This was the strongest possible support for Brandt's
hypothesis. Gazert's negative results were thrust aside, credited to defective technique; denitrifying bacteria did occur throughout the oceans and
could take their place honorably in Brandt's hypothesis.[80]

Nitrogen fixation too proved to be a process that occurred in the sea.
Joseph Keutner, a doctoral student in the Botanical Institute at Kiel began
work on marine nitrogen-fixing bacteria in 1902. In 1903, he and Wilhelm Benecke reported that the well-known nitrogen-fixing genera *Azotobacter* and *Clostridium* occurred in the sea. Keutner amplified these
results in 1905. He concluded that nitrogen-fixing bacteria were widespread all over the world in marine sediments, on attached algae, and in
the plankton of marine and freshwaters. They appeared to be the same
species of *Azotobacter* and *Clostridium* known from land; *Azotobacter*
in particular was tolerant of changes in salinity. Keutner's colleague Max
Keding provided more information on *Azotobacter* in the Baltic the next
year, showing that it occurred in algal slime, was tolerant of salt water,
and retained its activity in pure culture. These results, of only minor
interest to Brandt, played little part in the development of his ideas about
the nitrogen cycle because biological nitrogen fixation was believed to be
far less important than river inputs and atmospheric nitrogen fixation as
a primary source of nitrogen reaching the oceans.[81]

79. The cruise of SMS *Planet*, the ship, and its equipment are described by Lubbert 1909.
Only two scientists were aboard, Gräf and the physical oceanographer Wilhelm Brennecke. On
the results, see Gräf 1909; Brandt 1921a, pp. 183–184.
80. The young English biologist Harold Drew (1911, 1913) also found denitrifiers in the
open ocean both in the English Channel and the American tropics. He linked them to the
precipitation of calcium carbonate from solution in tropical regions.
81. Nitrogen fixation was regarded for many years as a minor process in the oceans. Work
begun in the 1960s has shown that significant amounts of nitrogen may enter the seas by

Although denitrification and nitrogen fixation (as processes confirmed by reputable biological work) fell into place quite neatly and were accepted as widespread in the oceans, other processes of the nitrogen cycle were less tractable. This was particularly true of nitrification, which played its role as a counter to denitrification in the sea. Observing in 1911 that there was an inverse relation between the amount of nitrate and temperature in seawater, Brandt speculated that when water temperatures fell in the autumn, denitrification would cease, whereas nitrification, which probably proceeded at lower temperatures, would begin to restore nitrate to the surface waters. In fact, from the beginning of his denitrification hypothesis more than a decade earlier, Brandt had assumed that nitrification must occur in the sea just as it did on land. But finding evidence that the process was widespread in the open sea, especially in the water column, proved to be extraordinarily difficult.

Among Baur's first discoveries for Brandt in 1900 was the detection of nitrifying bacteria in nearshore sediments at Kiel.[82] Earlier, H. L. Russell had noted the ability of marine bacteria to produce nitrate, as had Vernon in his ambiguous studies at Naples. Gazert, when searching for denitrifiers on *Gauss* in 1901–1903, had found no nitrifiers, but his results were suspect. More telling and harder to explain away were H. H. Gran's inability to find nitrifying bacteria except very close to shore in Norwegian coastal waters, reported in 1903, and Alexander Nathansohn's (see Chapter 3 near n. 71) statement, resulting from work in Naples in 1901–1903, that he could find no nitrifiers whatsoever there except where run-off from the city entered the sea, a result he attributed to the low ammonia content of the sea.[83]

At Brandt's instigation Peter Thomsen of the Botanical Institute at Kiel undertook a thorough study of nitrification in 1906, using samples of seawater and sediments from Kiel, Naples, and Helgoland. His results, when they appeared in 1910, added detail but did not clarify the basic problem of nitrification. Nitrite-forming bacteria could be detected in sediment samples from the three locations but not in seawater, on attached algae, or in plankton samples. They appeared to be ubiquitous in the sea. But nitrate-formers were absent away from shore; Thomsen found them only "in the greatest proximity to land,"[84] just as Gran and Nathansohn had done before him. Thomsen took the further step of

---

fixation, especially in the tropics. The main organism involved is a pelagic cyanobacterium, *Oscillatoria* (formerly named *Trichodesmium*). See Capone and Carpenter 1982, Carpenter and Capone 1983.

82. Brandt 1902b, pp. 72–73.
83. Nathansohn 1906b, pp. 357, 367–368.
84. Thomsen, 1910, p. 17.

isolating the nitrite and nitrate-forming bacteria from his samples. They proved to be morphologically and functionally identical to the well-known *Nitrosomonas* and *Nitrobacter* from land, and like them had high-temperature optima. His results supported the belief held by Gran and Nathansohn that nitrifying bacteria, especially nitrate formers, were carried to the sea from land, where they might survive close to shore as expatriate terrestrial organisms.

Between the time of Thomsen's work in 1906 and Brandt's last summary of his views on the nitrogen cycle in 1927 little changed, although interest remained high in the ways nitrate came to be in the open sea. The Russian microbiologist B. L. Issatchenko, in a series of studies published between 1908 and 1926, found nitrification only in sediments and near-bottom water of the Russian arctic coast, the Black Sea, and the Sea of Azov, while in Dutch waters F. Liebert failed to find nitrifiers in the open North Sea but found them close to shore and in the Zuiderzee, where, he suspected, nitrification might be a purely chemical rather than microbiological process.[85] Searching in the tropical Pacific and Florida waters, the American C. B. Lipman reported no nitrification in tropical near-shore water but found it in the calcareous sediments. At much higher latitudes in the Pacific, Cyril Berkeley had failed to find nitrifiers in the Strait of Georgia, British Columbia, in 1919. And at Plymouth, H. W. Harvey reported in 1926 that, when he enriched seawater samples from the English Channel with ammonium-nitrogen and stored them in the dark, nitrate appeared after a few days only in the samples taken from near the bottom. Nothing happened in poisoned samples. "This very important result" convinced Brandt that classical nitrification (bacterial conversion of inorganic ammonia to nitrate) occurred only near or in the sea bottom.[86] Gran's influential conclusion in 1912 that no nitrification occurred in the open sea appeared to be borne out by all the accumulating evidence during the next decade and a half, as it was indeed by research in the 1930s after Brandt's death.[87] As late as 1946 the American microbiologist Claude Zo Bell summarized knowledge of marine nitrification: "Although, in the light of circumstantial evidence, it is tacitly assumed by many oceanographers that nitrate is formed by bacteria at the bottom of the sea, and thence carried into the photosynthetic zone, there is no conclusive evidence bearing on either the mode or place of formation of nitrate in the sea."[88]

85. Liebert 1915; see also Brandt 1920, p. 219; Brandt 1927, p. 245; Waksman, Hotchkiss, and Carey, 1933, p. 140.

86. Brandt 1927, p. 246.

87. See esp. Waksman, Hotchkiss, and Carey 1933, pp. 138–146; Kreps 1934; Gilson 1937; and Carey 1938.

88. Zo Bell 1946, p. 154.

But the distribution of nitrate in the deep waters of the open ocean provided evidence, in Brandt's eyes, for some form of nitrification in the water column, despite the discouraging results from microbiology. J. Gebbing's analysis of the seawater samples from *Gauss* (1901–1903), which he carried out under Raben's direction in the chemistry division of Brandt's Internationale Meereslaboratorium at Kiel between 1904 and 1907, provided the first stimulus for a new way of thinking about nitrification in open water. Gebbing's results, combined with those of Gräf from *Planet* (1906–1907) showed that although the surface waters of the Atlantic, especially at low latitudes, had very little nitrate, the deep water was exceptionally rich.[89] Below a few hundred meters nitrate was abundant everywhere; there was even evidence of a maximum amount of nitrate at 800 meters followed by a slight decrease in deeper water.[90] Despite the evidence to the contrary, Brandt never abandoned the idea that nitrification must occur in the water column, but if it was not based on ammonia, which the experimental results indicated, there might be another source, which in his last monograph Brandt suggested could be organic nitrogen compounds.[91] Ammonia-oxidizing bacteria were indeed absent in the open sea, but the vertical profiles of nitrate in the sea could not be wholly attributed to diffusion from the bottom and transport by currents from coastal waters. According to his last results, compounds of dissolved organic nitrogen were abundant in the sea,[92] but no attempt had been made to see if they could be nitrified. If they could be, the problem of nitrification would be solved.

Time has not been kind to Brandt's attempts to account for nitrification in the sea, but, in fairness, his hypotheses stemmed from reasonable interpretations of the microbiological evidence available between 1899 and 1927. In fact, the problem of nitrification remained unsolved for another forty years, until nitrifiers were first unequivocally isolated from the open sea and when it became clear in the 1980s that their peculiar sensitivity to light had made their transformations of reduced nitrogen difficult to pinpoint earlier.[93] Denitrification, after all, was at the core of Brandt's scheme, not nitrification. He could afford to put aside the problems of nitrification in building a picture of the nitrogen cycle adequate for his purposes, which were to explain both regional variations in the abundance of plankton and the seasonal cycle.

89. Gebbing 1909, pp. 160–171; Gebbing 1910, pp. 59–62; Gräf 1909; Brandt 1915, pp. 33–48.
90. Brandt 1915, pp. 33–34.
91. Brandt 1927, pp. 246–248.
92. Brandt 1927, pp. 248–269.
93. See esp. Olson 1981 a, b; and Ward et al. 1982 for recent work and references to selected older work.

To solve the third problem in the development of his denitrification hypothesis—distribution of nitrogen-containing plant nutrients, plankton, and denitrifying bacteria—Brandt needed a suite of chemical techniques for measuring the dissolved nitrogen compounds in seawater in order to determine whether the distribution of nutrients in the sea was consistent with the denitrification hypothesis. When Brandt first presented his denitrification hypothesis in 1899, there were neither highly reliable techniques nor large amounts of data on nutrients in the sea which could be used to evaluate and bolster his theory.

The few trustworthy values available before 1904 originated with Konrad Natterer (1860–1901), Professor extraordinarius of chemistry at Vienna and director of its Second Chemical Laboratory, who had accompanied Austrian naval ships to the Mediterranean, the Sea of Marmara, and the Red Sea on six cruises between 1890 and 1896.[94] Natterer tested water and sediment samples from all depths with the established techniques—the Nessler reaction for ammonia, zinc iodide with starch for nitrite, and diphenylamine plus sulfuric acid for nitrate (a test that was suspect, for it responded to nitrite as well as to nitrate). Ammonia proved to be widespread at low levels throughout the water column, even abundant in some sediments, whereas nitrite (except for a few ambiguous traces) was absent everywhere. By contrast, nitrate was reasonably abundant in deep water (below a few hundred meters), though absent or sparse in the surface layers.[95] Although Brandt recognized the relevance of Natterer's results—and accommodated them to his denitrification hypothesis—he recognized that Natterer's techniques were being used at or below their limits of sensitivity. New more sensitive methods were imperative to allow reliable, large-scale surveys of soluble nitrogen compounds to be made.

These new methods and a large amount of new data on the distribution of nutrient salts became available during the first few years of the twentieth century because of Germany's membership in a new organization, the International Council for the Exploration of the Sea. The Kiel school's research profited directly from programs that originated in Scandinavian scientific and economic concerns during the late nineteenth century.

94. Natterer 1892 a, b; 1893; 1894; 1895; 1898. Most were reprinted in the *Monatshefte für Chemie* shortly after they appeared. Most of Natterer's work was based on cruises of the ship *Pola*, described by its captain Wilhelm Mörth in 1892.
95. Summarized by Brandt 1902b, pp. 55–62.

3 International Oceanography, the Kiel School, and the Fate of Brandt's Hypothesis

> Hydrographic researches ought to have as object the distinction of the different layers of water according to their geographical distribution, their depth, their temperature, their salinity, the gas content, [their] plankton and currents, with the aim of discovering fundamental principles, not only to determine the environmental conditions [affecting] useful marine animals but also for extended meteorological predictions in the interest of agriculture.
>
> RESOLUTION OF STOCKHOLM CONFERENCE
> 1899, FROM MAURICE 1928

German scientists were not alone in being interested in the quantitative study of the sea during the first few years of the twentieth century. The early development of Brandt's hypothesis coincided with a major program of collaborative marine science in northwestern Europe, originating in Scandinavia. This took institutional form in the International Council for the Exploration of the Sea (ICES), whose first meeting was held in Copenhagen in July 1902.

ICES originated in national concerns about the state of fisheries, about the links between the sea and the weather, and in the belief that the scale of these problems was so great that only collaborative programs by several nations could solve them. Germany, like Britain and the Scandinavian nations, had special interests in a broad study of the seas, for its fisheries were believed to be underdeveloped in the late nineteenth century, and certainly were far less scientifically based, Hensen and Brandt claimed, than agricultural production.

Brandt's hypothesis was explicitly an attempt to bring the principles of scientific agriculture to bear on the problem of production in the sea. Fortuitously, it was formulated just as ICES was formed; Brandt quickly took advantage of the special opportunities provided by the ICES program of research to amass information to bolster his hypothesis. As his ideas developed, incorporating new, more accurate data, the whole oceanographic context in which the denitrification hypothesis developed changed. The ICES program provided, directly and indirectly, a new climate of ideas about the chemical and physical environment in which the marine nitrogen cycle occurred.

The varied fortunes of two major fish stocks, Arctonorwegian cod and Atlantoscandian herring, provided the initial stimuli that led to the formation of the International Council for the Exploration of the Sea at the end of the nineteenth century. Especially in Scandinavia, cod and herring were the economic mainstay of coastal peoples, and both fisheries fluctuated violently, leading to economic and social distress of large proportions.

Arctonorwegian cod, which have been fished around the Lofoten Islands of Norway (especially in the Vest Fjord) since the early middle ages, spawn in that area from January to April. Their floating eggs develop into larvae that grow rapidly as they drift northward in the Norwegian coastal current. By June, immature cod have drifted north to the Svalbard shelf and the Barents Sea. There they occur at the boundary of the warmer currents and the cold arctic water mass, staying in relatively warm water of greater than 2°C. These nursery areas vary in position from year to year, depending on hydrographic fluctuations, but the principle holds that the young cod mature in the north (where they are fished off Spitsbergen and in the Barents Sea), then migrate south as adults to the spawning areas of the Vest Fjord, where many are captured by the traditional Norwegian "spring" fishery.

The second, historically important stock of fish on the Norwegian coast, Atlantoscandian herring, breeds at several locations between Bergen and the Lofoten Islands between February and March in water of 3 to 13°C. The immature herring, like the cod, drift north, then northeast, to the polar front between Jan Mayen and Iceland. The traditional Norwegian fishery occurs on spawning herring ("spring herring") in the south, along the Norwegian coast, and on young, supplemented by Icelandic stocks, off Iceland. Farther south, the Swedes too fish herring, when, at irregular intervals, herring enter the Skagerrak, Kattegat, and eastern Baltic allowing a fishery off Scania. This fishery, one basis of the wealth of the Hanseatic League, is probably on North Sea herring (rather than Norwegian ones), which are most abundant around southern Sweden in cold years when the Norwegian fishery fails.[1]

1. Information on the cod and herring fisheries is provided in Cushing 1975, and esp. Cushing 1982, pp. 9, 34, 53–55, 78–90. See also Cushing 1966 on Arctonorwegian cod.

Fluctuations in the abundance of cod and herring around Scandinavia have been noted for centuries, but until the nineteenth century little could be done to examine their causes. During the early years of that century Norwegian cod stocks were fished heavily, beginning just after the Napoleonic Wars, a period that coincided with the warming of European seas at the end of the Little Ice Age. The expanding fishery of the 1840s and 1850s was checked in 1860, when it inexplicably failed in the Vest Fjord. As a direct result, G. O. Sars was appointed in 1864 by the Norwegian government to investigate the fishery and to explain its fluctuations. Sars's studies led rapidly to his famous discovery that cod eggs, contrary to expectation, float, and are carried great distances in the coastal current.[2] Thus it might be necessary to look for the causes of fluctuations in the fishery far from the spawning grounds, where hydrographic conditions might affect the success of the larval and juvenile fish. For this reason, as Sars recognized, the study of cod had to move to the open ocean. He began to cooperate with the director of the Norwegian Meteorological Institute, Henrik Mohn, who suspected that there was a close link between hydrographic changes (especially sea temperature) and the climate of Norway.[3] Mohn's collections of temperature data compiled from observations made by sealers in the Norwegian Sea and around Spitsbergen were used by Sars in 1893 to show that herring catches were linked with warm water off the Norwegian coast.[4] A few years later Sars and Mohn organized a series of summer expeditions on the Norwegian ship *Vøringen* (the Norwegian North Atlantic Expeditions of 1876–1878), which showed, among many other significant results, that young cod from the Lofoten Islands were carried north to Spitsbergen.[5]

After the crash of 1860, the Norwegian cod fishery fluctuated for the next thirty years, until in the mid-1890s it reached a critically low level again; between 1893 and 1899 the catch was halved, coinciding with a lengthy period during which the herring fishery was also poor off the

2. Sars 1869, 1877. For a brief account of the background see Rollefsen 1966 and Solhaug and Saetersdal 1972, pp. 401–403. Sars's career is described by Nordgaard (1918), Calman (1927), and Broch (1954).

3. On Henrik Mohn (1835–1916) see Spiess 1935, Thorade 1935, Pedersen 1974, and Kutzbach 1979, pp. 240–241. He was director of the Norwegian Meteorological Institute in Oslo, 1866–1913, and professor of meteorology at the University of Oslo. Mohn was among the first to use mathematical methods (derived from meteorology) to calculate oceanic circulation.

4. On Mohn's early data see Helland-Hansen and Nansen 1909, p. 6; on Sars's use of them see Solhaug and Saetersdal 1972, p. 402.

5. See Solhaug and Saetersdal 1972, p. 403. Mohn used the physical oceanographic observations to develop the first mathematical model of the circulation of the Norwegian Sea (Mohn 1885, 1887). Sars's opinions that the fluctuations of the herring fishery were due to climate and that the sea affected the weather and thus the well-being of the country are quoted in Wille 1882, p. 4.

Norwegian west coast. By 1900, at the end of a cold period in northern Europe that had begun in 1878, there was famine in Norway as well as economic distress for farmers and fishermen. In 1900, Johan Hjort, who had succeeded Sars as research fellow in fisheries at the University of Oslo in 1894, was appointed Director of Fisheries for Norway. "It was in these circumstances," he said, "that the Norwegian Government requested me to attempt to find out the causes of these failures." A ship for fisheries research, *Michael Sars*, was provided in 1900; from then until 1914, under Hjort's direction, an increasing number of scientists (among them H. H. Gran) assessed the state of Norwegian fish stocks and carried out basic biological research on the marine biota of the northeastern Atlantic.[6]

Hjort neatly summarized the popular Norwegian attitude to marine science at the end of the nineteenth century:

> The main question for the Norwegian fisherman and perhaps the only scientific problem which could interest him was at that time whether science could enlighten him as to the puzzling changes in the occurrence of the shoals of fish. His experience and that of his forefathers through a thousand years had taught him that the large spawning shoals of herrings and cod in the spring months occur for a short period and are concentrated upon small areas, after which they disappear to unknown localities. Experience had moreover for hundreds of years shown the fishing population that the yield of the fishery fluctuated from year to year or from one group of years to another.[7]

Herring provided a problem even more significant than cod, for they were caught not only along the Norwegian coast, but for a few decades every century off the southern coast of Sweden, where their economic significance had been great since at least the fourteenth century.[8] As climate warmed during the mid-nineteenth century, herring appeared to be moving north along the Norwegian coast, permitting a fishery that peaked during the 1840s but remained significant, though variable, during the 1850s and 1860s. Farther south, the Scanian fishery failed after 1808 and remained insignificant during the following seven decades. After 1870 the Norwegian spring herring fishery began to decline, coin-

---

6. Solhaug and Saetersdal 1972, pp. 405–406. Lamb (1982, pp. 245–247) describes the cold winters at the end of the century. The tangled background of Norwegian fisheries research late in the nineteenth century is described by Rollefsen 1966 and Brattström 1967, pp. 26–28. The ship *Michael Sars* and its work was described by Hjort 1901, 1928.

7. Hjort 1928, pp. 188–189.

8. Cushing (1982, pp. 78–83) outlines the main periods of the Swedish herring fishery. These were ca. 1307–1362, 1419–1474, 1556–1587, 1660–1680, 1748–1808, and 1877–1896.

ciding with colder winters that began during 1878–1879. At the same time the Scanian herring returned, providing rich catches between 1878 and 1896, then somewhat lower ones until 1906.[9] The implications for Sweden were positive, for Norway negative; off Norway the herring fishery declined from south to north, until Norwegian fishermen were forced to fish off Iceland or to turn to other species, such as North Sea mackerel.[10]

The fluctuations of cod and herring that occurred during the late nineteenth century affected most of the fishing nations around the northeastern Atlantic to a greater or lesser degree. There was widespread agreement then in Germany that both the North Sea and Baltic fisheries were depressed, or at best underdeveloped, due to the lingering effects of the Napoleonic Wars, bad relations with Britain, and the instability of the German states before and during the early years of the Prussian hegemony. The economic collapse of 1873, the "Great Depression" that lasted (although with irregular and lessening intensity) until 1896, further depressed the fishery, in concert with other primary industries such as agriculture.[11] Both Victor Hensen's early political activities and the formation of the Deutsche Fischerei-Verein in 1870, which led to the formation of the Kiel Commission, were responses to the economic underdevelopment of the North German fisheries and to variations in fish stocks of unknown sizes and uncertain habits.

The situation in Scandinavia (and the Nordic countries in general) was far more serious. With low, scattered populations, the prevalence of primary industries such as forestry, agriculture, and fishing, and their dependence on export trade, the Nordic countries were perhaps the poorest in western Europe during the nineteenth century.[12] As a case in point, Norway was largely agrarian; its farmers, mostly coastal folk, depended on the cod and herring fisheries for a seasonal food supply. Between 1866 and 1900 about 10% of all adult agrarian workers were also fishermen, catching cod and herring, a fishing population that in the mid 1880s numbered between 120,000 and 130,000. Until 1870, when industrialization began to increase in Norway, fish exports made up 45% of the value of all exports, a figure that had decreased only to 35% at the

9. Solhaug and Saetersdal 1972, pp. 400–403.

10. Solhaug and Saetersdal 1972, p. 403. In fact, the results were not all bad, for Norway's high seas fishing fleet began to develop under the stimulus of failures in the coastal cod and herring fisheries after about 1879.

11. See Reincke 1890, Dittmer 1902, Henking 1910, and Illing 1923a, b for contemporary opinions of the fishery and responses to the problems. The depression had widespread consequences throughout Europe. Rosenberg (1976) discusses its effects on Germany, and Stone (1983) shows how the depression had its greatest effect on landowners and commodity prices.

12. The discussion is based on Jörberg 1973.

beginning of World War I. In Sweden, 72% of the population depended on agriculture and fisheries in 1870, 48% in 1910, although the importance of fishing alone was lower than in Norway during the nineteenth century because of the failure of the Scanian herring fishery between 1808 and 1810.[13]

Nordic populations seem to have been poised on the brink of famine and economic crisis throughout the nineteenth century. Major crop failures occurred in the 1860s; 20% of the population of some regions of Finland died during those years. Profound consequences also ensued in Sweden. The rate of increase of Norwegian population decreased during the same period; this coincided with the first major failure of the Lofoten cod fishery, soon to be followed by the failure of the spring herring fishery.[14] Agricultural failure was also frequent during the cold years of the 1880s and 1890s, exacerbated in Norway by the poor state of the cod and herring fisheries. These factors, combined with the depressed economy of Europe after 1873, led to a major emigration of Norwegians, Swedes, and Finns to North America late in the century. They also resulted in major investigations of the links between the oceans, climate, and the fisheries beginning during the 1870s. As late as 1909, Bjørn Helland-Hansen and Fridtjof Nansen expressed the Scandinavian hope, born in economic hardship and famine, that physical science could help to ameliorate conditions in Scandinavia, claiming that

> of more general interest [than hydrography alone] are perhaps the annual variations in the currents . . . and their relations to the variations in the climate of Norway, the variations in the fisheries, and also the variations in the harvests of Norway, the growth of the forests, etc. We have also been able to trace a certain relation between these variations and cosmic causes. We think that these discoveries give us the right to hope that by continued investigations it will be possible to predict the character of climate, fisheries, and harvests, months or even years in advance.[15]

This sanguine view had its origins thirty years before in Sweden, when the herring fortuitously returned to the Skagerrak in 1877–1878. Gustaf Ekman, who had worked on the physical oceanography of Swedish waters, was asked to investigate the causes of the striking fluctuations in the stocks in relation to hydrography[16]—a particularly Swedish preoc-

13. Jörberg 1973, pp. 375–376, 403–404, 432.
14. Jörberg (1973, pp. 380–381 [esp. diagram 1], 399) outlines the demographic results of the famine.
15. Helland-Hansen and Nansen 1909, pp. iv–v.
16. Pettersson 1930, p. 288.

cupation that came to have major implications for an international community of marine scientists two decades later.

The reasons for the beginning of physical oceanographic work in Sweden during the middle of the nineteenth century are obscure. What is clear is that the pioneering early efforts were those of Frederik Laurentz Ekman (1830–1890), professor of chemistry at the Stockholm Polytechnic, who, about 1869, began to study the temperature, salinity, and currents of Swedish coastal waters. He designed special equipment for these studies, including a new water-sampling bottle[17] and made major surveys, in addition to devising physically based but essentially qualitative theories of oceanic circulation. His major works, published in 1875 and 1876, described the relation between inflowing and outflowing estuarine waters and accounted for the general circulation of the oceans.

It was F. L. Ekman who brought his young relative Gustaf Ekman (1852–1930) into oceanography. The younger Ekman, while studying chemistry at Uppsala in 1876, working from a yacht with a friend, made temperature and salinity sections of the Skagerrak, Kattegat, and eastern Baltic.[18] In the following year the two Ekmans undertook a major hydrographic expedition sponsored by the Swedish Academy of Sciences. Using the naval ships *Alfhild* and *Graf Klint* they traveled 2100 nautical miles, making 1800 temperature and salinity determinations along transects of the Baltic Sea.[19] It was largely as a result of this work, unprecedented in scope, that Gustaf Ekman was invited to investigate the fishery when herring reappeared around southern Sweden in 1878.

Another collaborator entered the scene in the 1870s. Otto Pettersson (1848–1941), a physical chemist and later professor of chemistry in his turn (1881–1909) at the Stockholm Polytechnic, first met Gustaf Ekman in 1872 at Uppsala.[20] But Pettersson's entry into oceanography occurred a few years later, after he published a paper on the properties of ice which was seen by the arctic explorer Otto Nordenskiöld. At Nordenskiöld's request Pettersson worked on the hydrography of the *Vega* expedition through the Northeast Passage, a study that led eventually to his collaboration with the two Ekmans in a major Swedish hydrographic expedition in 1890. Despite F. L. Ekman's death that year, the study proceeded, using five steamers in the Skagerrak and Kattegat. During one week in

17. McConnell 1982, pp. 122–124.
18. Pettersson 1930, p. 288.
19. F. L. Ekman and Pettersson (1893) describe the work and Ekman's water bottles.
20. Pettersson's career, which was critical to the development of oceanography in western Europe, is outlined briefly by Thompson 1947 and Went 1972b, pp. 167–168. More detail is provided by Hans Pettersson 1923.

February Pettersson and Gustaf Ekman occupied 70 stations, taking 1000 water samples. These became the basis of a prize-winning paper by Pettersson on the hydrography of southern Swedish waters published in 1891.[21]

By 1892 the Danes too were involved in hydrographic work in the Baltic region and its approaches. Pettersson and Gustaf Ekman, because of their interest in the effect of climatic fluctuation on fisheries and agriculture, recognized that only international collaboration could provide enough information to establish connections among sea, climate, fisheries, and agriculture. The first step was to make a concerted attack on the sea, apportioning large regions to countries that could carry out detailed surveys. Such a program was proposed by Pettersson in 1892 when a group of Scandinavian scientists met in Copenhagen. The "Copenhagen Program," with Pettersson its dominating force, began in 1893 when Swedish, Norwegian, Danish, German, and Scottish scientists made quarterly hydrographic sections in the Baltic, Kattegat, Skagerrak, and North Sea based on a protocol established earlier by the Danish Hydrographic Office.[22]

The results were considerable: a large amount of new hydrographic data on the seas of northwestern Europe gathered in an international program that united the varied scientific and economic interests of five nations. But Pettersson's ambitions were greater. He wrote in 1894 that the Copenhagen Program

> is naturally only a *preliminary programme for an international hydrographic survey*. . . . I hope that the experience gained from this scientific cooperation will lead to an international agreement about the division of labour, and satisfactorily settle the question of methods and measures to be adopted in the course of future hydrographic survey.

Otto Pettersson wasted little time beginning a campaign for an expanded scheme of international marine research based on the Copenhagen Program. He outlined his ideas to the Sixth International Geographical Congress, which met in London in 1896, winning its support.[23] But the precise mechanism by which a collaborative program could be orga-

21. Pettersson 1894, p. 281; Thompson 1947, p. 122.
22. Pettersson 1894, pp. 282, 631–634. Pettersson, Gustaf Ekman, and August Wijkander were appointed by the Royal Swedish Academy of Science to direct the Swedish research in the Baltic for a year beginning in May 1893. The first cruise took place in November 1893. Pettersson's early interest in climate and the ocean (which probably went back to the 1880s) was expressed in a major paper in 1896 making the claim that fluctuations in the Gulf Stream had an effect on Scandinavian weather. See also his well-known papers of 1915 and 1926, as well as the lesser known one, 1905b.
23. Went 1972b, pp. 3–4.

nized took some time to work out, for Pettersson was not the only proponent of a major program of hydrographic and biological research.

In 1897 John Murray of Britain had agreed with Johan Hjort and Fridtjof Nansen of Norway that an international conference to organize a major program of research should be held soon, if not in Christiania then in Edinburgh. Nansen, writing to Pettersson in April, 1898, promised Norway's cooperation in an international program that might be organized in Britain, Norway, or Sweden. Later that year, Walther Herwig, H. Henking, and Friedrich Heincke of Germany and P. P. C. Hoek of the Netherlands met, agreeing to ask the German and Dutch governments to sponsor a conference to organize a program of international research. The situation was complicated by the deteriorating situation in South Africa, which led to the Boer War of 1899 to 1902. Anti-British feeling was strong in the Netherlands and Germany; moreover it was not certain that Britain would join an organization in which Germany and the Netherlands played a part.[24] This problem was resolved when the Swedish government, through its Department of Foreign Affairs, invited delegates from Germany, Denmark, Sweden, Norway, the United Kingdom, Holland, and Russia to attend a conference, beginning on June 15, 1899, in Stockholm, that would undertake the organization of "international exploration of the Arctic Ocean, the North Sea, and the Baltic, in the interest of fisheries."[25] This conference, nominally the idea of the philanthropic King Oscar II of Sweden,[26] bore the stamp of Otto Pettersson, Johan Hjort, and Fridtjof Nansen; as Hjort later expressed its beginnings, "The international cooperation in north European waters began on a very small scale and chiefly on the basis of personal friendship between scientists of the three Scandinavian countries."[27]

The program proposed at Stockholm in 1899 was for the foundation for at least five years of an International Council for the Exploration of the Sea, to which each participating country would appoint two delegates. The council would establish a central office and laboratory, coordinate quarterly research cruises by each nation, and collect statistics on fisheries. Its research activities would be twofold, hydrographic and biological; each nation would be responsible for work in a specially designated area of the northeast Atlantic Ocean or the Baltic Sea. The

24. Went 1972b, pp. 4–5. A conventional institutional history of ICES is given by Went (1972a, b) and other details are in Jenkins 1921. A sociopolitical analysis of ICES is needed.

25. Maurice 1928 (I, p. 4).

26. The role of Oscar II, whose greatest interests were in geographical exploration and religious philosophy, is only hinted at in eulogies like those of Böttiger and Nathorst (1928). The king, and his Department of Foreign Affairs, appear to have been figureheads for Pettersson and Ekman.

27. Hjort 1945, p. 3.

delegates at Stockholm proposed that the council and its activities begin formally on May 1, 1901.[28]

It was in fact 1902 before the council was established because of such politically delicate problems as the location of the central office and laboratory, the negotiation of financial contributions from each nation, and the need to assure that ships would be available for the work at sea.[29] Many of them were worked out, or had just been resolved, when, at the invitation of the Norwegian government, a second organizational conference took place in Christiania in May 1901 with the original seven nations plus Finland and Belgium attending. The delegates reiterated their wish to have a five-year program of research around Europe, appointed a slate of officers, and established the central office in Copenhagen and the international laboratory in Christiania (under Nansen's direction).[30] With minor changes the Stockholm program of 1899 was to go ahead, it was hoped, with much of the work at sea done on specially built steamers from each nation.[31]

ICES was inaugurated in July 1902 in Copenhagen, with delegates present from Germany, Great Britain, Denmark, Finland, Norway, the Netherlands, Russia, and Sweden. Within a few months, in August 1902, the first regular research cruises began (although Germany had conducted one in May using its new ship *Poseidon*). The practical orientation imposed by each nation on its delegates was clear, for according to Hoek, the general secretary of the conference, the main object of ICES "was the practical benefit of the fisheries and the obtaining of a scientific and economically correct basis for international legislation." Within this general framework, the main problems to be attacked were "the migrations of the principal food fishes of the North Sea, especially of the herring and cod, [and] the question of the overfishing of those portions of the North Sea, the Skager Rak and Kattegat mostly fished over by trawlers, and especially with regard to the plaice, sole and other flat-fish

28. Maurice 1928 (I, pp. 6–19); Anonymous 1899; Went 1972b, pp. 6–8.

29. See Went 1972b, pp. 8–9.

30. A good deal of the early difficulty in getting ICES going was the problem of where to put the central office and laboratory. This was complicated by Norwegian nationalism, which was at its peak just before Norway left the Swedish Crown in 1905. Nansen, among others, believed that the laboratory, which was to standardize instruments and carry out original research, should be close to deep water. Eventually a compromise was reached, under pressure from Norway and Denmark, by which the central office was in Copenhagen and the laboratory (paid for by the Norwegian government) was in Christiania with Nansen as unpaid director. There it remained until 1908, when it was disbanded; some of its functions (e.g., the production of standard seawater) were transferred then to Martin Knudsen in Copenhagen. On the negotiations and the laboratory, see Went 1972b, pp. 8–12, 147–153.

31. Maurice 1928 (I, pp. 19–23); Went 1972b, pp. 10–12. The steamers and personnel of each nation during the early years are listed in Anonymous 1905.

as well as the haddock."[32] Each nation used its specific concerns to justify membership in ICES. Britain, which was worried by the decrease of demersal fish off its shores and by the depredations of foreign trawlers outside its three-mile limit, told its delegates to "propose that the scientific investigations shall be accompanied by a practical exposé of the steps to be taken in order to bring the exercise of sea-fishing more in accord with the natural conditions regulating the growth and increase of fish, and thus permanently increase the supply of fish in the markets of the countries adjoining in the North Sea."[33] The British delegation reported that "the delegates of the other nations we found imbued with the feeling that the International Survey would be discredited before Europe if in a brief number of years it did not give some striking proof of the assistance that science could render to the fishing industry." Walter Garstang, one of the delegates, made it clear to a parliamentary committee that scientific investigations were absolutely necessary for fisheries control because the Germans, who controlled the haddock and plaice nursery areas in the eastern North Sea, insisted on scientific evidence that trawling in those areas could harm the populations of adult fish.[34] These general British concerns were reinforced by specific Scottish demands that traditional British trawling grounds along the east coast of Scotland be closed to foreign trawlers for the protection of Scottish line-fishermen.[35]

Rather different from the practical demands of the British delegates were the concerns of the Norwegian and Swedish delegates to develop a program that would allow them to predict the fluctuations of cod and herring. The Norwegian Prime Minister, J. Steen, opening the conference at Christiania in 1901, reflected Scandinavian interests when he said that "the goal . . . is . . . to help the folk who sail the sea to become acquainted with and to choose the area of their activity, just as does the farmer his field and his meadow."[36]

These aims, which were readily agreeable to the German delegates, in fact pervaded the ICES program from the beginning. Pettersson's and G. Ekman's long-held interest in the effect of climate and fisheries are evident in the aims stated at the Stockholm conference of 1899, aims incorporated directly into those of ICES when it was formed in 1902. Pettersson's hand is clear in the prescription that ICES hydrographic research should not only indicate the environment affecting animals, but should also be useful "pour les prédictions météorologiques de périodes

32. Hoek 1903, pp. xxvi–xxvii.
33. Went 1972b, p. 6; see also Borley 1928, pp. 152–153.
34. Moncrieff and Thompson 1903, pp. 48, 94.
35. Anonymous 1928, pp. 171–172, 175.
36. Maurice 1928 (I, p. 21).

étendues dans l'intérêt de l'agriculture." The biological work, for its part, should be concerned with "détermination des variations périodiques dans la présence, l'abondance, et la taille moyenne des poissons utiles, et leurs causes."[37] Pettersson was the secretary general of the Stockholm conference, so it is hardly surprising that its communiqué bears his style and concerns. But his interests in climate, the oceans, and fluctuations in the fisheries were assented to, at least in part, because each nation saw the developing ICES program as at least compatible with, if not fully favorable to, its commercial and economic interests in the fisheries of northern Europe. Thus ICES was born by a clever compromise out of Scandinavian concerns that could be shaped to their own interests by the other nations.

Viewed more than eighty years later, it is clear that ICES, which has long outlived its nominal five years, was inordinately significant in the development of marine science after 1900. First, it encouraged systematic observations of biology and hydrography throughout the northeastern Atlantic using standard methods. As a result, large amounts of new data of unprecedented accuracy accumulated, allowing a detailed analysis of seawater chemistry, plankton, and the fisheries from year to year. New techniques and instruments were developed either in the International Laboratory during its short career, or at home in the laboratories of participating nations. And in response to the formation of ICES a series of courses for marine scientists of all nations was taught between 1903 and 1913 at the Bergens Museum by eminent marine scientists such as Johan Hjort, H. H. Gran, Adolf Appellöf, and Bjørn Helland-Hansen.[38] By this means the latest techniques and discoveries passed rapidly to a large scientific community. Both because of the courses and through its yearly meetings, ICES was responsible for the rapid origin and development of an international community of marine scientists, many of whom considered themselves oceanographers, who maintained regular scientific and personal contact. It was also responsible for the origin of another new profession, that of the nationally based technicians who did much of the routine work at sea and in the laboratory.

Norway and Sweden, the founding countries of ICES, had the most to gain from its early programs of scientific work, but Germany was among the first to respond to the call for internationally coordinated research. Karl Brandt, in particular, used the German work for ICES to expand his own research in ways that would have been difficult if not impossible without the equipment and personnel provided by his government for the international investigations.

37. Maurice 1928 (I, pp. 12, 16).
38. Brattström 1967, pp. 29–30.

Germany quickly responded to the formation of ICES by establishing the Deutsche wissenschaftliche Kommission für internationale Meeresforschung, which had representatives from the Deutsche Seefischerei-Verein (its president Walther Herwig and secretary H. Henking), the Biologische Anstalt Helgoland (Friedrich Heincke its director), and the Kiel Commission (Brandt and Otto Krümmel). At the beginning, Herwig and Henking were responsible for fishery statistics, Heincke for the biology of fishes in the North Sea and Baltic, Brandt for general marine biology, and Krümmel for physical oceanography.[39] After 1908, when the regular quarterly cruises ceased, Helgoland took over the plankton and benthic work in the North Sea (keeping the North Sea fisheries), while Brandt took over responsibility for the commercial fish of the Baltic, meanwhile carrying on his work on plankton. Most important of all, the German Internationale Meereslaboratorium for the ICES investigations found a home at Karlstrasse 42, a few streets from Brandt's Zoological Institute. There, in a rented house, it shared quarters with the Kiel Commission's offices and library. In the cramped laboratories shared by the two institutions the biological and hydrographic work of both was carried out; in addition, students came there for lectures and practical work on marine science.[40]

Brandt had hoped for six scientific positions to do the German ICES work but got only three at first. He appointed Emil Raben (1866–1919) to do the all-important chemistry, Carl Apstein to work on plankton, and Johannes Reibisch to do benthic investigations. They were aided by a small corps of assistants supplemented on occasion by doctoral students or assistants from the Zoological Institute. These made up for a chronic lack of personnel in handling the mass of samples coming in every three months.[41]

As early as 1895 the Kiel scientists complained of the difficulty of getting research time on ships. Otto Krümmel, Brandt's colleague in the university and the chief hydrographer of the Internationale Meeres-

39. The German contribution in 1910 is described by Kofoid 1910, pp. 218–221. Brandt 1902a, pp. 290–291; Brandt 1920, pp. 263–273; and Brandt 1921a outline the formation and work done at Kiel under the umbrella of ICES.

40. Brandt 1908, p. 17; Kofoid 1910, p. 220; Reibisch 1933, p. 5; Brandt 1928. The group occupied Brunswickerstrasse 12 until 1905, then moved to larger quarters in Karlstrasse. Karlstrasse 42 existed and was used for teaching as late as 1974 (Gabriele Kredel, personal communication, May 1984). Even the street has now disappeared because of the expansion of the Kiel medical school.

41. Emil Raben (1866–1919), who was born in Hadersleben, southern Denmark, studied pharmacy in Munich and Kiel, practiced as a pharmacist, then in 1899 began to study chemistry in Kiel. After receiving his Ph.D. in 1902, he was hired by Brandt to develop the chemical methods necessary for his plankton investigations. On Raben's early career see Raben 1902. See Brandt 1905a and Brandt 1920, pp. 263–265, on his personnel and the difficulties they faced.

laboratorium, commented then that a full understanding of physical oceanography required studies on the open oceans, yet he had great difficulty getting time on ships except by specially negotiated arrangements with the Kiel Commission or the German Navy. With the advent of ICES, a research vessel, *Poseidon*, was built in 1901–1902 by the German government specifically for the ICES investigations (Figure 3.1).[42] Both Krümmel and Brandt saw this as one of the golden opportunities provided by Germany's membership in ICES. But, although the new ship brought new opportunities, it was much sought after, tied up too much in routine cruises, and underfunded. Even after 1908, when the quarterly cruises stopped, the ship was less available than Brandt had hoped. Nonetheless, by combining work on *Poseidon* with short cruises on other vessels, and by having samples gathered for him on others' expeditions (even the voyage of the *Graf Zeppelin* to Spitsbergen in 1910), he amassed more than 2400 quantitative determinations of nutrients between 1902 and 1917.[43] This collection of analytical data resulted in large part from ICES cruises and from Raben's development of new methods for nitrate, ammonia, silicate, and phosphate analysis. It allowed Brandt, step by step, to erect a complex theory of the nitrogen cycle in the sea which went well beyond mere agreement with the predictions of the denitrification hypothesis.

Brandt's elaboration of his denitrification hypothesis into a theory of the nitrogen cycle in the sea occurred in three stages. Between 1899 and 1905, before much information was available on dissolved nutrients, he developed an outline of the analogy between terrestrial and marine systems, postulated that denitrifying bacteria had a critical role, suggested that dissolved nitrogen was a Liebigian limiting nutrient, and outlined a program of fieldwork that would verify his hypotheses. From 1905 to 1910 the Kiel group was hard pressed to collect and analyze the ICES samples taken between 1902 and 1908. Thus Brandt presented the most detailed accounts of his ideas later, between 1911 and 1920, basing his interpretations on the accumulation of data from the deep oceans all over the world and on a large amount of information about the plankton cycle near Kiel. Then in 1927, after his retirement, when the Kiel group of the earlier decades had dispersed, Brandt began a synthesis of his ideas on the

42. Krümmel (1895, p. 82) complains of the difficulties. *Poseidon* cost Germany 330,000 marks for construction and equipment.

43. Brandt 1925d, pp. 81–82. By far the most important were the Kiel group's own samples from the North Sea and Baltic and Gräf's from the voyage of SMS *Planet*, 1906–1907, part of which he worked up at sea, the rest in the Internationale Meereslaboratorium in Kiel after the voyage. Brandt never willingly acknowledged that the nutrient analyses from *Gauss* 1901–1903 done by Gebbing in the Kiel laboratory were especially useful, but he used them freely with the values from *Planet* to describe the nitrogen cycle in the open ocean (see esp. Brandt 1915, 1927).

FIGURE 3.1 The German research vessel *Poseidon*, built in 1901–1902 for ICES investigations and other work on marine science. From a glass negative in the Institut für Meereskunde, Universität Kiel, courtesy of Heye Ruhmohr; published with permission of the institute.

nitrogen cycle, attempting to incorporate information from North America, Germany, and the rest of Europe into a series of monographs that would give a definitive account of the factors governing the seasonal and geographical abundance of plankton. True to his thorough training as a systematist and comparative physiologist, Brandt set out a program of research in the early years, followed it with nearly obsessive attention, then developed its implications, adding complexity to a framework of theory that actually changed very little from its beginning late in the 1890s.

The first theoretical formulation of Brandt's ideas was quite simple.[44] By analogy with land, nitrogen was a limiting nutrient in the sea. Overall, plant abundance in the sea depended on the input of nitrate from land and its removal by denitrification. Nitrogen was demonstrably a limiting nutrient, for it was scarce in the sea relative to its abundance in organisms (although admittedly silicon was likely to govern diatom abundance).[45] And denitrification did occur in the sea; proof positive had been obtained by Baur, Gran, and Feitel (see Chapter 2 near n. 74) that denitrifying bacteria were widespread there.[46] As far as the evidence went, which was only as far as Konrad Natterer's data on nutrients (see Chapter 2 near n. 94) could be extended, the distribution of ammonia and nitrite in the sea was consistent with the denitrification hypothesis. If the Mediterranean, a warm sea, had more dissolved nutrients than the Baltic, the denitrification hypothesis would be disproved.[47] On the basis of the abundance of ammonia, this was manifestly not the case. In addition, the temperature relations of the bacteria agreed with the hypothesis; between February and August 1903 Brandt and Feitel discovered denitrifying bacteria throughout the water column at Kiel, but in all cases their temperature optima were high. They could be fully active only in summer.

When Germany's ICES investigations began in 1902 Brandt outlined a program for its work. Basic observations of hydrography, nutrients, and plankton were essential, supplemented by benthic studies that would show the role of the sediments—whether as a chemical system or as a habitat for organisms—in governing the decomposition of organic matter and its release as soluble inorganic nutrients to the water column.[48] A newly developed method of Emil Raben's for ammonia and nitrite plus nitrate (also his method for dissolved silicon) analysis had just begun to produce results by 1905, enough to show that nitrite plus nitrate (they

44. See Brandt 1899a, b; 1901; 1902b; 1904; 1905b.
45. Brandt 1902b, pp. 53–72.
46. Brandt 1902b, pp. 46–53.
47. Brandt 1904, pp. 395–396; Brandt 1905b, p. 9.
48. Brandt 1902b, pp. 74–79.

were measured together) dropped in summer and that silicon decreased after the spring diatom bloom at Kiel.[49] The first results of Raben's analyses were consistent with the denitrification hypothesis: as Brandt said in 1905, "The great diminution in the nitrates and nitrites during the warm period of the year 1904 I consider to stand in relation to the great decomposition of the nitrates and nitrites by the so-called denitrifying bacteria, which are more active in warmer than in colder water."[50] Phytoplankton, he believed then, could not or did not deplete nutrients; it was the bacteria that accounted for the disappearance of nitrate during the summer. Progress beyond this could be accomplished in two ways: first, by experiments on the nutrient requirements of phytoplankton,[51] second, and most important in Brandt's view, by increasingly detailed surveys of nutrients and plankton during the seasons in as many different environments as possible.

The ICES investigations between 1905 and 1911, combined with Gebbing's data from SMS *Gauss* and especially Gräf's from *Planet*, yielded all-essential data on nutrients. From the Baltic and North seas alone Raben provided 400 analyses of ammonia and nitrate. The expeditions added about 300 more samples documenting the pattern of nutrients at all depths in portions of the Mediterranean Sea and the Atlantic, Southern, Indian, and Pacific oceans. The fully developed theory emerged then in two installments, the first concerned with the distribution of nitrate in the oceans, the second with the interrelation of nutrients, the law of the minimum, and the seasonal course of the plankton cycle in temperate latitudes.[52]

Brandt's 1915 monograph *Über den Nitratgehalt des Ozeanwassers und seine biologische Bedeutung* put his ideas into a fully oceanographic context. It incorporated the hydrographic information from the cruises of *Gauss* and *Planet* and was inspired, in part, by the criticisms that Brandt's hypothesis had evoked (which are described later in this chapter). In it, Brandt attempted to defend the view that denitrification, combined with oceanic circulation, was a global force controlling plankton production. The first indication that this might be possible appeared in 1911 when Brandt noted, using information from the *Gauss* and *Planet* samples, that the nitrate content of ocean water and its tempera-

49. Brandt 1905b, pp. 7, 10–11.
50. Brandt 1905b, p. 7.
51. Brandt 1905b, p. 11. Brandt began culture experiments with phytoplankton in 1905 or earlier, but apparently never carried them far. His early results indicated that phosphate could control the summer plankton. This, combined with criticisms of his use of the law of the minimum, is probably the origin of the multifactor nutrient hypothesis that appears in later papers (esp. Brandt 1920, 1927).
52. Brandt 1915, 1920, 1921b. A short preliminary account appeared in 1911.

ture appeared to be inversely related, just as one would suspect if denitrifying bacteria were the factor controlling nitrate.[53] The detailed examination of this relationship is the subject of *Über den Nitratgehalt*. It depended entirely on Raben's distillation technique for ammonia and nitrite plus nitrate, developed in the Internationale Meereslaboratorium, which was described in 1905, but had been used for at least a year previously.[54] The basic premises on which Brandt's argument proceeded were that nitrate is present everywhere in seawater, but in very low concentration, especially in warm water. Moreover, "denitrifying bacteria are represented in every cubic centimeter of open sea surface water. Feces of animals and corpses provide organic nutriment. On or in suitable substrata small groups existing nearby can reproduce themselves and break down minute portions of the nitrate supply. *Extraordinarily numerous very small effects are summed in the course of time to yield a perceptible result.*"[55] Gräf's results from *Planet* had shown that denitrifiers were diverse and abundant in the open ocean. There, in their isolated microenvironments, they could be expected to cause profound effects on the distribution of dissolved inorganic salts of nitrogen.

As Brandt had suggested in 1911, when the nitrate values in the *Gauss* and *Planet* samples from all over the world were arranged by temperature, there was a strikingly inverse relationship between the two.[56] But nitrate and temperature could not be directly related causally, an idea Brandt called "vollkommen unverständlich"; denitrifiers must be the intermediary. If the relationship between temperature and nitrate was used empirically as an indication of the varying role of denitrification, further tests could be made using the data to demonstrate that it was more likely that a bacterial link, rather than physical processes, was responsible for the correlation. First, Brandt asked, does the same nitrate content occur in the same region at different depths but the same temperature? If so, the effect of depth alone can be ignored. Twenty-four samples from *Gauss* taken in the Antarctic at different depths but similar temperatures had a relatively narrow range of nitrate content, indicating that effects of depth were negligible.[57] Furthermore, one could also exclude the effect of depth on nitrate content and of processes unique to specific locations in the ocean by checking the data to see if at similar depths but different temperatures there was a reciprocal relation between temperature and nitrate content. With minor exceptions this proved to be

53. Brandt 1911, p. 78.
54. Additional information was published by Raben (1910 and 1914).
55. Brandt 1915, p. 10.
56. Brandt 1915, pp. 23–24.
57. Brandt 1915, pp. 28–29.

the case, although the uptake of nutrients by algae in the surface waters and the complexity of hydrographic processes near shore made interpretation difficult under those special circumstances.[58]

The vertical distribution of nitrate showed Brandt in a striking way that physical processes (such as pressure) alone could not be responsible for the patterns of nutrients in the open ocean. If nitrate were dependent on pressure or temperature, one would expect to see a smooth increase of nitrate with increasing depth. But the data showed that nitrate reached a maximum at about 800 meters, and declined above and below that depth.[59] The deep ocean was a great storehouse of nitrate; in the surface phytoplankton (and near shore the fixed algae) kept nitrate abundance low, while in the depths other processes increased it in an unexpected way.

Part of the explanation for the peculiar distribution of nitrate lay in the circulation of the oceans, according to Brandt. W. Brennecke, who had analyzed the hydrographic results of the *Planet* expedition, had concluded that water from the deep ocean upwelled near the equator;[60] this was borne out by the nutrient values, for higher values of nitrate occurred near the surface there than at higher latitudes. But Brandt was convinced that nitrification must be involved too. Although Brandt could not solve the problem of nitrification, the anomalous distribution of nitrate was apparently a stimulus that led him to look for patterns of abundance relating nitrate and its presumed source, ammonia.

Ammonia too showed puzzling patterns of distribution. Nowhere truly abundant, ammonia proved to be more abundant in cool than in warm water, but, when closely examined, had no particular pattern of vertical distribution or relationship with temperature. What it did appear to show at some locations was a reciprocal relationship with nitrate; although total nitrogen frequently remained nearly constant from sample to sample, high ammonia and low nitrate or low ammonia and high nitrate frequently occurred together. And Brandt concluded that in the open-ocean nitrate maximum near 800 meters ammonia was lower than average. These circumstances were just as would be expected if nitrification of ammonia to nitrate occurred in the water column.[61] Solid empiri-

58. Brandt 1915, pp. 29–32.

59. Brandt 1915, pp. 34–36. The revelation that nitrate was abundant in the deep sea dealt a final blow to the old but widely held view that ammonia was the main form of combined nitrogen in solution in the sea.

60. Brennecke 1909. His scheme of circulation was a two-cell one in which water sank at high latitudes then ascended to the surface on either side of the equator before returning to high latitudes at or near the surface. Such a two-cell scheme was not unique to Brennecke; it dates at least to the 1820s (see Wüst 1968; Deacon 1971, chapters 10, 11; Mills 1973).

61. Brandt 1915, pp. 44–48.

cal evidence showed that nitrification occurred only in or near the bottom, but the nutrient data accumulating through Raben's and Brandt's efforts allowed Brandt to postulate open-water nitrification to account for the mid-water maximum of nitrate.

Thus by 1915 Brandt was using information from microbiology, physical oceanography, and nutrient analyses to develop a global theory of marine denitrification and nitrification. He used this same information to explain the seasonal plankton cycle (see Chapter 4) in his third monograph titled *Ueber den Stoffwechsel im Meere*, written during enforced leisure during the war and published in 1920. Brandt did not return to the nitrogen cycle per se until 1927, when the first of a projected but unfinished series of monographs summarizing his life's work was published. But it only added detail to a theoretical structure that had been intact in 1915.

Brandt's final monograph, *Stickstoffverbindungen im Meere*, published four years before his death, concentrated again on his opinion that "above all, the formation of nitrite and nitrate in seawater requires explanation."[62] But it was written in a wholly new scientific environment by a man whose active scientific career had ended, with the mobilization of his students and young colleagues early in the war. In 1927 he took the opportunity to present the Kiel group's long unpublished information on nutrients in sediments, showing that recycling of nutrients through decomposition in the bottom and the release of inorganic nutrients such as nitrate was a far more important source of nitrogen to the water column than the influx of nitrate from rivers.[63] Dissolved organic nitrogen, which Raben had measured for years (mainly as the difference between total nitrogen and the inorganic forms) proved to be abundant; it, suggested Brandt, could be the basis of nitrification in the water column, since according to Harvey's results oxidation of ammonia did not occur by bacterial action in surface waters.[64] But most of the monograph is a review of the old work at Kiel and a defense of Raben's analytical methods against new, more precise but untested methods. Brandt was on the defensive, not for the first time. He still claimed in 1927 that "even now there are no convincing reasons to reject my hypothesis that denitrifying bacteria break down the excess of nitrogen compounds, and that it is they that maintain the existing equilibrium in nature."[65] His theoretical formulation of the nitrogen cycle, incorporating the denitrifi-

---

62. Brandt 1927, pp. 274. The content of the unfinished monographs is hinted at in Brandt 1929, which is mainly an abstract of his earlier work.
63. Brandt 1927, p. 224.
64. Harvey 1926. See discussion in Chapter 2 near n. 86.
65. Brandt 1927, pp. 204–205.

cation hypothesis, remained intact and little changed since before the war, but his hypotheses and methods were under attack.

W. A. Herdman, professor of oceanography at Liverpool, in his address to the British Association for the Advancement of Science at Cardiff in 1920, expressed a widespread view of Brandt's denitrification hypothesis.

> Brandt first attributed the poverty of plankton in the tropics to the destruction of nitrates in the sea as a result of the greater intensity of the metabolism of denitrifying bacteria in the warmer water; and various other writers since then have more or less agreed that the presence of these denitrifying bacteria, by keeping down to a minimum the nitrogen concentration in tropical waters, may account for the relative scarcity of the phytoplankton, and consequently of the zoo-plankton, that has been observed. But Gran, Nathansohn, Murray, Hjort and others have shown that such bacteria are rare or absent in the open sea, that their action must be negligible, and that Brandt's hypothesis is untenable.[66]

Attacks of greater or lesser weight began shortly after publication of Brandt's rectoral address, "Ueber den Stoffwechsel im Meere," in 1899 (see Chapter 2 at n. 16). They centered on nitrogen fixation, on the role of denitrification, and on Brandt's use of the law of the minimum. The first published attack, an alternate hypothesis to replace Brandt's, was presented in 1903 by Johannes Reinke (1849–1931), Professor ordinarius of botany at Kiel. Reinke's argument was based on the incontrovertible observation that inorganic nitrogen compounds were scarce in the sea. This, according to Reinke's line of reasoning, was not because denitrification removed significant amounts of nitrogen but because the supply to the oceans was small. Plants in the ocean received their nitrogen either from decomposing organic remains in the bottom sediments or by nitrogen fixation. Reinke suggested that atmospheric nitrogen reached the sea only in very small amounts in rain and snow, and that these sources were most important at high latitudes, where plankton abundance was high. But because the total amounts were so low there must be another source of fixed nitrogen. That source was nitrogen fixation by marine bacteria, demonstrated the same year by Benecke and Keutner (see Chapter 2 near n. 80). Nitrogen-fixing bacteria such as *Azotobacter* and *Clostridium* living on the surface of algal cells or macroalgae could fix atmospheric nitrogen and pass it directly to the plants; thereby the plants received

---

66. Herdman 1920, p. 14. His mention of Murray and Hjort probably refers to their text, published in 1912, in which Gran attacked the denitrification hypothesis. Murray and Hjort themselves had little or nothing to say about it.

nitrogen for metabolism, but the fixed nitrogen never appeared in the water.

Brandt described Reinke's idea as an unoriginal conjecture.[67] But he devoted 20 pages to refuting it, basing his argument on the idea that low amounts of nitrogen in seawater did not necessarily mean a low supply. Brandt's calculations indicated that a large supply reached the sea from the land and that the important regulating process of the nitrogen cycle had to be removal of superabundant nitrogen, especially in the warm seas, not its addition by fixation. If nitrogen fixation were truly important it should function most effectively in the tropics at high temperatures, just the regions where plankton was impoverished. Only denitrification could explain this distribution, so anomalous if only nitrogen fixation was considered. Reinke was wrong in his assessment of how much nitrogen reached high and low latitudes in precipitation; more important, he had wrongly assessed the total nitrogen supply of the sea. Nitrogen fixation must be rejected as a process that significantly affected the abundance of plankton.

Reinke's conjectures were a lightweight attack on Brandt's conception of the nitrogen cycle. The serious criticisms came from those who directly attacked the significance of denitrification, especially H. H. Gran, Alexander Nathansohn, and to a lesser extent J. Gebbing. Their criticisms centered around the physiology of denitrifying bacteria and the question of whether or not a chemical, rather than a microbiological, explanation existed for the pattern of plankton abundance and of inorganic nitrogen compounds in the sea.

Gran had been skeptical from the beginning of his microbiological work (published in 1902) that there was enough organic material in the open sea to support denitrification. He was skeptical too, as he pointed out in 1912, that denitrifying bacteria would use nitrate or nitrite for oxygen, because molecular oxygen was abundant throughout the oceans. Gran had been among the first to find nitrifying bacteria only very near shore. He doubted that nitrifiers occurred in the open sea, thus that nitrate could have a significant role in the nitrogen cycle of the open sea. He summarized his criticisms of Brandt's hypothesis by saying that "most of the combined nitrogen in the sea occurs as organic compounds or as saline ammonia, neither of which can be reduced by denitrification."[68] Clearly, if nitrate was not present in the open sea then Brandt's hypothesis was unsupportable. Marine denitrifying bacteria existed—Gran himself, Baur, later Drew, had shown that they were widespread—but

67. Brandt 1904, p. 401.
68. Gran 1912a, p. 370.

Gran would not concede that they played a major role in the nitrogen cycle.

Brandt's reply to Gran both denied and accepted the force of his criticisms. He pointed out that Gran's own experimental results (like those of Baur) showed that the denitrifying bacteria he had isolated from seawater off Texel were actually inhibited by low oxygen rather than preferring anoxia, and that Gran had suggested that well-aerated conditions might also favor denitrification in the open ocean.[69] Gran's criticism that because nitrate was extremely low in tropical seas and organic matter was low everywhere denitrification could not occur was a more serious problem. Brandt countered it by suggesting that denitrifying bacteria found localized sources of nitrate in the decomposing feces and corpses of planktonic organisms. Such tiny point sources of denitrification, summed throughout the tropical oceans, resulted in the low nutrient content of the tropical waters and the abundance of denitrifying bacteria throughout the water column observed by Gräf during his work on SMS *Planet*. Thus by postulating many small, organic-rich, probably anaerobic, microenvironments, Brandt linked denitrification to the well-known abundance of organic remains in the water column and blunted the force of Gran's criticism.

The most important attacks on Brandt's hypothesis were those mounted by Alexander Nathansohn and J. Gebbing between 1906 and 1910. Each presented a new model of the nitrogen cycle in which denitrification had no part. In each, chemical rather than biological processes maintained the low level of inorganic nitrogen in the sea.[70] Nathansohn based his attack on the results of his search for bacteria of the nitrogen cycle at Naples, while Gebbing based a similar assault on the denitrification hypothesis on the results of the *Gauss* expedition of 1901–1903.

During his work at Naples between 1901 and 1903 Nathansohn found denitrifying bacteria but was unable to detect nitrification in seawater except very close to shore, even when he enriched cultures with ammonium chloride.[71] He noted that nitrate entered the sea via the sewers of Naples but quickly disappeared, especially where the shore algae grew luxuriantly. Aware that macroalgae stored nitrate in their cells,[72] Nathansohn believed that the limited amounts of nitrate entering the sea were rapidly absorbed by algae, which outcompeted denitrifying bacteria

69. Brandt 1915, pp. 9–10, cited in Gran 1902b, p. 16.
70. Nathansohn's main attack was 1906b (but see shorter versions in 1906a, c); Gebbing's was 1909 and 1910.
71. Nathansohn 1906b, p. 366.
72. Nathansohn 1906b, pp. 367–368.

for that nutrient. There could be no question of an accumulation of nitrate in open water, for the only reservoir was on land, and that supply was bound up by algae near shore. No nitrifying bacteria could be found in the open sea, thus denitrifiers, which depended on nitrification, were not responsible for the low amounts of inorganic nitrogen found in the sea. But why were there no nitrifiers—and thus no denitrifiers—in the open sea? Nathansohn returned to the concept the Schloesings had suggested in the 1870s, out-gassing of ammonia from the sea to the atmosphere (see Chapter 2 near n. 47). The decomposition of nitrogen-containing algae would result in the release of ammonia to seawater; from there it would be released rapidly to the atmosphere by the dissociation of ammonium carbonate (which Nathansohn believed was the major form of ammonia in seawater). This "ständiges Überdestillieren vom Meere" of ammonia carried ammonia to the continents from near-shore waters and the open sea. There it was rapidly absorbed by the soil, resulting in the low ammonia content of the air that critical observers had noted since the time of Boussingault and Lawes. Thus by combining the absence of nitrification with chemical-physical processes that remove ammonia from the oceans, there was, as Nathansohn stated, "no logical necessity" for Brandt's denitrification hypothesis.[73]

When Nathansohn wrote in 1906, there was no evidence that nitrate was abundant in deep ocean waters. That evidence was provided by J. Gebbing, whose analyses of the samples from the voyage of SMS *Gauss* (1901–1903) were done in Brandt's ICES laboratory at Kiel, and by Gräf using even more geographically widespread samples from the cruise of SMS *Planet* a few years later. By 1909, Gebbing had to contend with the discovery that nitrate was not scarce but actually very abundant, at least in deep ocean waters. He was face to face with the problem of nitrification. But Gazert, who had also worked on *Gauss*, had found no evidence of nitrifying bacteria in the open ocean and precious little evidence of denitrification.

The distribution of nutrients also spoke against Brandt's hypothesis, according to Gebbing's reasoning.[74] If denitrifying bacteria were widespread and temperature-dependent, as Brandt believed, there should be a symmetrical distribution of inorganic nutrients in the sea, low at the equator and high toward each pole. This was manifestly not the case, as the results from *Gauss* showed, for nitrate was much higher in the Southern Ocean than in the North Atlantic. This could be accounted for best, according to Gebbing, if nitrate was carried from some source in the south, rather than produced at all latitudes by nitrification and removed

73. Nathansohn 1906b, p. 365.
74. Gebbing 1909, pp. 177, 181–182; Gebbing 1910, pp. 59–63.

by denitrification in the tropics. One of Brandt's own tests of the hypothesis could be turned against him. Gebbing's conception of the nitrogen cycle was largely chemical, like Nathansohn's, but unlike Nathansohn, Gebbing extended chemistry to nitrification (for nitrate was evidently abundant in the sea) as well as to the loss of ammonia from the sea surface. Gebbing convinced himself that Nathansohn had been correct about the Schloesings' hypothesis by measuring experimentally the loss of ammonia from an alkaline solution of ammonium chloride maintained in an airstream at 25 to 27°C. After 33 days nearly 60% of the nitrogen in solution had been lost, direct evidence for the out-gassing of ammonia from the oceans.[75] The process was temperature-dependent, and, as Gebbing noted, would probably be accelerated by the strong winds of the tropics. Nitrification too could be a purely chemical process, mediated by a host of potential catalysts in the ocean. Here Gebbing used evidence from chemical laboratories rather than his own experiments. His conclusions ran directly counter to Brandt's developing views of biological nitrification: "Indeed one can assume with considerable certainty that this oxidation of ammonia to nitrite and nitrate may also take place in the sea by purely chemical means. Therefore one would be able to dispense with the hypothesis of nitrifying bacteria as essential for the cycle of nitrogen in the sea."[76]

On this basis, Gebbing outlined a theory of the nitrogen cycle that differed in significant ways from Brandt's. According to his scheme, inorganic nitrogen compounds (ammonia, nitrate, and nitrite) entered the sea from land, from the atmosphere, and by volcanic activity. Plants used inorganic nutrients, which were passed on to grazing animals and the rest of the food chain. By decomposition of organisms ammonia was produced, much of which was lost by chemical means to the atmosphere. Sinking organisms released ammonia, which if not lost could be nitrified anywhere in the water column by purely chemical oxidation catalyzed by a potential host of agents in seawater. In the surface water, nitrate was absorbed immediately by plants, whereas in deep water it accumulated, giving rise to the distribution evident in the results of the vertical sections done on *Gauss*. Nitrate in deep water could be, and was, carried by horizontal and vertical currents, welling up near the equator, as the results from *Gauss* indicated. The main source of nitrate lay in the south; from thence nutrients were carried at least to the tropics. And in general, nitrogen was so abundant in the sea that it could not be limiting.[77] One problem remained. If nitrification occurred everywhere throughout the

75. Gebbing 1909, pp. 192–194.
76. Gebbing 1909, pp. 191–192.
77. Gebbing 1909, pp. 190–194; Gebbing 1910, pp. 63–65.

water column (modified only by absorption of nitrate by plants at the surface) why was nitrate more abundant in the Southern Ocean than anywhere else? Gebbing's conception of the nitrogen cycle did not accommodate this fact well. He evaded any direct explanation, satisfied that he had dealt adequately with Brandt's hypothesis on other grounds.

The variety of speculations about the marine nitrogen cycle in the early twentieth century is illustrated further by Hans Gazert's attempt to account for the high levels of nitrate in the Southern Ocean that Gebbing had found but could not explain. Relatively few "empirical" elements—the well-known nitrogen cycle on land, conflicting evidence about marine nitrifying and denitrifying bacteria, the distribution of inorganic nutrients—could be molded into yet another plausible theory of the nitrogen cycle, joining those promulgated by Brandt, Nathansohn, and Gebbing. Each asked, as Gazert did, "What do bacteriological and chemical investigations tell us of the nitrogen cycle in the sea?"[78] The answer depended on the weight given to sketchily known marine microbiological or chemical processes.

Gazert was willing to agree with Brandt that denitrification occurred in the sea (albeit in limited amount, according to his results) and that nitrogen salts limited plant production. These salts were derived from two reservoirs, the Southern Ocean and the deep sea. Away from each, inorganic nutrient salts (notably nitrate) diminished; for example, along the axis of the Benguela Current (which originates in the Southern Ocean and transports nutrients from it) nitrate decreased to the north because of uptake by phytoplankton everywhere and bacterial denitrification near shore.[79]

It was the origin and fate of nitrate and ammonia in deep water that concerned Gazert the most. If nitrification occurred, the nitrate content of the deep sea should exceed that of the Southern Ocean, since nitrate should slowly form as Southern Ocean water moved north, filling the deep sea basins.[80] But the nitrate contents of both were the same, and Gazert had not been able to detect nitrification during his incubations on *Gauss*. Thus it was unlikely that nitrification occurred in the open ocean. But, if nitrification did not occur, the constant decomposition of organisms by bacteria should increase the ammonia content of the sea. Then why was it so low and uniform?

The answer to this problem lay, Gazert believed, in special conditions governing the formation of ammonia in the deep sea and Antarctic. If increasing ammonification of Antarctic water did not occur as it moved

78. Gazert 1912, p. 283.
79. Gazert 1912, pp. 284–285.
80. Gazert 1912, p. 286.

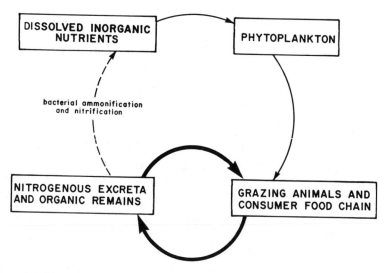

FIGURE 3.2 The marine nitrogen cycle as described by Gazert in 1912. Heavy arrows show the abbreviation of the cycle that Gazert believed was responsible for the low level of ammonia in the oceans and the absence of nitrification (broken arrow) by bacteria.

north, perhaps ammonia was not formed at all. Gazert speculated that animals might produce excreta not suitable for bacterial decomposition (he suggested urea), or that under the special conditions of the Antarctic and the deep sea, excreta and decomposition products were absorbed directly by simple multicellular organisms that competed with ammonifying bacteria for nitrogen-containing organic remains.[81] At low temperatures ammonifying bacteria could not compete successfully with simple animals for nitrogen-rich organic remains, thus ammonia was not formed and the classical nitrogen cycle was short-circuited. By this means, Gazert suggested, the nitrogen cycle could be closed without postulating either nitrification or ammonification (Figure 3.2).

Even with this ad hoc modification of the nitrogen cycle, Gazert expected very low levels of ammonification to occur, for the bacteria of decomposition were widespread (though sparse) in the open ocean. Deep ocean water should slowly increase in ammonia content, an expectation not borne out by his measurements. Accordingly, nitrification must occur throughout the sea, and to counter it, there must be denitrification of the small amounts of nitrate formed from ammonia.[82]

A more serious problem was the origin of the nitrate in the deep sea and the Southern Ocean. Deep-sea nitrate was carried there by deep currents

81. Gazert 1912, p. 287.
82. Gazert 1912, p. 288.

from some southern source that Gazert found hard to pinpoint. Rejecting terrestrial origins, fluxes from the atmosphere, and volcanic gases as adequate sources, he opted for a biotic origin of nitrate, associated in some way with the ice edge where he had found high levels associated with diatoms. Seeking the origin of nitrate under what he called "specifically polar conditions," Gazert viewed the diatom-rich ice edge as an analogue of coastal sediments where nitrification had frequently been observed. Issatchenko's observations that nitrification could occur in arctic sediments gave Gazert confidence that this hypothesis was correct. The ice edge and the cold waters adjacent to it were a special system where nitrate formed and was provided to sinking deep water, and where lower animals out-competed bacteria for nitrogenous organic remains. The result was water of low ammonia content but very high nitrate content moving north along the bottom. Pressing further beyond empirical results, he suggested that nitrogen fixation was likely to be the process enriching the ice edge, the ultimate source of the seemingly paradoxical richness of nitrate in Antarctic waters.[83]

Nathansohn, Gebbing, and Gazert directly challenged Brandt's conception of the nitrogen cycle by criticizing its details and by presenting alternative models of the cycle. Rather than directly challenging the models, Brandt responded by showing that denitrification (and nitrification, which played an increasingly greater role in his ideas) could better account for the detailed observations of plankton abundance and nutrients provided by the Kiel investigations and the cruises of *Gauss* and *Planet*. The observations were seldom in doubt; it was interpretation and basic assumptions that varied from theorist to theorist. Brandt's 1915 monograph, *Über den Nitratgehalt*, was intended to be a decisive refutation of his critics' views. It was also an expression of his faith in his intuitions about the nitrogen cycle.

The *Gauss* and *Planet* cruises and their results provided two crucial kinds of evidence used equally by Brandt and his critics. First, both studies revealed that the deep sea contained large amounts of nitrate. And despite Gazert's modest success in finding denitrifying bacteria during the earlier cruise, Gräf showed decisively on the later one that denitrifiers were widespread in the sea. Gebbing's belief, based on Gazert's results, that denitrifiers were unimportant in the open ocean, was decisively refuted.[84] Furthermore, Gebbing had claimed that the asymmetrical distribution of nitrate with latitude was strong evidence against denitrifica-

83. Gazert 1912, pp. 289–290. Gazert must have been aware that nitrogen fixation results in compounds of reduced nitrogen (e.g., ammonia) being formed, not nitrate. If so, he left his speculative scheme of the Antarctic nitrogen cycle incomplete.

84. Brandt 1915, pp. 8–10.

tion. But, as Brandt took pains to point out, Gebbing's conclusion that nitrate was low in the northern hemisphere was based on only five samples, two of them in the Gulf Stream, which, because of its tropical origin, could hardly be expected to have high amounts of nitrate. Even if Gebbing's conclusions were correct, Brandt indignantly rejected the implication that he had assumed that hydrographic processes were uniform north and south of the equator and that a symmetrical distribution of nutrients should occur.[85] Far more important were the detailed relations that he could discern between temperature and nutrients and between ammonia and nitrate.

Both Nathansohn and Gebbing had denied that any clear relationship existed between temperature and nutrients or temperature and plankton abundance, as the denitrification hypothesis required. They also denied that dissolved nitrogen was scarce enough to be limiting. The great value of the observations made on *Gauss* and *Planet* was that general impressions could be tested using large amounts of data. Brandt did just that, showing that, except under very complex coastal conditions, nitrate (but not ammonia) and plankton abundance varied inversely with temperature in a way consistent with increased denitrification at high temperatures.[86] Equally, because plankton and nitrate covaried so closely from place to place (as well as seasonally in the Baltic and North seas) Brandt convinced himself that nitrogen was a limiting nutrient.[87] The Schloesings' hypothesis, accepted by Nathansohn and experimentally tested by Gebbing, could also be rejected using the data from *Gauss* and *Planet*, which indicated that ammonia was virtually uniform in the oceans, not low in the tropical surface waters as Gebbing's results would have suggested. Moreover, Gebbing's experiment was unrealistic, for combined ammonia was released only if the pH was unusually high, outside the range of normal seawater.[88] All the evidence available to Brandt then and later indicated, in the words of Kurt Buch, that "the sea therefore would be regarded as a great consumer of nitrogen which practically never releases the consumed material."[89]

The most striking difference between Brandt's view of the nitrogen cycle, based on the data from *Gauss* and *Planet*, and the views of Nathansohn and Gebbing was that Brandt accepted some form of nitrification. Even though nitrifying bacteria could not be isolated, Brandt used indi-

85. Brandt 1915, pp. 25–27, 53–54.
86. Brandt 1915, pp. 23–32.
87. Brandt 1915, pp. 37–38, 41.
88. Brandt 1915, p. 31; Raben 1914, p. 211; see also Brandt 1920, p. 331; Brandt 1927, p. 205; Whitfield 1974.
89. Brandt 1927, pp. 205, 277. The reference is to Buch 1920.

rect evidence—the presence of a nitrate maximum at 800 meters (where ammonia was low) and the reciprocal abundances of ammonia and nitrate both geographically and on smaller local scales—to infer that nitrification must occur.[90] Ammonia varied in a way consistent with nitrification; nitrate varied in a way consistent with both nitrification and denitrification. He rejected the evidence used by his critics, absence of nitrifying bacteria, because the consistency of his theory required nitrifiers as well as denitrifiers. Even H. W. Harvey's evidence (in 1926) that nitrifying bacteria (using ammonia) did not occur in the water column did not permanently shake Brandt's faith in nitrification. Instead he postulated nitrifiers using dissolved organic nitrogen rather than ammonia, thus retaining the explanatory completeness of his scheme.[91] His theory required nitrifiers; there was indirect evidence (but no direct evidence) for nitrification, therefore nitrifiers of some kind must exist.

Brandt's treatment of his critics was not even-handed, for he ignored Gazert's speculations about the "specifically polar processes" giving rise to high levels of nitrate in deep water. Instead he directed withering fire on Gebbing, whose "completely erroneous view" of the nitrogen cycle in the sea was the main focus of attack in 1915. Gebbing not only summarized and extended the views of Nathansohn, he appeared to be a traitor to the cause, for he had been trained in Brandt's laboratory and had enjoyed its hospitality for three years (1904–1907) after the cruise of *Gauss*. Like Lohmann, whose work Gebbing aided by chemical analyses while he was in Kiel, Gebbing seems to have found difficulty agreeing with an authoritarian figure like Brandt; Brandt in turn used him as the target of his maneuvers as he defended, amplified, and extended the denitrification hypothesis between 1902 and 1915.

Despite the importance of denitrification, its implications, and its elaboration into a complex theory of the control by limiting nutrients of plankton abundance in the sea, it was only one important aspect of Brandt's theory between 1899 and 1927. The second major theoretical idea was the law of the minimum, credited originally to J. Liebig.[92] Brandt's use of the law of the minimum, like his faith in denitrification, was criticized by Nathansohn and Gebbing, among others. Their criticisms forced Brandt to change the context into which he fitted the law, though his basic view changed very little from start to finish.

90. Brandt 1915, pp. 44–47.

91. Brandt 1927, pp. 246–248.

92. The life and work of Justus Liebig (1803–1873), the great organic and agricultural chemist of Giessen, is outlined by Holmes 1973. What came to be called the law of the minimum was formulated first by Liebig in the third edition of his textbook *Organic Chemistry in Its Relation to Agriculture and Physiology*, 1843 (Rossiter 1975, pp. 43–44).

Brandt's earlier reference to the law, in 1899, stated merely that "the growth of plants is governed by the amounts of the nutrient that is provided to them in the lowest quantity."[93] More definitively, three years later he established the context in which it applied to the plankton.

> Plants produce as much organic material as is possible under the general conditions of life under which they exist and according to what is made possible by the inorganic nutrients at their disposal; *but thereby they are subjected to the Law of Minimum. If one alone of the indispensable plant nutrients is present in relatively smaller amounts the production is also sparse.* . . . As a result, the intensity of production depends upon the amount of that indispensable plant nutrient which is relatively most scantily replaced.[94]

This definition, embodying the idea of plants' direct response to limited nutrients while governed in a general way by the surrounding physical, chemical, and biotic variables, was accepted by Brandt throughout the evolution of his ideas between 1899 and 1927.[95] But his critics forced him to describe the context in which the law applied, to specify which nutrients were minimal, and to defend its applicability in the sea.

Alexander Nathansohn's original and disquieting criticisms of Brandt's approach appear to have been rooted in views that he credited to Hensen and Schütt. He stated that "single individuals have the same nutritive-physiological relationship to one another as the parts of the body," and that this organismic unity allowed a productive approach to the general metabolism of the sea, just as Hensen had suggested.[96] Equilibrium and the adaptation of organism to organism were cardinal points in Nathansohn's approach to the world of the plankton. In detail, it applied as follows.[97]

The law of the minimum, as defined by Brandt, applied only to agricultural systems. Nathansohn believed that it could apply only where exhaustion of nutrients occurred, a situation that was impossible if the crop remained to fertilize succeeding generations of plants. The removal of agricultural crops on land allowed the law of the minimum to operate on land, but in the sea, grazers and decomposers "self-fertilized" the phytoplankton by providing ready nutrients either in recycled form or as

93. Brandt 1899a, p. 5.
94. Brandt 1902b, pp. 45–46. A similar but not identical formulation is in Brandt 1905b, p. 5.
95. See for example the definition in Brandt 1920 (p. 231) and the discussion in Brandt 1925b (pp. 67–76).
96. Nathansohn 1906b, p. 361. On Schütt's holism see Chapter 1 near n. 93 and Chapter 4 near n. 10
97. Most of the argument summarized here is in Nathansohn 1908; see also Nathansohn 1909 a, b.

excreta.[98] Grazers in the sea (for example, copepods cropping phytoplankton) were the critical regulating factor, causing "dynamic equilibrium of plankton abundance" by reducing the plants to levels at which exhaustion of nutrients could not occur, and by fertilizing them simultaneously.[99] Thus when grazing was intense (the usual situation in the sea) single limiting factors were replaced by a complex of limiting factors, including, according to Nathansohn, various nutrients, dissolved carbon dioxide, inhibitory substances produced by the organisms, species-specific preferences of the phytoplankton for nutrients and vertical circulation of the water.[100] Single nutrient factors, such as nitrogen, were seldom if ever limiting, for two reasons. First, the zooplankton should always crop phytoplankton to levels at which nutrients would be superabundant. Second, there was evidence from Brandt's work that nitrogen (as nitrate) and silicate never completely disappeared from the Baltic, even during the spring bloom. Nathansohn believed that every apparent case of nutrient depletion cited by Brandt could be accounted for by some other cause, usually a change of water masses.[101] Equilibrium at levels obviating the operation of the law of the minimum was paramount; it operated through grazing, rapid transfer of nutrients from organism to organism, and through the overall quality of the water, a complex of biotic, chemical, and physical factors. All of these affected the production of marine organisms (the balance of growth and consumption), not just the maximum productivity, which Nathansohn believed Brandt considered to the exclusion of all else.[102]

The multifactorial approach to the control of production in the sea championed by Nathansohn was eagerly accepted by Gebbing, who agreed pointedly that "there is no single cause in nature."[103] Using Liebig's concept of environmental resistance (the sum of factors impeding the growth of a plant), Gebbing redefined the law of the minimum in those terms: "The yield of a soil is dependent upon the nutrients available in minimum and the resistance which opposes their assimilation by the plant." Like Nathansohn, he believed that the broadening of the law to include several environmental factors working for and against production converted Brandt's static conception of nutrient control to a dynamic one in which rate of production (that is turnover) rather than mere amounts was considered.[104] Applying this directly to the sea, phyto-

98. Nathansohn 1908, pp. 38–39.
99. Nathansohn 1908, p. 1.
100. Nathansohn 1908, pp. 60–64.
101. Nathansohn 1908, pp. 41–50.
102. Nathansohn 1908, pp. 60–61.
103. Gebbing 1910, p. 66. See also his longer discussion in Gebbing 1909, pp. 184–190.
104. Gebbing 1909, p. 186.

plankton production was likely to be limited by several nutrients, not one, and maintained at a constant level by the equilibrium between phytoplankton growth and grazing.

Brandt's reply came in 1920, when he had time because of the war to write complete accounts of the Kiel group's work. His comments were direct, sometimes scathing, about what he perceived as misunderstandings of his long-held beliefs about the control of production. He claimed not to have ignored the complex of factors affecting marine production, but he emphasized that the complex of environmental factors (such as light, temperature, salinity, gases, and so on) could only set the general bounds of biological production; proximate control had to be sought in nutrients, as a host of experiments on land showed.[105] Most important of all, the law of the minimum must apply to all systems, not just cropped agricultural ones. The sea was not unique in being grazed and "self-fertilized'; so too were even the most highly managed agricultural areas, where insects grazed plants and left their droppings among the plants as fertilizer. Grass meadows, which responded to fertilizers as would be expected under the law of the minimum, were just like the sea because the plants there were also grazed, and experienced competition for space, light, and nutrients. The law of the minimum applied there, just as in the sea; so too it applied in ponds and lakes, which, might have even more favorable environmental conditions than the ocean. Growth of the plants would soon compensate for the removal of material caused by grazing and would result in lowered nutrient levels; then some environmental factor, usually a nutrient such as nitrogen, would prove to be limiting despite the interplay of other factors.[106]

Despite his critics' claim that nitrogen was never scarce enough to be limiting, Brandt was convinced that other substances necessary for plant metabolism, such as carbon dioxide, were seldom if ever limiting. On land, winds would easily provide more than enough for photosynthesis. In the sea, where the absolute abundance of carbon dioxide was higher than on land, it was unlikely to be in short supply. But nitrogen could be limiting, despite the fact that it was never completely exhausted from seawater, as long as there was a lower limit below which plants could not absorb inorganic nitrogen compounds from seawater. Evidence from plant physiology supported this hypothesis, that plants could never remove the vestiges of nutrients from seawater;[107] moreover if the various species had different lower limits there was a physiological basis for the seasonal succession seen in nature.

105. Brandt 1920, pp. 223–225.
106. Brandt 1920, pp. 241–245.
107. Brandt 1920, pp. 239–241, 245–246.

Brandt was sarcastically critical of the distinction that Nathansohn and Gebbing had seen in his work between maximum productivity and rate of production of material, a balance between growth and consumption. He called Nathansohn's criticism a mere play on words, because productivity, the realized rate of growth, must depend on nutrients. Productivity is governed by growth rate, growth rate by limiting nutrients. To claim otherwise was to confuse concepts, to employ empty verbiage.[108]

Throughout the years between 1899 and 1927 Brandt's arguments repeated the same themes—law of the minimum, nitrogen a limiting nutrient, nitrogen cycle following a course in the sea similar to the better known terrestrial one. Each significant restatement by Brandt, in 1902, 1905, 1915, 1920, 1925, and 1927, affirmed his long-held views, emphasizing that new data amplified and extended the original basic ideas. But hidden in these monographs are significant changes forced from him by his critics. To maintain the status quo he enriched the conceptual structure of his model in several significant ways, particularly by being forced to concentrate on multiple external factors, as Nathansohn and Gebbing had done. This step provided Brandt with a mechanism for the seasonal succession of phytoplankton species. And because carbon dioxide was postulated as a limiting nutrient Brandt was forced to examine in increasingly greater detail the evidence that nitrogen was usually the most significant limiting factor and that its variations in nature were consistent with the denitrification theory and with plant physiology. The paradoxical absence of ammonia-oxidizing nitrifying bacteria in the open ocean directed his thoughts toward unorthodox pathways of nitrification based on the distribution of organic nitrogen. Finally, he was forced to contend with the possibility that vertical and horizontal currents might play some controlling role in marine production.

The main proponent of the idea that oceanic circulation had a significant role in controlling oceanic production was Alexander Nathansohn, whose monographic studies between 1906 and 1910 substantially weakened the force of Brandt's argument that denitrification was the major controlling process affecting phytoplankton abundance. Nathansohn (1878–1940) was born in Brzezany, East Prussia (now in Poland). In 1884, his parents moved to Leipzig, where Nathansohn was educated, eventually studying natural history with Carl Chun, K. G. F. Leuckart, and Wilhelm Ostwald, among others. His doctoral studies, completed in 1900, were on amitotic nuclear division in plants; his Habilitationschrift (1902) dealt with uptake of nutrient salts by plant cells. He became a Privatdozent in Leipzig in 1904 and was Professor extraordinarius from

108. Brandt 1920, pp. 246–248.

about 1910 through 1917. Nathansohn worked at Naples during 1901–1903 on bacteria of the nitrogen and sulfur cycles. Early in 1907, he began work with H. H. Gran on the plankton cycle in Christiania Fjord. Later that year, he went to Monaco, where, for periods of several months until early January 1909, he studied the relationship of hydrographic change to plankton abundance.

Nathansohn's later career is less well known. In 1918–1919, he was Professor (extraordinarius?) in Vienna; in 1920, he gave his address as Berlin Lichterfelde–West. During this period, into the 1930s, he worked on applied botany in the Kaiser-Wilhelm-Institut für physikalische Chemie. Both in 1933 and in 1938, while he was an industrial chemist in Torino, fearful of German oppression, Nathansohn attempted to return to his early work on plankton in Monaco. Thereafter, his fate is unclear. Unlike his colleague at Monaco, Mieczyslaw Oxner (1879–1944), an assistant director of the Musée océanographique, who was killed at Auschwitz, Nathansohn died early in World War II in Torino. His last contributions to the problems of plankton dynamics, influential ones, were in 1910.[109]

Nathansohn had strong opinions about how research on the biology of the oceans should be conducted. He was convinced that the role of expeditions was finished; the future lay in new foundations such as the Prince of Monaco's Musée océanographique at Monaco, where long-term research could be done at permanent sampling sites.[110] Most of all, he was an exponent of the marriage of physics and biology in oceanography, for he believed that the control of production could be explained mainly by the physical circulation of the oceans, not by the interplay of bacterial metabolism and nutrients.

Early in his career Nathansohn had been convinced by Brandt's first publications that the nitrogen cycle was critical to plant production. At Naples he searched for denitrifying bacteria with success, but was unable to find any evidence that nitrifiers were present anywhere but very near shore. He began to reexamine his instinctive assent to Brandt's ideas. In the course of this reexamination, physics replaced bacteriology as the most significant governing force influencing plankton abundance.[111]

Without more detail of Nathansohn's early career I have been unable

---

109. A brief outline of Nathansohn's early life is given in the preamble to his doctoral dissertation, 1900. Documentary information was provided by the Canada Institute for Scientific and Technical Information, the Staatsbibliothek Preussischer Kulturbesitz in Berlin, the Bezirksamt Steglitz von Berlin, and the Centro di Documentazione Ebraica Contemporanea, Milano. I am grateful to Christiane Groeben, Archivist of the Stazione Zoologica, Napoli, and to Jacqueline Carpine-Lancre, Librarian of the Musée océanographique de Monaco, for copies of letters giving information on Nathansohn's work and life between 1901 and 1940.

110. Nathansohn 1909b, p. 624.

111. Nathansohn 1906b, pp. 357, 367.

to establish how he first began to see links between oceanic circulation and variations in the abundance of phytoplankton. He knew and was helped by Carl Chun at Leipzig; perhaps Chun's work on deep-water plankton and food supplies reaching the deep sea predisposed Nathansohn to look for physical explanations of planktonic events.[112] It is more likely, though, that Nathansohn's views had a Scandinavian source, for the dominating work on physical oceanography at the turn of the century originated in Sweden and Norway. Nathansohn's publications cite the most recent work on physical oceanography by Fridtjof Nansen, Otto Pettersson, Martin Knudsen, Bjørn Helland-Hansen, and J. W. Sandström, as well as the older work of F. L. Ekman. By 1906, when Nathansohn's publications on physical processes had begun to appear, these Scandinavian physicists had worked out the physical structure and current systems of the North Atlantic (concentrating on the Norwegian Sea) in considerable detail, based on the distribution of temperature and salinity. This work provided the basis for the first theoretical determinations of current flows made by Henrik Mohn, V. F. K. Bjerknes, J. W. Sandström, and Bjørn Helland-Hansen.[113] Most important, it was widely known, because after 1902 ICES was the forum in which their work was widely discussed and from which it was disseminated in a flood of reports, bulletins, and semi-official manuscripts. Nathansohn's friendship with Gran, who until 1905 was an official of the Norwegian Fisheries Department under Johan Hjort, probably gave him access, via ICES or in person, to the Scandinavian physical oceanographers and their work.

Nathansohn based his divergent views of the control of plankton production firmly on physical circulation, that is, on what he had learned of Scandinavian physical oceanography. Each area of high production in the oceans could be accounted for by a physical mechanism in which the coasts had little role (contrary to Brandt's early ideas about nutrient supply). On what Nathansohn called "compelling logical grounds" vertical water circulation took the place of the nitrogen cycle and the law of the minimum, providing the control of plankton abundance under highly varied hydrographic conditions and in widely separated regions.[114] Denitrification could be discounted on other grounds too, according to Nathansohn, for if ammonia was lost from the sea to the atmosphere there would be no need for denitrification. And because ammonia was lost, and thus was scarce in the sea, nitrifiers too could be dispensed with. Admittedly denitrifying bacteria were present in the sea—Nathansohn

112. On Chun's work see Mills 1980, 1983.
113. See especially Mohn 1885, 1887; Bjerknes 1898; Sandstrom and Helland-Hansen 1903, 1905.
114. Nathansohn 1906b, p. 411.

himself had found them—but they were merely facultatively dependent on nitrate, and could switch their metabolism to operate, presumably, as aerobic heterotrophs. Even where nitrate was abundant, close to shore, at river mouths and around the sewer pipes of large cities, large attached algae would out-compete denitrifying bacteria for nitrate and sequester it in their cells. Finally, completing this chain of reasoning, Nathansohn postulated that once the role of denitrifying bacteria had been discounted, temperature became relatively unimportant in phytoplankton abundance. Paradoxical exceptions to Brandt's hypothesis became less puzzling. Nathansohn referred here to the high plankton volumes discovered in the warm waters of the South Equatorial Current during the Plankton Expedition (see Figure 1.4), which would be inexplicable by Brandt's hypothesis if temperature-dependent denitrifying bacteria had stripped nitrate from the water. Thus both temperature (controlling denitrification) and the coasts (providing nitrogen to the sea) were unnecessary features. They had no part in a general theory that could explain plankton abundance under a wider range of circumstances than could Brandt's hypothesis.

Nathansohn made his theory of vertical water circulation explicit in 1906, four years after he had read Nansen's speculations on the reasons for the high fertility of arctic waters where they entered the North Atlantic circulation southwest of Greenland and off Iceland. During the remarkable drift of his ship *Fram* across the Arctic Ocean between 1893 and 1896 Nansen made temperature observations and plankton collections, but he had not been able to do water chemistry.[115] He noted in particular that plankton was sparse in the high Arctic, although farther south, where the Arctic water met warmer North Atlantic water, plankton and fish were abundant, giving rise to the fisheries at Iceland, off Newfoundland, in the Lofoten Islands, even in locations much farther north such as Denmark Strait, Jan Mayen, and the White Sea. This distribution could be explained if the Arctic Ocean was rich in nutrients and acted, in Nansen's words, "as a kind of lungs in the circulation of the ocean."[116] He speculated that the Arctic Ocean was enriched by nutrients (especially nitrogen, phosphorus, and silicon) carried to it by the great Asian rivers draining north but that ice cover and low temperatures prevented plankton production in the north. As the Arctic waters flowed south and their temperature rose, freeing them of ice, their nutrients were the basis of rich oceanic production and the well-known northern fisheries.

Nansen's suspicions that Arctic waters had unrealized potential for

---

115. The background of the expedition is described in Neatby 1973, pp. 113–176.
116. Nansen 1902, p. 423.

production were reinforced when he read Brandt's second paper (1902) on metabolism in the sea and Gran's on denitrifying bacteria. The first made it clear that denitrifying bacteria were temperature-dependent, the second that they required organic matter. Nansen concluded that

> it will be easily seen that both these assumptions will lead to the conclusion that the surface water of the North Basin (the true Polar water) must possess specially favourable conditions for higher organic life. On the one hand the very low temperature (generally about −1.5°C) practically prevents the decomposition by the bacteria of the nitrates and nitrites received from the rivers. On the other hand the amount of constituents containing carbon, necessary for the activity of these bacteria, is probably very small in the waters of the North Polar Basin, even on its coasts, because there is relatively little organic life to produce them (e.g., the mud from the Siberian coast contains very little organic matter), and the low temperature greatly retards the decomposition of the organic substances. The result must be that as the thick ice-covering prevents the development of plant life, the nutritive constituents containing nitrogen must be stored up in the surface-water of the North Polar Basin, until this true Polar water divests itself of its ice-covering in lower latitudes and light obtains access to the water.[117]

The flaws in this appealing hypothesis were spotted by Nathansohn. If nutrients in the polar water, accompanying it from the Arctic Ocean, were the source of production where Arctic and Atlantic waters met, the bloom at the juncture of the two should decline as the summer progressed because of exhaustion of nutrients. But observations contradicted this— the zone of mixing was productive (that is, it had high plankton volumes) all summer long. Some other mechanism must be operating, one that would also explain the abundance of plankton in the Antarctic, where there were no continental rivers to enrich the polar waters with nutrients.[118]

Nathansohn found "in the vertical motion of the water a principle of the most general significance for the productive capacity of the sea."[119] The importance of vertical water movement was demonstrable, according to Nathansohn, by examining a hypothetical case in which it was absent, an isolated, current-free marine basin. There plankton produced at the surface would sink slowly to the depths, nourishing a sparse mid-water fauna and a richer benthic one. At the bottom, nutrients would be regenerated, but they would diffuse only slowly, if at all, into the surface waters. Thus "motive power and building materials are spatially sepa-

---

117. Nansen 1902, p. 426.
118. Nathansohn 1906b, pp. 371–372.
119. Nathansohn 1906b, p. 372.

rated and for that reason cannot find use in the synthesis of living bodies."[120]

But "motive power and building materials" could be united if vertical currents carried the accumulated nutrients from the bottom to the surface waters. Nathansohn regarded the Mediterranean as a good example of his paradigmatic isolated basin; Konrad Natterer's results showed that the nutrient-poor surface waters were also impoverished in plankton, yet, where vertical mixing was intense, for example in the Strait of Messina and off the Algerian coast, the plankton was rich.[121] Similarly, vertical currents could explain the richness of polar waters, of the mixed region where polar water met Atlantic water, and of the equatorial regions (where Hensen had observed paradoxically high plankton volumes).

In general, Nathansohn ascribed vertical mixing to cooling and sinking of saline surface waters, which caused their replacement by nutrient-enriched deep water.[122] The temperature of the underlying water would determine the depth of vertical mixing, for surface water, if cooled, would sink to a depth where it met water of the same specific gravity. In the open temperate and tropical oceans, Nathansohn expected that cooled surface water would sink only to the depth of the seasonal thermocline, 25 to 140 meters in the Atlantic, perhaps 110 to 180 meters in the Pacific, according to profiles from the German Deep-Sea Expedition (1898–1899), and that surface production would be low. But south and west of Iceland, in the Irminger Sea, the temperature of the water was virtually the same right to the bottom, according to Otto Pettersson's observations.[123] In that region, surface cooling would cause exchange throughout the whole water column—and that was just where Hensen had observed high plankton volumes (see Figure 1.4). Nathansohn linked increasing depth of vertical circulation to increased amounts of plankton production, for it was the accumulated nutrients near the bottom, in general, that fertilized the surface waters most effectively. Thermal vertical circulation of this kind was not common, but when and where it occurred he believed it to be extraordinarily important in increasing plankton production at the surface.

Nathansohn was not modest in his application of the (mainly) Scan-

120. Nathansohn 1906b, p. 373.

121. Nathansohn 1906b, pp. 374–375. The Strait of Messina was famed for the rich deep-water plankton carried to the surface by turbulence; Lohmann, among many others, visited the region to study it. The importance of the Algerian coast was mentioned and explained by Adolf Puff (1890) in a little-known dissertation at Marburg. He accounted for the rich plankton by wind-induced upwelling.

122. See Nathansohn 1906b, pp. 376–385, for an extended discussion of "thermal vertical circulation."

123. Pettersson 1900, pp. 64–65.

dinavian studies of physical oceanography to oceanic circulation and plankton production; he ambitiously described the possibility of linking global circulation patterns and production.[124] Cold surface currents, the zones of water mixing, and stable subtropical-tropical gyres could all be accommodated in his causal model of plankton production. Cold surface currents (such as the "Polar current" discussed by Nansen) were extraordinarily fertile because of the vertical circulation induced by ice, an idea Nathansohn borrowed from Otto Pettersson (Figure 3.3).[125] Such a local circulation around icebergs (or the ice shelf of the Southern Ocean), which enriched the surface water with rich bottom water, could account for the fertility of the Southern Ocean and for the rich nutrients of the East Greenland Current (Nansen's "Polar water").

Where polar and Atlantic waters mixed, Nathansohn suggested another mechanism, based on Otto Pettersson's observations of the Irminger Sea and on Henrik Mohn's calculations of the counterclockwise circulation of the Norwegian Sea. In both regions, "aspiration" of deep water to the surface should occur (a phenomenon now known as *doming*, the upward displacement of water at the center of cyclonic eddies). Using the same principle, upwelling in areas of low pressure, Nathansohn suggested that the "cold tongue," a region of cool temperatures and abundant plankton where the North Equatorial and Guinea Currents diverged, could be explained by increased vertical circulation.[126] Upwelling of this kind, in cyclonic eddies or divergences, accounted for the steady season-long plankton bloom of the northern North Atlantic, whereas the seasonal mixing of the frontal regions where polar and Atlantic waters met would allow blooms only in spring and autumn; at other seasons circulation would cease when warm saline Atlantic water lay on the surface (in summer) or cold brackish polar water replaced it (in winter).

The sparseness of plankton in anticyclonic eddies like those of the Sargasso and Mediterranean seas was easy to explain according to Nathansohn. In those regions water accumulated and never circulated vertically. But at the southern edge of the Sargasso Sea near the equator, where cold deep water upwelled, completing the circulation of water along the bottom from high latitudes, plankton was abundant, in sharp contrast to the situation in the stable tropical gyres.

Vertical circulation could be invoked for far more localized phenomena than the subtropical gyres and the equatorial current systems, ac-

124. Nathansohn 1906b, pp. 386–413.
125. Pettersson 1905a, esp. figure 1.
126. The "cold tongue" was discussed by Krümmel (1893), who analyzed the physical observations of Hensen's Plankton Expedition.

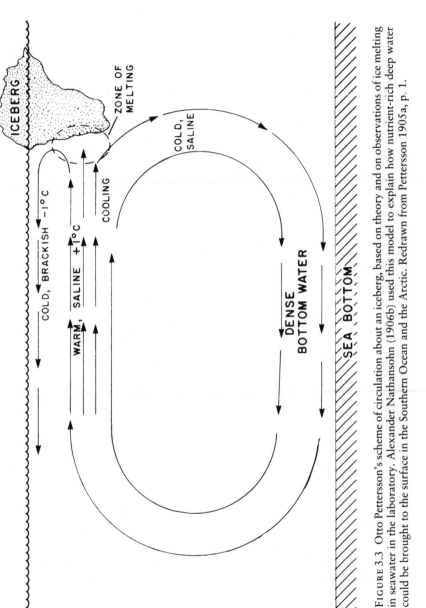

FIGURE 3.3 Otto Pettersson's scheme of circulation about an iceberg, based on theory and on observations of ice melting in seawater in the laboratory. Alexander Nathansohn (1906b) used this model to explain how nutrient-rich deep water could be brought to the surface in the Southern Ocean and the Arctic. Redrawn from Pettersson 1905a, p. 1.

cording to Nathansohn.[127] He accounted for coastal upwelling by suggesting that offshore winds drove surface water from the coast, causing its replacement by nutrient-rich deep water. High plankton production ensued.[128] The spring bloom in the Norwegian coastal current could also be explained by vertical circulation, for, Nathansohn believed, the current would cause an inward and upward movement toward shore of deep Atlantic water, bringing nutrients, just as F. L. Ekman had suggested was the case in estuaries.[129] When the Baltic Current, the source of the Norwegian coastal current, strengthened in spring, vertical mixing by this mechanism should increase. Nathansohn noted that, just as expected, the bloom occurred when the Baltic outflow increased in spring. The current, causing vertical circulation, caused an infertile water mass to be replaced by a fertile one. Similarly, the spring bloom off shore in the open Atlantic (which H. H. Gran had found began later than the bloom near the coast) could be explained if a south-flowing current of polar water increased in late spring, increasing the mixing of nutrient-rich deep water to the surface.[130]

By the time his second monograph on plankton production had been published in 1908 Nathansohn had a well-articulated theory of global dimensions explaining variations in plankton production. "The regions of the sea and the times of year with more intensive water mixing are the most plankton rich, the rest are more or less plankton poor."[131] This could be applied worldwide, to circumstances as varied as those in the equatorial currents, the Sargasso Sea, or the Norwegian Coastal Current. In general, the deep waters were like fallow fields, rich in nutrients from the decomposition of sinking organisms. If they were mixed upward at some optimal rate—not too quickly, not too slowly—rich plankton production must result. But the exact conditions for enhanced production had yet to be investigated in nature. This Nathansohn did next by examining the hydrographic control of plankton abundance at Monaco between November 1907 and December 1908.

At Monaco, aided by Jules Richard, the director of the Prince of Monaco's Musée océanographique, Nathansohn could work in deep oceanic water close to shore (Figure 3.4). Moreover he knew that the

127. Nathansohn 1906b, pp. 414–420; Nathansohn 1908, pp. 48–50.

128. Nathansohn 1906b, p. 414; based on Puff (1890). In fact it is *longshore* winds that cause coastal upwelling. The significance of V. W. Ekman's (1905) demonstration that water moved at an angle to wind direction seems to have escaped Nathansohn's notice.

129. F. L. Ekman 1875a, b; F. L. Ekman 1876. This phenomenon, now called *entrainment*, results in in-flowing deep water being mixed up into the out-flowing fresher water.

130. Nathansohn 1908, p. 66; Gran 1902a, pp. 117–118. It puzzled Gran that the bloom did not begin first in the warm Atlantic waters but began in the cold Norwegian current.

131. Nathansohn 1908, pp. 65–66.

FIGURE 3.4 Alexander Nathansohn (left) aboard *Eider,* a small research vessel of the Musée océanographique de Monaco on February 26, 1908. His companions include (left to right) Jules Richard, Jules Rouch, Jacques Liouville, and Mieczyslaw Oxner. Photograph Jules Richard (collections of the Musée océanographique de Monaco Rj-1574-107). With permission of the Musée océanographique de Monaco (reproduction Alain Malaval).

waters of the western Mediterranean were relatively uniform in temperature and salinity, so that marked changes of temperature, salinity, and plankton abundance would be easy to observe when they occurred. Sampling weekly using open and closing nets, determining temperature and salinity, and observing rainfall and wind, he arrived at a picture of plankton production off Monaco fully consistent with the hypothesis of control by vertical circulation.[132]

Nathansohn's observations of the yearly cycle, which have not been greatly changed by more recent studies,[133] ran as follows. Cooling of the surface water between October and December mixed it deeper and deeper—at least to 75 meters by December. In December a late autumn bloom occurred, mainly of Atlantic neritic diatoms from local waters, probably caused, Nathansohn speculated, by carbon dioxide–rich rain and river water being mixed into the sea. By January, the winter cooling caused mixing to 100 meters. The early bloom at the surface ended, but was replaced by an abundance of plankton at 60 to 120 meters, where he believed surface water and deep water were mixing. During February this bloom diminished; plankton became sparse throughout the water column because the water column had stabilized. During March, periods of intense cooling, caused by evaporation at the surface, resulted in deep mixing. Each cooling event corresponded with a phytoplankton bloom. As the month proceeded, phytoplankton, though abundant, slowly decreased, while grazing zooplankton increased. Periodic blooms also occurred in April, apparently as a result of fresh water bringing carbon dioxide. By May the water column had stabilized again, above a thermocline at 50 to 75 meters, which persisted until October; phytoplankton was scarce, although occasionally small increases occurred. Thus the cycle at Monaco involved a long period of low plankton abundance during the summer (occasionally broken by evaporative cooling and mixing), productive periods during March–April and December which coincided with the end and beginning of winter circulation, and throughout the winter, when the water was isothermal to at least 200 meters, variable amounts of plankton, much of it in deep water.

Quite reasonably, Nathansohn believed that his theory of vertical circulation had power and generality. It was also opportune, for he emphasized the role of physical circulation when physical oceanography was growing, whereas Brandt's was based mainly on older biological explanations of regional and seasonal variations in plankton abundance.

---

132. See Nathansohn 1909a, b; Nathansohn 1910.

133. See Bernard 1937, 1938, 1939, on the plankton cycle at Monaco. The hydrographic events described by Nathansohn were certainly real; his interpretations, especially about the role of carbon dioxide, have less plausibility.

During the first decades of the twentieth century, when physical explanations of ocean circulation were developed rapidly and publicized by means of ICES, Nathansohn's views were widely read, discussed, and accepted. Brandt's theory was quietly set aside as idiosyncratic and out of fashion.

Brandt reacted to this discouraging climate of opinion in 1920 by attempting to show in historical perspective that he had recognized the role of currents and vertical circulation in plankton production very early.[134] In fact, his publications say very little about circulation; Brandt had been preoccupied with the microbiology of the nitrogen cycle. So, although he attempted to divert the force of Nathansohn's theory, Brandt was forced onto the defensive.

The denitrification hypothesis never recovered from the attacks of its critics and from insights originating in Nathansohn's theory of vertical circulation. The beginning of its end lay in Nathansohn's recognition that physical circulation could replace denitrification as a sufficient explanation of plankton production. When physical oceanographic processes were married fully to the theory of the plankton bloom after Nathansohn's time, during the last years of the 1920s, the demise of the denitrification hypothesis was assured. But until that time Brandt's hypothesis had provided the focal point from which a theory of the plankton cycle developed. It diverted the attention of the Kiel school from the nitrogen cycle to a complex of chemical and physical factors controlling the abundance of plankton throughout the year.

134. Brandt 1920, pp. 201–206.

4 "The Water Blooms": The Discovery of the Spring Bloom and Its Control

Now the question is: on what does the peculiar position of plankton organisms' maxima and minima depend? Why is summer, which on land uniformly represents a production maximum, plankton poor, and why are spring and autumn, which represent transitional times on land, plankton rich?

THEODOR BÜSE 1915

The anomaly that Brandt's hypothesis explained was the unexpected abundance of phytoplankton at high latitudes compared with the tropics. Thus Brandt was concerned at the beginning of his studies on plankton dynamics with geographical patterns of plankton distribution. Between 1899 and 1920 the attention of the Kiel school—namely Brandt, Hensen, their professional associates, and students—shifted from geographical patterns of plankton abundance to seasonal changes in abundance. During this period the concept of the spring bloom was born, accompanied by a chemical theory of its control that was married, by degrees, to ideas of physical circulation favored by Alexander Nathansohn and physical oceanographers.

A textbook might suggest that the spring bloom was discovered and explained during a few decades of the late nineteenth and early twentieth centuries. In reality this "discovery" was more complex and is more adequately explained as a change of viewpoint, a cognitive shift, than as a sudden revelation of an unexpected scientific fact.[1] Concomitantly, as the viewpoint changed, Brandt's denitrification hypothesis, one of the par-

---

1. My approach is similar to that of Nicolson (1983, p. 313), who considers the "cognitive interests" of early plant ecologists, that is, conceptual shifts within the scientific community, as a social entity, rather than the provision of discoveries to a waiting world.

ents of the new viewpoint, became progressively more and more isolated and irrelevant.

Barry Barnes's lapidary statement that "to speak of 'discovery' is to abet a form of collective self-forgetfulness" and Ralph Jewell's comment that "a theory is a compact historical document" refer to the complexity of cognitive change that occurs when a shift of scientific viewpoint takes place.[2] The discovery of the spring bloom was not a unitary event, but a collective development, summarized after two decades in a chemical theory of the control of plankton abundance. A reorientation of thought occurred, under the influence of physiology and agricultural chemistry, from a mainly taxonomic approach to the phytoplankton to one much less concerned with individual species. The new approach emphasized the controlling role of the chemical and physical environment. In this chapter, I concentrate on the spring bloom as a newly discovered event and as a phenomenon explained increasingly as the outcome of chemical and physical fluctuations in the marine environment at temperate latitudes. Plankton blooms, during the first decade of the twentieth century, were reified; the concept became the expression of a new and influential approach to the biology of the seas.

A phenomenon as striking as the sudden appearance of phytoplankton cells during spring in temperate and high latitudes should have been noted very early, perhaps even incorporated into fishermen's folk-wisdom. Yet there is little or no mention of phenomena that in modern terms would be called the spring bloom in the scientific literature of the early nineteenth century.

Without doubt the most acute and experienced observer of the seas during that period was William Scoresby, Jr. (1789–1857), who combined a whaler's practicality with the acumen of a trained naturalist. Yet Scoresby has little to say in his great work *An Account of the Arctic Regions with a History and Description of the Northern Whale-Fishery*, published in 1820, about seasonal changes occurring in the sea. During his many voyages to the Arctic, particularly on those between 1807 and 1818, Scoresby made many scientific observations in addition to collections of marine organisms. Yet in that work he noted only "long bands or streams" of green water, often of great length, where whales preferred to feed. In the green water

> on observing that the water was very imperfectly transparent, insomuch, that *tongues* of ice, two or three fathoms under water, could scarcely be

2. Barnes 1982, p. 45. I quote Jewell from a conversation in Bergen, Norway, in July 1982. The complexity of scientific "discovery" is thoroughly explored by Brannigan 1982.

discerned, and were sometimes invisible, and that the ice floating in the olive-green sea was often marked about the edges with an orange-yellow stain, I was convinced that it must be occasioned by some yellow substance held in suspension by the water.

Whereupon he melted some of the colored ice, examined the sample with a microscope, and noted "a great number of semi-transparent spherical substances, with others resembling small portions of fine hair." There, perhaps for the first time, he observed what we now know to be under-ice diatoms.[3] Yet despite his presence in the Arctic over and over again between April and July, when the peak of the Arctic spring bloom occurs, Scoresby did not note (at least in his published works) the onset of the bloom, only its presence as an unnamed spatial phenomenon. He held the sophisticated view that plankton organisms (all of which he believed were animals) must be the food of more advanced ones such as whales, but had no reason to expect seasonal variations in their abundance.

Scoresby's lack of attention to what is now called the spring bloom was not unusual, even much later in the century when quantitative plankton biology first developed. Victor Hensen himself was more interested in the production system of the seas and in the possibility of quantitatively sampling a regularly distributed plankton than in the details of local or temporal abundance. Hensen's viewpoint, summarized in his monograph *Über die Bestimmung des Planktons* in 1887, was that plankton must be part of the metabolism of the sea, and that production in the sea was analogous to that on land.[4]

That monograph summarized a series of 34 sampling cruises in the Baltic conducted for Hensen by Franz Schütt between August 1883 and August 1886. From 1883 through 1885 they took place in every month of the year, providing an unprecedented series of samples allowing the taxonomic composition and abundance of the plankton to be assessed throughout the seasons. Hensen's careful compilation of the results shows seasonal peaks of various diatoms and dinoflagellates in spring and autumn, corresponding closely with modern knowledge of the plankton cycle off Kiel.[5]

There is no doubt that Hensen was aware of these changes, indeed he

3. Scoresby 1820, pp. 175–181. In the 1969 reprint Alister Hardy discusses the significance of Scoresby's biological observations.
4. Hensen was aided in his first comparisons of the yield in the sea compared with that on land by Hermann Rodewald, who had come to Kiel in 1881, where he directed the seed-testing station of the Agricultural Institute. Rodewald's influence on Brandt was mentioned in Chapter 2.
5. Hensen 1887b, pp. 23, 71–72, 83–89, 96, plus Fangverzeichnisse (summary tables). On current knowledge see Lenz 1981 and Smetacek 1985.

asked, "Welche Rolle die Jahreszeiten bei der Produktion des Meeres spielen?"[6] But, in practice, he was more concerned with the adequacy of his sampling and the application of physiology to marine production than with the details of how and why individual species varied in abundance. To Hensen there appears to have been no obvious link between plant production and short-term changes, nor was there any a priori reason to expect pulses of abundance. In estimating the annual production of the Kiel plankton, Hensen gave no extra weight to production occurring during the spring bloom, instead he calculated annual production based on simple average values throughout the year.[7] His belief in the validity of averages did not encourage a close look at variation, much of which Hensen attributed to sampling error.[8] Thus seasonal changes, although many were obviously real, were of secondary interest to Hensen, ranking far behind the ideal of production estimates and the problem of the *Gleichmässigkeit* (regularity of distribution) of plankton organisms, on which his system of investigation depended.

Franz Schütt, a botanist and plant physiologist, was the first evangelist of Hensen's methods, describing them for a general scientific audience in his *Analytische Plankton-Studien.*[9] Schütt's approach differs in interesting ways from Hensen's, indicating the variety of intellectual approaches brought to bear on the quantitative study of plankton during the early years of Hensen's work.

Schütt, unlike Hensen, was not preoccupied with production, but with the total vegetation of the earth, a viewpoint apparently based on August Grisebach's *Die Vegetation der Erde nach ihrer klimatischen Anordnung* of 1872, which was very influential during the last decades of the nineteenth century.[10] The taxonomic viewpoint is strong in Schütt's two books, *Analytische Plankton-Studien* and *Das Pflanzenleben der Hochsee*, both published in 1892, but it is combined with a mechanistic holism

6. "What role does the time of year play in production of the sea?" Hensen 1887b, p. 102.
7. Hensen 1887b, p. 96.
8. See Chapter 1 near n. 82.
9. Franz Schütt (1859–1921) was born in Woldegk, Mecklenburg. His Ph.D. was granted at Heidelberg in 1883. In the same year, he became an assistant in the chemical laboratory of the Technische Hochschule in Darmstadt, then in 1884 an assistant in the plant physiology institute at Göttingen. He came to Kiel in 1885 as assistant in the Botanical Institute, then from 1887 Privatdozent, and in 1894 Professor titular (an unpaid, largely honorary position). Besides conducting Hensen's cruises between 1883 and 1886, Schütt spent the winter of 1888–1889 in Naples doing quantitative plankton hauls for Hensen and Brandt and was a member of the Plankton Expedition in 1889. He was the first of the Kiel school to leave, becoming Professor ordinarius of botany at Greifswald in 1895. See Volbehr and Weyl 1956, p. 212, and Borriss 1956.
10. Nicolson 1983, chapter 1, esp. pp. 54–56. Schütt appears to fit neatly into the Humboldtian tradition of plant ecology described by Nicolson.

most fully expressed in his rectoral address at Greifswald in 1904. In these books he rationalized old ways while following what he believed to be the new path blazed by Hensen.

According to Schütt, "Hensen's question is briefly summarized as follows: what living things, microscopic or macroscopic are present at each location in the ocean, and how much of each form is present?"[11]

Individual species must be studied because their *qualities* (not just their quantities) were different, just as the seeds of a variety of species, available for agriculture, varied in their potential for the farmer. In a curious analogy, Schütt described Hensen's methods as being like the screening carried out at seed-testing stations which separated "indifferent" components from those of greater value.[12] Just as seeds were of different quality so too were plankton organisms; for example, copepods were of different qualitative significance from dinoflagellates. The nature of the organisms, their abundances, and the physical environment were emphasized by Schütt because he regarded them all as properties of a superorganism with emergent properties.

> The life of the sea as a whole is the product of very many individual factors. But these factors are not self-sufficient and independent of one another, the individual relationships do not occur without coordinated interrelationships, thus it is not a matter of the simple sum of phenomena in the sea but of a product, in which each single factor influences all the others, as a function of very many factors, which are all correlated with one another, and which reciprocally supplement, restrict and interlock with one another like the gear-wheels of a clock.

In other words,

> life in the sea appears to be a superorganism [*grosser Gesammtorganismus*] in which each individual organ has its own unique function, and which furthermore activates, hinders, furthers, regulates all the other organs and influences their functioning, thereby also exerting influence upon the totality of living phenomena, upon the metabolism of the whole [super] organism.[13]

The significance of species to the superorganism was likely to be proportional to their abundance, a variable that Hensen's methods resolved especially well; counts were necessary to allow assessment of the inter-

---

11. Schütt 1892a, p. 8.
12. Schütt 1892a, p. 13. Herman Rodewald's influence is evident; at the time he was director of the university's seed-testing station.
13. Schütt 1892a, pp. 9–10.

relations (*Zusammenwirken*) of organisms, amplifying the information gained from taxonomic discrimination of the species involved.

To the incautious reader, Schütt's views might seem like a late-nineteenth-century version of Naturphilosophie. He was explicit in renouncing such a view, claiming that his science was based on "positive observations" and "exact investigations," a viewpoint made clear in his rectoral address in 1904 when he showed how color of the water, a property that might have been beloved of Goethe or Schelling, was the product of its physical structure and the physiology of the organisms that lived in it, including denitrifying bacteria. He foresaw a positive, experimental cosmology that would investigate the properties of natural systems as self-renewing, homeostatic entities.[14]

It was clear to Schütt from the results of Hensen's cruises between 1883 and 1886 that there was a seasonal cycle of the plankton off Kiel. He noted that the main species peaked at different times: "For each species there proves to be a specific time at which its occurrence reaches a maximum."[15] The diatom *Chaetoceros* reached a maximum in March, its relative *Rhizosolenia* in June–July, whereas the dinoflagellate *Ceratium* was most abundant in October. Not only was there a succession of peaks, the individual peaks occurred at the same time each year.[16] This was an insight only possible after several years of investigation—Hensen's cruises and those carried out by Carl Apstein for Brandt beginning in 1888. Schütt described the phenomenon in a characteristically colorful analogy.

> One form appears, grows and vanishes yet again from the surface waters and makes way for another form, which now asserts its dominance for its own time, yet again to fade away, and this play repeats itself year after year with the same regularity as every spring the trees turn green and in autumn lose their leaves; with just such absolute certainty as the cherries bloom before the sunflowers, so Skeletonemas arrive at their yearly peak earlier than the Ceratiums.[17]

But it was the periodicity and succession of species as an assemblage, not the bloom per se, that interested Schütt in 1892. His view of the events occurring throughout the year was conditioned by his interest in the interactions of species, their contribution to the *Gesammtleben* (life of the assemblage of plankton cells). The factors that unified the plankton

14. Schütt 1904, pp. 23–25. Stripped of their *naturphilosophisch* terminology Schütt's views resemble some brands of holistic ecology or Lovelock's (1979) Gaia hypothesis.
15. Schütt 1892a, p. 93.
16. Schütt 1892a, pp. 94–97.
17. Schütt 1892a, p. 95.

assemblage interested him more than did the reasons for striking seasonal events in the plankton. Although he speculated about the causal factors involved in periodicity and succession, his interests were elsewhere.

The formative years in the development of a concept of the spring bloom occurred between 1896 and 1905. A variety of events happened then, beginning with the work of Carl Apstein, who had begun a five-year series of plankton collections for Brandt in 1888 (see Chapter 2 near n. 7). He had also, since 1891, been carrying out a series of investigations on the lakes of Schleswig-Holstein near Plön. Apstein's 1896 book *Das Süsswasserplankton* documented the increase of lake plankton which reached a peak in October. Moreover, the amount of plankton in each lake appeared to be a direct function of the amount of ammonia and nitrate present in the water.[18] Clearly the Kiel school realized (and was discussing) the general role of nutrients in 1896, but it was not ready to view them as determinants of seasonal fluctuations. Even in 1899, Hans Lohmann, discussing his plankton collections at Messina in 1896–1897 (see Chapter 2 near n. 66), concentrated mainly on the systematics of the species, although he speculated that the seasonal increase of appendicularians was due to an increase of their food, namely diatoms and dinoflagellates. The factors causing the appearance and disappearance of the phytoplankton were unknown.

Karl Brandt himself, summarizing in 1897 Hensen's, then Apstein's, cruises, was still preoccupied with species relations—the "biocoenotischen Verhältnisse" beloved of Schütt and other plant ecologists. But the 300 collections then available made it clear that two significant questions could soon be asked and perhaps answered: how much plankton was present at various times of the year, and what was its chemical composition? Both were directed toward Hensen's quest for a valid comparison of production on land and in the sea.[19] Increasingly after 1897 Brandt became concerned with the reality, then, later, the causation, of the evident seasonal changes of the amount of plankton off Kiel. In 1902, with the denitrification hypothesis well established as the central concept of his investigations, Brandt hinted that local studies, rather than geographically widely scattered ones might yield the most significant clues to the processes governing the abundance of plankton.[20] Seasonality of production itself did not much interest Brandt in 1902; he saw it as a microcosm of the regional differences first reported by Hensen as a result of the Plankton Expedition. But his conversion to studies of the factors governing seasonal production followed swiftly.

18. Apstein 1896, pp. 86–87, 102–106.
19. Brandt 1897, pp. 28–31.
20. Brandt 1902b, pp. 76–77.

Germany's involvement in a series of regular cruises sparked the detailed study of local production (something Hensen had espoused first in the 1880s). The German quarterly cruises for ICES, beginning late in 1902, made it possible to achieve Hensen's ideal of studying the gains and losses of populations at short intervals so that their annual production could be assessed accurately.[21] And the rapidly increasing volume of data on nitrate and silicate levels, made possible by Raben's new methods of chemical analysis, soon showed that the plankton *and* nutrients (especially nitrate and silicate) fluctuated in concert. By 1905 Brandt was turning increasingly to a chemical explanation of the fluctuations of plankton he had first documented, without much interest, in 1897. Inadvertently, the spring bloom had become a major concern of the Kiel school. When Victor Hensen described to a general scientific audience in 1906 how "das Wasser blüht" and that in spring "a cycle of the reproduction and the assemblage of species takes place" he expressed the Kiel school's transition from preoccupation with the biology of species, through interest in the determinants of geographical differences in abundance, to a broader concept, in which the phenomenon of the spring bloom (and a lesser autumn one) had a major place.[22]

This German work and attitude were influential in England, where in 1907 W. A. Herdman and his colleagues had begun to study the plankton of the Irish Sea. The Liverpool group was especially impressed by Hensen's methods, which J. T. Jenkins described first in English in 1901. In 1908 W. J. Dakin discussed the work of Hensen, Brandt, and Lohmann up to 1906, and James Johnstone gave a lengthy account of the work at Kiel in his book *Conditions of Life in the Sea*. Dakin had first-hand knowledge of current thought at Kiel, having accompanied Lohmann and Apstein on one of the ICES cruises by *Poseidon*, probably in 1906. His attention too was shifting from species to the causes of plankton maxima, for he noted that the peaks of phytoplankton and copepods occurred at virtually the same time at Kiel and in the Irish Sea. He speculated that "the development of the Copepoda is therefore influenced by . . . development of the phytoplankton or both are influenced by some other conditions hydrographical or otherwise at present unknown which cause a coincidence in the maxima of the two groups."[23]

In Norway too, the bloom, rather than taxonomically based research, soon became a major focus of attention. There, H. H. Gran, in his doctoral thesis, gave the clearest early expression of this newly appreci-

21. Brandt 1905b, p. 4.

22. Hensen 1906, pp. 364, 373. On Apstein's views of the first year's collections and the seasonal variations of the plankton see his report dated 1908, esp. pp. 42–43, 44–46, 49–50.

23. Dakin 1908b, p. 781.

ated phenomenon, basing his analysis on collections made during the first cruises of the new research vessel *Michael Sars* in 1900–1901.[24] Apparently independently of the Kiel school (though he knew of their work by direct contacts and through ICES), Gran observed clear peaks of diatoms in the Norwegian Sea in March–April, then again between September and November. In addition, there were many small peaks near the coast, apparently influenced by land.[25] "This yearly bloom," as he called the spring event, had a number of striking characteristics: it always began in shallow water; it was brought about by neritic, spore-forming diatoms; it was dominated by a succession of arctic or northern species; it reached a maximum between March and May, depending on latitude, usually just after the water reached its lowest temperature; and it disappeared one to two months after it began.

Gran's is the first explicit statement of the nature of the spring bloom, clearly articulated a few years before the Kiel group achieved its conceptual reorientation to local studies of recurring phenomena in the sea. Apparently quite independently of the Germans he realized that the spring phytoplankton outburst was "such a universal phenomenon that it must have universally acting causes."[26] Unencumbered by the need to verify the denitrification hypothesis (although he accepted it for a time) and by the hydrographic (and thus biological) complexity of the Baltic, Gran turned his original mind to the control of the bloom in a series of ingenious and original studies of processes in the water column. As I show in Chapter 5, it was his insights during the next thirty years that led eventually to a synthesis of biological and physical oceanographic thought about the control of the bloom.

By the second decade of the twentieth century, a "moderne Planktonforschung" (Hans Lohmann's phrase[27]) had developed, quite different

24. Haakon Hasberg Gran (1870–1955), born in Tønsberg on the Oslo Fjord, studied botany in Oslo. He became a research assistant in the Norwegian fisheries department under Johan Hjort, participating in the early cruises of *Michael Sars* beginning in 1900, and a botanist at Bergens Museum in 1901. His Ph.D. was awarded at Oslo (then Christiania) in 1902. Gran studied plant physiology in Leipzig with Wilhelm Pfeffer in 1896–1897 (when he may have met Nathansohn) and worked with Beijerinck at Delft and Texel on marine bacteria in 1901. In 1905 he was appointed professor of botany at Oslo, a position he held until 1940. During 1931 and 1932 Gran led a study for the Canadian government of the hydrography and plankton of the Bay of Fundy. Biographical details are given by Braarud 1956a, b, and Holmboe 1921. The first cruise of *Michael Sars*, with details of the ship, is described in Hjort 1901.

25. Gran 1902a, pp. 112–113.

26. Gran 1902a, p. 114.

27. See Lohmann 1912. He distinguished a "Gestaltungsproblem" from a "Bevölkerungsproblem" in plankton biology. The former involved functional morphology (problems of cell shape, flotation, and comparative physiology), the latter community and population biology, including the control of abundance. Lohmann's career embodied the evolution of a distinctive new brand of plankton ecology. His early career was as a taxonomist; later he espoused

from the taxonomically based studies that had dominated the literature at the time of Hensen's early work. It evolved as a result of Hensen's physiological approach, Brandt's application of agricultural chemistry to a problem that was originally geographically drawn rather than local and temporal, and Gran's acute observations, perhaps originating in his training as a plant physiologist, that "general causes" must exist for such a striking and widespread phenomenon as the spring diatom outburst in the Norwegian Sea. Gradually the concept of the bloom, latent in Schütt's ideas, explicitly expressed in Gran's accounts ten years later, took center stage in plankton research. Hand in hand with it, a chemical theory of the control of plankton abundance evolved. Its intellectual and practical roots lay in agricultural chemistry, which, throughout the nineteenth century had emphasized the roles of nitrogen and phosphorus in plant growth. Both Hensen and Brandt, working at Kiel in the agrarian landscape of Schleswig-Holstein, in close contact with their colleagues in the Agricultural Institute, viewed scientific agriculture, a chemical discipline that had moved far past its Liebigian origins in Germany, as the appropriate model for quantitative plankton research.

This was both a practical standpoint and an act of faith for Brandt. Scientific agriculture was demonstrably effective, but it could be applied to the sea only if adequate data were available. Hensen in 1887 had provided a few chemical analyses of organisms; later Brandt himself did the same,[28] but reliable chemical analyses of seawater were extremely scarce until those done by the ICES nations in 1902 began to accumulate. Trustworthy analyses depended on new, quantitative techniques for the primary nutrients such as nitrate, silicate, and phosphate. Raben's newly developed gravimetric and colorimetric techniques eventually produced the data Brandt required to support his ideas about the control of production in the sea, but before 1904 any chemically based theory had to be purely hypothetical.

Franz Schütt, shortly before he left marine research in the late 1890s, became a convert to Brandt's hypothesis. But his early belief, summarized in *Analytische Plankton-Studien*, was that the plankton cycle in the sea was controlled by the same factors that imposed seasonality on land, namely temperature and light.[29] Gran, also a plant physiologist, was

---

Hensen's quantitative views to make significant advances in plankton sampling and the estimation of production. Although Lohmann was the gadfly of the Kiel school, criticizing both Hensen's and Brandt's methods or ideas, the three men agreed intellectually on the need to study plankton using fully quantitative techniques. Their approach was more than methodological— it was an article of faith.

28. Brandt 1898.
29. Schütt 1892a, pp. 111–117.

equally impressed by the role of temperature and light, but his first major work on the annual cycle soon showed him that a more complex theory of control was necessary.

When Gran's doctoral research began on the first cruises of *Michael Sars* in 1900 and 1901 the hydrographic complexity of the Kattegat-Skagerrak area and the southern Norwegian Sea had just been revealed by the hydrographic work of the Swedish physical oceanographers Otto Pettersson and Gustaf Ekman.[30] About 1893 Pettersson and Ekman began to collaborate with Per Theodor Cleve (1840–1905), professor of chemistry at Uppsala, whose hobby was diatom taxonomy.[31] Noting that the complex water masses of the Skagerrak each had a distinctive group of phytoplankton species, Cleve initiated a more ambitious study of the distribution of phytoplankton in the Baltic and Norwegian seas. Each current and distinctive water mass proved to have a distinctive phytoplankton assemblage that could be used as a diagnostic marker if the origin of a water mass could not be determined unambiguously on physical oceanographic grounds. Collaborating at times with C. W. S. Aurivillius, Cleve developed these ideas into major biogeographic compendia for the North Atlantic region.[32] It was in opposition to Cleve's belief that seasonal changes of the plankton represented changes in the pattern of currents that Gran's doctoral research originated.[33]

As Gran observed in 1902, if Cleve's assertion that variations in the plankton represented only hydrographic changes were correct, Hensen's quantitative studies would be pointless. Gran believed intuitively that Cleve's view was too one-sided; biological changes as well as hydrographic variations must be responsible for seasonal changes of the plankton. To establish the relative significance of biology and currents a lengthy study was necessary; as he said, "It is essential to adequately study single species during their life cycles and their distribution at various times of year."[34]

Gran's results, which showed that a yearly succession of phytoplankton species occurred, and that there was a characteristic spring diatom

30. See esp. Pettersson 1894.

31. Cleve, born in Stockholm, received the D.Phil. at Uppsala in 1863 and became professor of chemistry there in 1874. He had a distinguished career as a chemist and became a member of the Nobel Committee in 1900. His longstanding interest in diatoms probably became oceanographic when he went to sea with Pettersson, Ekman, and C. W. S. Aurivillius on the west coast of Sweden in 1893. See esp. Gran 1912a, pp. 337–338, and Cleve 1896a, b; also Cleve-Euler 1928.

32. See esp. Aurivillius 1896, 1898, and Cleve's papers from 1896 through 1902; also Gran 1902a, pp. 71–75, 98–106.

33. Gran 1902a, pp. 7–9, 71–75, 98–106.

34. Gran 1902a, p. 9; also his discussion pp. 71–106.

bloom that began in March off the southern Norwegian coast and progressively later as he went north, could not be reconciled with Cleve's
hypothesis. But accounting for the control of the seasonal cycle was a
difficult problem; there could be several "specific causes." As Gran realized, light might well account for the origin of the diatom bloom, but it
could not account for its sudden decline when daylength was still increasing. Temperature too might be involved, especially if the diatoms were
cold-adapted, which he believed they were. But because temperate-water
and arctic species declined simultaneously, temperature alone could not
be responsible. Gran turned to nutrient supply.[35]

Accepting Brandt's hypothesis as the only one available to account for
changes in plankton abundance, Gran commented that "the relative
richness of the polar seas, which ultimately is based upon the increase of
pelagic diatoms, must have the same causes as the immense increase of
our neritic diatoms during the spring." In spring, as light increased,
diatoms that had overwintered could grow rapidly in the nutrient-rich
waters. Under these conditions, "the proliferation of diatoms is so intense
that one can readily suppose that the supply of nutrients will soon
become depleted."[36] The nutrient-starved diatoms would then decrease
until nitrifying bacteria became active during the summer, when a second
bloom could occur.

Although this adaptation of Brandt's hypothesis to the seasonal
changes of the plankton off Norway explained these phenomena in a
general way, Gran was aware of some troublesome problems. Why, for
example, did the bloom begin near the coast in March, but begin in the
open sea only in May? Perhaps, he suggested, the timing of open-sea
production was linked to the mixing of nutrient-rich polar water with
Atlantic water, especially along the zones of contact (fronts) between the
two.[37]

Gran's view of the seasonal plankton cycle in 1902 was a prescient one.
Schooled in plant physiology and Scandinavian physical oceanography,
he was prepared to apply Brandt's hypothesis to the Norwegian Sea,
invoking nutrient supply and demand as well as physical processes that
could change the vertical distribution of the nutrient salts. But when he
wrote of the bloom in 1902 not a single reliable nutrient analysis had
been done using water from the Norwegian Sea. Gran provided the first
inspired but unsubstantiated model of the annual plankton cycle.

Two years after Gran's thesis was published, the first nutrient analyses

35. Gran 1902a, pp. 113–116.
36. Gran 1902a, p. 115.
37. Gran 1902a, pp. 116–120.

TABLE 4.1

Emil Raben's first nutrient analyses (1904), which were used as the basis of Karl Brandt's first model of the seasonal cycle

		Baltic Sea					North Sea		
	Date	NH$_3$-N (mg/l)	(N)	NO$_2$/NO$_3$-N (mg/l)	SiO$_2$ (mg/l)	(N)	NH$_3$-N (mg/l)	(N)	NO$_2$/NO$_3$-N (mg/l)
1902	August				1.037	(4)			
	November				1.260	(3)			
1903	February				1.450	(2)			
	May				0.65	(1)			
	August				0.83	(1)			
	November				1.084	(6)			
1904	February	0.068	(13)	0.199	1.015	(6)	0.063	(12)	0.216
	May	0.065	(13)	0.170	0.655	(2)	0.065	(15)	0.217
	August	0.057	(13)	0.095	0.926	(2)	0.061	(13)	0.079

Note: N = number of samples. Ammonia and NO$_2$/NO$_3$ were determined with the same samples. Underlined numbers, according to Brandt, showed the depletion of silicate by diatoms in the spring, and summer denitrification by bacteria.
Source: Based on Brandt 1905b, pp. 7, 10.

came from Emil Raben's chemical laboratory in the Internationale Meereslaboratorium in Kiel (Table 4.1). Using samples from the quarterly cruises in February, May, and August 1904, Raben showed that ammonia was virtually constant throughout the year, nitrate was lowest in August, whereas in samples from August 1902 through May 1904, silicate was very low in May, more abundant in summer, and most abundant in November.[38]

To Brandt, this evidence yielded a clear circumstantial picture of the bloom. As temperature and light increased in the spring, the diatom *Chaetoceros* grew rapidly in the silicate and nitrate-rich surface waters. By May the bloom was ended by the depletion of silicate, which was the Liebigian limiting nutrient for diatoms. During the summer, silicate was regenerated, causing an autumn bloom of the diatom *Rhizosolenia*. Simultaneously nitrate was removed by denitrifying bacteria. The dinoflagellate *Ceratium* also reached a peak in autumn before the entire phytoplankton population decreased to winter levels. Accounting for its abundance so late in the year (September–October) was difficult; Brandt speculated that it was governed by phosphate, which, judging from the limited evidence of Raben's few analyses, was highest in the autumn.[39]

38. Brandt 1905a, pp. 15–16; Brandt 1905b, pp. 7, 10.
39. Brandt 1905b, pp. 9–12. In 1905 Brandt was growing phytoplankton in the laboratory; his early results showed that only phosphate stimulated the growth of the autumn plankton. The complete results of this work were never published.

FIGURE 4.1 Hans Lohmann photographed as professor of zoology and dean, University of Hamburg, probably about 1930. In earlier life, Lohmann was the *enfant terrible* of the Kiel school. He disputed the accuracy of Hensen net collections and did a year-long study of the plankton cycle in Kiel Fjord (1905–1906) which threw doubt on Brandt's ideas of chemical control. Photograph from Klatt 1935, courtesy of Gabriele Kredel, Institut für Meereskunde, Universität Kiel. Reprinted with the permission of the editors of *Mitteilungen aus dem Hamburgischen Zoologischen Museum und Institut.*

But the analyses were so few that he was forced to conclude that "it is not yet explained how the cessation of the second diatom maximum occurs, nor how it is replaced by the luxurious growth of *Ceratium*."[40]

Despite its shortcomings, this early model, presented to and published by ICES in 1905, provided the skeletal framework of all Brandt's later work on the seasonal cycle of plankton. From it, the Kiel school, which had begun its work by attempting to explain spatial patterns of plankton production, moved steadily toward a broad theory of the seasonal cycle, impelled at first by the inexorable ticking of the ICES clock, its quarterly cruises, then by Brandt's obsession to show that the denitrification hypothesis was not superseded by, but lay at the root of, the theory of seasonal plankton dynamics. The result was a complex theory in which nitrogen played only a part, and to which other members of the Kiel school made major contributions.

Although Karl Brandt was the great motivator and organizer of research at Kiel between 1888 and World War I, the most energetic and original member of the Kiel school was Hans Lohmann (1863–1934) (Figure 4.1). Lohmann, who was born in Hannover, began his studies in Göttingen in 1885, then came to Kiel in 1886 to study natural science and

40. Brandt 1905b, p. 11. Later work showed that the dinoflagellates usually peaked before the diatoms.

medicine. After spending a clinical year at Greifsburg, apparently his last contact with medical studies, he completed a doctorate (1889) on marine mites at Kiel. His Habilitationschrift (1893) dealt with the results of the Plankton Expedition. At the instigation of Hensen and Brandt, Lohmann collected plankton at Messina between April 1896 and February 1897 and at Syracuse in 1900–1901. In 1902 he took the *Von Podbielski* to New York, collecting open-ocean plankton along its route. When he returned, Lohmann became assistant to the Kiel Commission, a post he held until 1911.[41]

During his study of appendicularians from the Plankton Expedition, Messina, and Syracuse, Lohmann became aware that very small cells, which he called "nannoplankton," were common in the sea (see Chapter 1).[42] In two studies published in 1901 and 1903 Lohmann showed that these small cells were abundant at Kiel and in the Mediterranean, by using filter paper or centrifugation to concentrate the plankton rather than a Hensen net, and that they appeared to be the food of many planktonic animals.

If nanoplankton were widespread and abundant, no study of the seasonal cycle of the plankton using only Hensen nets could give a satisfactory quantitative picture of its changes, or of the production of planktonic organisms. That was Lohmann's belief when in 1905 he set out to document fully the yearly plankton cycle at Kiel using his new methods. He had two aims, to compare losses from Hensen nets with the total content of the water column, and to show the importance of small cells in the metabolism of the sea, if indeed they were underrepresented by a factor of 5 to 100 in plankton hauls, as he believed.[43]

Lohmann's study is still one of the most complete ever done on planktonic organisms. Once a week between April 1905 and August 1906 he left Kiel at 7:00 A.M. in a rented motorboat for an hour-long ride to a station just west of the ship channel off Laboe, at the mouth of the Kiel Fjord. There he took vertical plankton hauls with Apstein's version of a Hensen net, pumped 54 liters of water (Figure 4.2) (it was preserved, then filtered in the laboratory), and took samples with a Krümmel water battle at 0, 5, 10, and 15 meters for centrifugation and salinity and nutrient analyses. By 10:00 A.M. he was back in the laboratory, where filtration, centrifugation, identification, and counting of all the organisms from

41. See espeically Hentschel 1933 and Klatt 1935. In 1911, Lohmann went south on Filchner's German Antarctic Expedition. He left Kiel in 1913 to become director of the hydrobiological section of the Zoological Museum in Hamburg, then director of the museum (1914); Lohmann became the first Professor ordinarius of zoology at the University of Hamburg when it was founded in 1919.
42. Lohmann 1909a, p. 233. The spelling "nanoplankton" is etymologically correct.
43. Lohmann 1908b, p. 134.

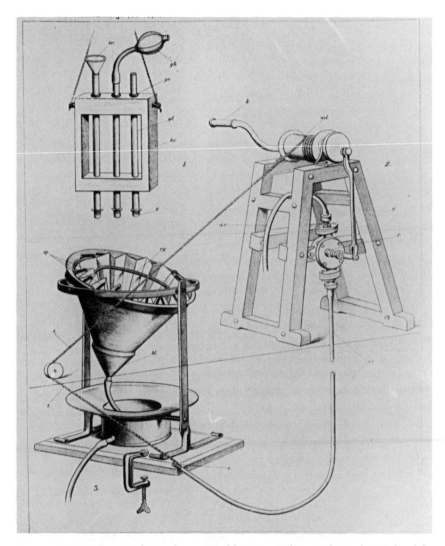

FIGURE 4.2 A pump similar to the one used by Hans Lohmann during his study of the plankton of Kiel Fjord in 1905–1906. From Lohmann 1903, Tafel III.

bacteria to macroplankton took until night or longer. His friend Gebbing did the chemical analyses. Hydrographic information was available from the Kiel Lightship, in addition to measurements of nutrients in Kiel Bight from Raben's and Brandt's quarterly ICES cruises.[44]

The volume of Lohmann's results, their unprecedented completeness and detail, confirmed in a general way what was known already by the

44. On his methods and sampling arrangements see Lohmann 1908b, pp. 137–140, 217, 232ff.

Kiel school, that the plankton was depauperate in winter but rich in the warmer months due to the rapid growth of a succession of plants and animals. The detail also allowed Lohmann to estimate the production of the plankton on a month-by-month basis, using very rough-and-ready assumptions about the growth of the algae and the grazing or predation rates of the consumers.[45]

Superficially, the plankton cycle that Lohmann observed at Laboe was much like that described by Brandt in 1905 and by Gran before him. During the short dark days of winter, plankton volume was very low. Production increased in spring when the diatoms bloomed; later, in the late summer, diatoms bloomed again and the "peridinians" (dinoflagellates, especially *Ceratium*) also reached their peak. Grazing animals were most abundant during the spring diatom bloom and late in the summer coinciding with the autumn bloom. After September the whole plankton community decreased rapidly to winter levels.[46] But Lohmann's account of the causal factors at work contained a number of significant heresies.

First, and most distressing to Brandt, Lohmann found it hard to match nutrient abundances with the changes of plankton stock, especially the blooms. Admittedly, silicate and phosphate were low in spring and summer, then increased in autumn and winter. But there was no clear annual periodicity of nutrients such as nitrate. In fact, the year-to-year variations were greater than seasonal fluctuations. The summer depletion of nutrients that Brandt had seen appeared to be a fiction; so too did the close link he had suggested between nutrient abundances and the blooms.[47]

Lohmann postulated that the spring bloom, as well as the overall seasonal abundance of plankton, resulted from the interaction of temperature and light. Correlating L. Weber's figures describing hours of sunlight at Kiel[48] with the temperature cycle known at Kiel, he suggested that light was the most significant factor at the beginning of the spring bloom, then as the season advanced there was a close link between the high volume of summer plankton and the product of temperature and light (in effect, degree-days).[49] These physical factors, not the law of the minimum, ultimately controlled plankton abundance.

45. Lohmann 1908b, pp. 344–347, esp. the table on p. 347. This table contains errors (or was modified by Lohmann in some way difficult to reconstruct now), some of which were discovered by H. W. Harvey, who printed a corrected version (Harvey 1928a, pp. 165–166). I have been unable to recalculate some of Lohmann's figures using the data in his monograph.

46. Lohmann 1908b, pp. 332–343, 349–353.

47. Lohmann 1908b, p. 233–236; see esp. figure 17, p. 236.

48. Weber (1889, 1895) used a specially designed photometer that compared sunlight with the intensity of an amyl acetate lamp at a fixed distance from the observer's eye. He tabulated daily and monthly means of sunlight and the maxima and minima at various wave lengths, as well as hours of sunshine per day throughout the year.

49. Lohmann 1908b, pp. 233, 332–333.

Light could not explain the rapid decline of the spring diatom bloom. Rejecting the law of the minimum as well as the role of light, Lohmann believed the spring diatom bloom declined when brackish eastern Baltic water reached the western Baltic in early summer. Then euryhaline, temperature-tolerant small cells replaced the diatoms until Atlantic water was restored to the western Baltic late in the summer, bringing about the fall bloom of *Ceratium* and diatoms.[50] This in its turn declined as the daylength decreased in autumn.

At the beginning of his work on the plankton cycle Brandt had been uncertain whether or not plankton actually declined in abundance during the summer.[51] By 1905, basing his ideas on Hensen's net collections and chemical analyses, he was convinced that the decline was real and that it was brought about by nutrient depletion.[52] Now Lohmann suggested that this was largely an artifact of net collections; indeed diatoms did decrease, but they were replaced by a host of nanoplankton cells, both colorless and pigmented, and by dinoflagellates, all important components of marine production systems.[53]

Lohmann's views could not be dismissed as merely wrong-headed. He had carried out the most thorough study of yearly changes in the plankton yet done, using the most advanced techniques of collection and analysis, all of them espoused by Brandt. He was acknowledged to be in the front rank of plankton researchers, with years of experience all over the Atlantic. As a result, the refutation of his heretical views was likely to require equally rigorous and long-lasting studies. These were undertaken by some of Brandt's students between 1911 and 1916, leading to re-evaluation of Lohmann's observations and a synthesis of his most significant insights with Brandt's chemically based theory. Brandt's work on the plankton cycle, which had begun to stagnate under the weight of work for ICES, was renewed shortly after Lohmann's ideas were published, as a direct result of their implications. The most significant question about Lohmann's results, according to Brandt, was whether they were representative of the sea as a whole, or if they represented some

50. Lohmann 1908b, pp. 328, 334–336, 349–351.
51. Brandt 1898, p. 90.
52. At first (1905b) Brandt believed that the summer minimum was attributable to removal of NO_3 by denitrifying bacteria. Very quickly he accepted the idea that nutrients declined because of uptake by the phytoplankton rather than because of bacterial action.
53. Lohmann 1908b, p. 332 and p. 335, figure 23. Lohmann's view of a complex reality was correct in essence. Lenz (1981, figures 7, 10) shows that at Kiel plankton biomass, although variable, is high from March through October and that primary production (^{14}C uptake) remains high through the summer. Chlorophyll *a*, on the other hand, often peaks in March–April and the autumn, probably reflecting the two diatom blooms. Regarding nutrients, Smetacek (1985) believes that Brandt's and Lohmann's views differed because Kiel Fjord was more eutrophic than Kiel Bight due to sewage and ships' wastes.

anomalous situation in Kiel Fjord. Thus he believed it was necessary to verify Lohmann's observations; equally important, they had to be placed in a broader context. Lohmann's heresies therefore led to a major synthesis of the Kiel school's work on the plankton cycle, published in 1919–1920 as the third of Brandt's monographs, *Ueber den Stoffwechsel im Meere.*

In 1910, Brandt's students and assistants in the Internationale Meereslaboratorium began a series of new studies of the annual plankton cycle. Among these was Theodor Büse, who worked up a year-long series of weekly plankton and hydrographic samples collected by the captain of the Fehmarnbelt Lightship between January 1910 and March 1911.[54] Beginning in March 1912, Brandt's assistant, Karl Müller, and a doctoral student, Alfred Wulff, restudied Lohmann's Laboe station, taking samples every three weeks for the next 2¼ years.[55] In addition, a series of six cruises by *Poseidon* between December 1910 and June 1911 provided an abundance of new information on seasonal changes of nutrients in the Baltic.[56] Uniting the new data with the old, which went back to Hensen's first cruises in 1883, by 1919 Brandt had available nine yearly series of plankton collections and at least 2200 determinations of nutrients, especially nitrate, ammonia, organic nitrogen ("albuminoid ammonia"), silicate, and phosphate.[57] This abundant information, combined with new ideas about physical circulation, enabled Brandt to unify his ideas about the control of the annual plankton cycle, to explain the richness of coastal waters compared with the open sea, and to account for the poverty of tropical seas. His synthesis was a physically based multinutrient hypothesis significantly different in detail from his ideas in 1905, in which the denitrification hypothesis played a relatively small, though still significant, part.[58]

Lohmann's claim that changes of salinity and temperature, rather than nutrient depletion, were the most important factors controlling the plankton blooms was a major stimulus to Brandt's reassessment of the old data; it also was the reason for Müller and Wulff's lengthy new study at Laboe. Brandt soon concluded that Lohmann was wrong, misled by

54. See Büse 1915, pp. 231–233.
55. W. Busch 1916, pp. 27–30. The samples were analyzed by Werner Busch, who received a Ph.D. in 1916. His publication of the results was completed in collaboration with Brandt when Busch, an army physician, was on furlough late in 1915. Karl Müller was killed during the war.
56. Brandt 1920, pp. 340–343.
57. The first lengthy series of phosphate analyses was published late, in 1916, when Raben had refined a gravimetric technique for that nutrient salt. Brandt's first lengthy discussion of phosphate is in Brandt 1920, pp. 273–304.
58. Brandt 1920, pp. 198–200; Brandt 1925, pp. 78–79. Brandt's assessment of the controlling role of silicate, nitrate, and phosphate changed significantly between 1905 and 1919.

TABLE 4.2

Monthly nutrient levels (in milligrams per cubic meter) in the North and Baltic seas, from 1191 analyses by Emil Raben between 1902 and 1914

	Phosphate		Nitrate		Nitrogen Ammonia		Organic		Silicate	
Month	Baltic Sea	North Sea	Baltic Sea	North Sea	Baltic Sea	North Sea	Baltic Sea	North Sea	Baltic Sea	North Sea
Feb.	152	154	**114**	157	60	64	—	—	860	738
March	180	139	200	—	34	—	92	693	—	—
April	—	125	21	—	71	—	650	—	—	—
May	105	75	129	144	67	65	159	110	**660**	658
June	104	67	189	132	39	45	130	192	800	1030
July	—	104	—	60	—	50	—	112	—	—
Aug.	157	123	**83**	85	66	60	—	—	740	848
Sept.	—	155	132	200	87	40	160	153	—	905
Oct.	—	118	—	105	—	43	—	118	1000	—
Nov.	144	161	**80**	89	62	59	135	85	1020	864
Dec.	115	—	138	—	43	—	130	—	—	—

Note: Figures in boldface type represent changes of nutrients associated with the spring and autumn blooms, according to Brandt. — = no data.
Source: From Brandt 1921, p. 23.

unusually brackish conditions in the early summer of 1906. In fact, based on a reassessment of data from Laboe and other sites in the western Baltic, Brandt was able to show that the spring diatom bloom invariably declined even if the salinity did not.[59] Thus Lohmann's claim that the spring bloom was ended by the advent of warm, brackish Baltic water and that the fall bloom coincided with the replacement of Baltic by Atlantic water could not be maintained.

Throughout the year, and from year to year, contrary to Lohmann's conclusion that nutrients did not vary in significant ways with the plankton, Brandt could see significant patterns of change that he was convinced were the cause of fluctuations in plankton abundance and species composition (Table 4.2). Especially striking was the decline of silicate, then of nitrate and phosphate, as the spring diatom bloom came to an end (Figure 4.3). Although the data on phosphate were limited, Brandt claimed that it too had a major role in limiting the bloom, since even though nitrate and silicate began to increase again in May and June, the diatoms did not. Only phosphate, which remained low until later in the summer, could account fully for the decline of the diatoms.[60] Later, after the summer minimum (which at least applied to the larger cells such as

59. W. Busch 1916, pp. 30–31.
60. Brandt 1920, pp. 273, 358, 422.

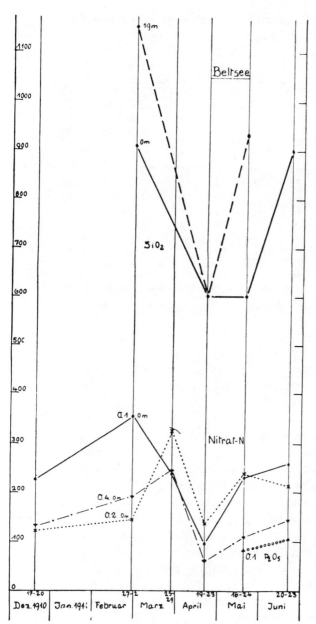

FIGURE 4.3 Changes of the major nutrients in the Beltsee off Kiel, spring 1911, expressed in milligrams per cubic meter. Station numbers accompany the months. Note the striking decrease of silicate and nitrate, which coincided with the decline of the spring diatom bloom. From Brandt 1920, p. 343.

diatoms, if not Lohmann's nanoplankton), regeneration of nitrate and phosphate from the bottom accounted for the fall bloom of diatoms and *Ceratium.*[61] Clearly, based on the array of data before him, the plankton was controlled by several nutrients, each obeying the law of the minimum. In addition, if the various phytoplankton species required different optimal amounts of each nutrient or had different lower limits of uptake, the seasonal succession of species as well as the general patterns of abundance could be explained by nutrients alone.[62]

Although Lohmann had been wrong in attributing the decline of the spring bloom to vertical water movements, Brandt was willing now to concede that vertical water movements had a profound effect on the seasonal plankton cycle. In this, he gave little credit to Nathansohn, whose ideas had been widely accepted, but which Brandt despised. In his 1919 monograph Brandt attempted to show that he and Otto Krümmel had recognized the role of vertical circulation as early as 1893 and that it had played a significant part in his ideas of the factors controlling plankton production from then on.[63] In truth, until 1916 Brandt's publications say very little about vertical circulation. When his interest in vertical circulation arose it was undoubtedly in response to Nathansohn's ideas, increasingly influential since 1906, but the immediate cause was his discovery of the work of the American aquatic biologist, George C. Whipple.[64]

Whipple, who in 1890 was appointed biologist in charge of the Boston Water Works laboratory, was concerned with the effect of diatom blooms on water quality. His work on a number of reservoirs in Massachusetts showed that there were two peaks of diatoms a year in deep lakes (just like the spring and fall blooms in the sea) which were closely associated with renewal of nutrients from deep water to the surface when vertical circulation took place in spring and fall. Whipple suggested that vertical circulation had two effects, first to carry sinking diatoms or their spores back to the surface, second to bring nutrients (originating from decomposition of organic matter in the sediments) back into the lighted upper waters. Under stratified summer conditions, diatoms sank below the productive zone and nutrients could not be regenerated.

Whipple's studies, published beginning in 1894, were summarized in

---

61. Brandt 1920, p. 278; Brandt 1921b, pp. 25–26.

62. Brandt 1920, pp. 239–40, 302, 374–375; see also Chapter 3 near n. 107.

63. Brandt 1920, pp. 191–221.

64. George Chandler Whipple (1866–1924) became director of the Mount Prospect Laboratory, Department of Water Supply, Brooklyn, New York, in 1894; later he was Gordon McKay Professor of Sanitary Engineering at Harvard.

his book *The Microscopy of Drinking Water* in 1899.[65] Apparently, they were not known in Europe until 1910, when Adolf Steuer mentioned them in his book *Planktonkunde*, a review of plankton studies. H. H. Gran, either independently or through reading Steuer's book, quickly recognized the significance of Whipple's observations.[66] Brandt was slower; sometime between 1916 and 1919, he read Steuer's book and realized that Whipple's hypothesis about the regeneration of nutrients could be applied to the plankton cycle at Kiel.[67] Once linked with the wind, vertical circulation of nutrients could be used to explain some of the anomalies in Lohmann's results; in addition it explained seasonal events and year-to-year regularities of nutrient supply to the surface waters.

Even before reading Whipple's work, Brandt had begun to examine the role of the wind in regenerating nutrients. Commenting on his student Werner Busch's analysis of the Kiel Fjord samples taken in 1912,[68] he stated that a major fault of Lohmann's work had been its neglect of current-caused changes of nutrients within the fjord. Noting that sea level there was strongly dependent on wind direction, Brandt suggested that west and southwest winds increased nutrients in the surface waters by displacing surface water seaward; it was replaced by nutrient-rich water from the bottom. The fall bloom in the fjord could be accounted for in this way, since it occurred in mid-September following the onset of westerly winds in late August. Although this was a general phenomenon in the Baltic, as at Laboe, there were many wind-induced local effects at Laboe that were not characteristic of Kiel Bight or of the Baltic as a whole. Lohmann had been misled by the hydrographic complexity of Kiel Fjord and had neglected the effect of wind in such enclosed waters.[69]

Whipple's work made it clear that two-bloom pattern of seasonal production was a general one and that both nutrients and wind must be invoked to give a fully coherent explanation of the patterns observed in nature. Brandt now concluded that, in general, blooms occurred only when nutrients were stirred into the surface waters. Moreover, the abundance of diatoms and wind strength should be directly related because

65. See Whipple 1894; 1895; 1896; 1897; and 1899, chapter 7. The spring and fall blooms in lakes were first noted by G. N. Calkins in 1892, published in 1893.

66. Gran 1912a, pp. 377–380; Gran 1915, pp. 135–137. A brief remark by Gran (1912a, p. 372) suggests that he may have read *The Microscopy of Drinking Water* in 1907.

67. Brandt 1920, p. 198.

68. W. Busch 1916, p. 35. See also n. 55 above.

69. W. Busch 1916, pp. 39–40. Smetacek (1985) makes the point that Brandt's and Lohmann's views about the nature of the cycle may be reconciled by recognizing that Kiel Fjord was heavily polluted by effluents from ships and the city at the turn of the century. As a result, nutrients varied less there than in the open waters of Kiel Bight.

greater stirring would result in higher levels of nutrients for plankton production. To support this idea, Brandt showed that the same months in years with different wind strengths had different-sized blooms, (for example, stronger winds in March led to bigger blooms). Even the timing of the spring bloom could be affected by differences in the onset of windy periods.[70] Wind direction, too, was important, especially in the Kiel region; westerly winds moved surface water off shore, causing vertical "reaction currents" that brought nutrients from the enriched near-bottom region.[71] The beginning of the fall bloom, in particular, could be explained in this way, for demonstrable nutrient enrichment of the surface water and the bloom followed a week or two after the seasonal shift of summer winds at Kiel to the west or southwest. Once the winds changed, Brandt's nutrient data showed a decline of ammonia and nitrate in the sediments which coincided with their transfer to the surface waters.[72] Shortly thereafter the fall bloom began, originating in shallow water and spreading seaward, because the nutrient supply increased near land first as a result of stirring by the wind.[73]

Nutrients (and thus indirectly the wind) were also implicated in the control of dinoflagellate abundance according to Brandt's synthesis. Lohmann, like many before him, had observed the summer-long increase of *Ceratium*, which reached a peak in early autumn just before the second diatom bloom. Lohmann provided strong evidence that temperature and light were involved in the increase, but it had not been clear what role, if any, nutrients played, nor was it easy to explain why *Ceratium* and the autumn diatoms reached their peaks at different times. Brandt's explanation, largely an ad hoc one of significantly greater complexity, went beyond Lohmann's temperature–light correlation to include nutrients and grazing.[74]

By calculating the growth rate of *Ceratium* in addition to its abundance, Brandt showed that growth rate was closely linked to light and temperature, but that the population reached its peak three months later than the light-temperature peak (Figure 4.4). This, he believed, was explicable only if grazing (an unmeasured factor) kept the numbers of *Ceratium* low initially, then decreased; thereupon the fall dinoflagellate bloom occurred.[75] Nutrients too were implicated, for once the diatoms

70. Brandt 1920, pp. 348–357; Brandt 1921b, pp. 26–28; Brandt 1925b, pp. 89–97.
71. Busch 1916, p. 36; Brandt 1920, p. 353.
72. Brandt 1921b, pp. 28–29; Brandt 1920, p. 424.
73. Brandt 1920, p. 357.
74. Brandt 1920, pp. 376–393; Brandt 1921b, pp. 16–18. Lohmann (1908b, pp. 335–336) had suggested that *Ceratium* was favored by high light and temperature whereas diatoms were not; once the temperature decreased in the fall *Ceratium* also decreased.
75. Brandt 1920, pp. 380–382; Brandt 1921b, pp. 15–18, esp. figure 1.

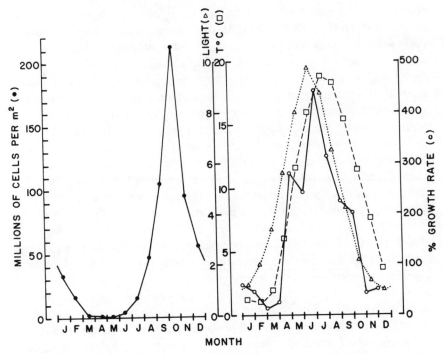

FIGURE 4.4 The relationship between the growth rate of the dinoflagellate *Ceratium* and its seasonal peak of abundance. Growth rate (right), expressed as percent increase per month, is closely correlated with light (in relative units) and temperature, which are greatest in June–July. The population (left) peaks later, in October. Brandt accounted for this by suggesting that *Ceratium* was kept in check, despite its very high growth rate, by high levels of grazing in midsummer. Modified from Brandt 1921b, p. 17.

began to increase, *Ceratium* decreased.[76] There was also a significant effect of wind; in years when the wind was stronger, causing more vigorous regeneration of nutrients, diatoms dominated in autumn, whereas in years with light, variable winds dinoflagellates were favored.[77]

In summary, the main features of Brandt's mature theory were as follows.[78] In spring, diatoms increased, primarily as a result of the regeneration of nitrate and phosphate caused by bacterial action and wind stirring. Increasing light and temperature promoted the spring bloom but were not the primary causes. The bloom was ended by nutrient depletion; usually phosphate was the limiting nutrient, but it might be silicate, nitrate, or even organic nitrogen. Diatoms remained low during the

76. Brandt 1920, pp. 383, 403–404.
77. Brandt 1920, pp. 404–405.
78. Brandt 1920, pp. 189–190, 421–426.

summer, except when local, wind-induced nutrient regeneration occurred in the inner fiord, but *Ceratium* and small flagellates, capable of
growing at low nutrient concentrations in strong light and at high temperatures became abundant. Phosphate slowly increased, but species
capable of using it at significant rates were not present. Nutrient regeneration proceeded in the bottom waters, but except under unusual
circumstances the nutrients remained below the surface waters.

By the end of the summer, phosphate and especially nitrate had increased in the surface waters. In direct response, both the dinoflagellates
(*Ceratium*) and diatoms increased, resulting in an autumn bloom and a
rapid decrease of the major nutrients. This was complicated by competition between diatoms and dinoflagellates for nutrients; diatoms were
favored by high nutrient levels, but at low levels *Ceratium* could, in some
unknown way, be more productive than the diatoms. Nitrate depletion
brought to an end the autumn diatom bloom; *Ceratium* by contrast
decreased in response to the decrease of daylength and temperature.

During winter these populations remained low because of low light,
temperature, and nutrients; Brandt believed that it was primarily low
nutrient levels that prevented the bloom beginning when the days lengthened in January and February.[79] Only when nitrate increased in late
winter could the spring bloom begin, renewing the cycle and reestablish-
ing the complicated interplay of nutrients and phytoplankton.

Brandt's response to Lohmann's original and unorthodox conclusions
added much new detail to the Kiel school's theory of plankton dynamics.
Brandt acknowledged that "the picture that I give will not only be
supported by wider investigations but will be corrected in many ways. It
could not be otherwise with such complicated problems."[80] Nonetheless, by 1920 it was a coherent picture, based on chemistry and simple
physical oceanography, that appeared to account for all the features of
the yearly cycle. The Kiel school's theory of seasonal plankton dynamics
in 1920 reflected twenty years of work by Brandt and his colleagues. It
had grown from a microbiological explanation of Hensen's paradoxical
observations into a sophisticated biological, physical, and chemical theory of seasonal changes in plankton abundance based on the insight that
local seas examined at all seasons were a microcosm of the more distant
open oceans. Characteristically, Brandt used inductive correlational evidence—such as the association between decreases of nitrate or silicate
and the decline of spring diatom populations, or the lag between *Ce*-

79. Brandt (1920, p. 426) admitted that nutrient levels in winter were very poorly known
since the seasonal ICES cruises were in November and February.
80. Brandt 1920, p. 421.

ratium growth rate and its peak population—as evidence for causation without being able to exclude other hypotheses. The theory was first and last *observational*, despite Brandt's early attempts to give firmer foundation to his ideas using culture experiments.

Although the ultimate weakness of the Kiel school's theory in 1920 was its lack of empirical content, its inductive rather than deductive structure, there were practical weaknesses as well, noted by biologists who worked on a larger geographical scale in regions with simpler hydrographic structure than the Baltic. According to Brandt's model, seasonal plankton production should be controlled mainly by nutrient levels, they, in turn, linked to enhanced vertical circulation from the bottom. Yet as Gran had noted as early as 1902, the bloom did not develop uniformly in regions that appeared to be homogeneous physically or biologically.[81] Why, for example, did the spring bloom develop first close to the Norwegian coast rather than farther seaward? How could regional differences develop when temperature, light, and nutrients appeared to be uniformly distributed from place to place? And why did a summer minimum occur even when hydrographic conditions appeared ideal for high levels of production?

The solution to these problems was achieved when the Kiel school's synthesis was united with a yet more sophisticated application of physical oceanography to the biology and chemistry of the bloom. Vertical circulation, which came late to the theory of plankton dynamics, proved to be the key to the major local and temporal anomalies of plankton production. As its application to the theory of the bloom broadened, the anomalies apparent in nature became explicable, but there was a casualty—the denitrification hypothesis received its final, fatal blow.

---

81. Gran 1902a, pp. 113, 117–118. By 1915 Gran subscribed to the Whipple-Brandt view of the control of the bloom; see his discussion then, Gran 1915, pp. 134–138.

5 Hydrography and the Control of Plankton Abundance: Solving the Problem of Plankton Blooms

> Nathansohn has clearly formulated that . . . renewal can take
> place in two ways: either by a supply from land or by circula-
> tion, and circulation can . . . occur either by convection,
> when the surface strata are cooled, or by rising currents
> which bring great quantities of water up from the bottom to
> the surface. . . . An extensive material of observations is,
> however, required in order to enlighten us as to the manner
> in which the renewal of nutrition takes place in the various
> regions of the sea and conditions their constantly fluctuating
> productivity.
>
> H. H. GRAN 1928

During the 1920s and 1930s the Kiel school's model of the control of the plankton cycle, based on the uptake and regeneration of nutrients, was widely accepted. This synthesis provided the focus for field and experimental studies that resulted in opinions, such as that of Gran, that "the theory of BRANDT—that phosphates and nitrates can be and often are the limiting factors of the production of the sea—and that of NATHANSOHN—that these salts accumulate in the depth of the sea and circulate by upwelling of deep water and in winter by the temperature convection—have been confirmed by a series of newer investigations."[1] Among the field studies were those of W. R. G. Atkins at Plymouth, of H. H. Gran and Birgithe Ruud Føyn in the Norwegian Sea, and of Sheina Marshall and A. P. Orr in the Clyde Sea. Each showed that the plankton varied in synchrony with fluctuations in nutrient supply, particularly the abundance of the phosphate ion. Marshall and Orr's work, in particular,

---

1. Gran 1930, p. 6; see also his comments in Gran 1928, p. 197.

a detailed weekly study in 1926, was widely cited as showing the role of phosphate in limiting diatom growth during the spring bloom.[2]

Despite the widespread acceptance of chemical control of pelagic production by all the major students during the first three decades of the twentieth century, direct evidence for the controlling role of nutrients such as phosphate, nitrate, and silicate was rare. Brandt in 1905 referred briefly to unpublished experiments showing that phosphate could stimulate the growth of the autumn plankton.[3] The work of E. J. Allen, begun in 1905, showed that diatom growth was proportional to the amount of nitrate in the medium, and that phosphate was frequently important. In 1907–1908 Gran added ammonia and nitrate to phytoplankton suspended in glass bottles in the sea; most species increased when their nutrients were increased. Later, in 1916 and 1917, Torbjørn Gaarder and Gran did a series of ingenious in situ experiments in Oslo Fjord which indicated that ammonia, nitrate, and phosphate could increase phytoplankton growth, although their results were not always consistent or easy to explain. During experiments in the spring of 1922 Gran found that nutrients had the greatest effect when the bloom was declining, presumably because nutrient depletion was beginning, and that two added nutrients might be more effective than one.[4] Most influential of all were the careful experiments of Ernst Schreiber at Helgoland, using bacteria-free phytoplankton cultures, which showed the controlling roles of light, temperature, and particularly nutrients in minimum quantities. He was able to estimate the "productive power" of North Sea water by growing laboratory cultures of algae in freshly collected water; there was close correspondence between the growth of the algae and the nitrogen and phosphorus content of the water.[5]

Throughout this period experimental studies were far less important and influential than inductive, correlational accounts of the control of annual cycles in such locations as Kiel Bight, the Norwegian Sea, Oslo Fjord, or the Clyde Sea. In each location the general features of plankton control by nutrients were evident, but local variations, such as those noted by Gran as early as 1902, preoccupied each investigator. Why, for example, did the bloom begin so late in the apparently more temperate

2. Marshall and Orr 1927, pp. 842–855.

3. Brandt 1905b, p. 11.

4. Gaarder and Gran's studies might have been more influential had they been published earlier. They did not reach a wide audience until 1927. For a brief description of the first experiments see Gran 1912a, pp. 376–377. Later experiments were done near Bergen in 1922 (Gran 1927, pp. 29–34).

5. Schreiber 1927, esp. pp. 22–29, 29–31. In addition, Schreiber (1929) showed that algal cultures could be used as bioassay devices to verify the results of new nutrient methods, such as those proposed by Atkins and Harvey.

nutrient-rich open Atlantic but early in the cold inshore waters of the Norwegian coastal current?[6] Nathansohn's view that hydrographic variations had a major role in the plankton cycle, outranking nutrient depletion in importance, and explaining local variations in production, had been clearly expressed: "Variations of the plankton flora are attendant phenomena to hydrographic variations, and if we find sparse vegetation in a place in summer which was rich in spring, it indicates that at present the mixing conditions, under which the development of the surface waters takes place, have become unfavourable for plant production."[7]

Gran, who worked with Nathansohn for a time, readily accepted the idea that vertical circulation restored nutrients to the surface waters. Even Brandt, reassessing his observations of the annual cycle at Kiel in the light of Whipple's theories, agreed by 1920 that vertical circulation was an essential component in the resupply of nutrients to productive surface waters.

On the premise that vertical circulation was essential to resupply nutrients to surface waters, mixing played a positive role in the theory of plankton dynamics accepted by Brandt, Gran, and many others until the early 1930s. Mixing and high levels of production went hand in hand; production declined when and where the water column was stabilized, shutting off the nutrient supply. This reasonable, simple view was replaced early in the 1930s, mainly as a result of studies in Norway. A new view of vertical circulation replaced Nathansohn's; ironically, before it brought explanatory light to many apparent anomalies of the plankton cycle, application of the old view completely eliminated the need for Brandt's denitrification hypothesis. By fully accepting the importance of vertical circulation as a source of nutrients to the surface waters, Brandt was forced to reject the hypothesis that had been the basis of his research for three decades.

The Norwegian studies of plankton dynamics that contributed significantly to the revision of Brandt and Nathansohn's ideas in the early 1930s began much earlier with Gran's doctoral research in 1900–1901, in which he examined the plankton of the Norwegian Sea on the first cruises of the fisheries research vessel *Michael Sars* under Johan Hjort, the Norwegian director of fisheries. As discussed in Chapter 4, Gran (Figure 5.1) noted that the spring bloom off Norway began near the coast, then spread seaward, occurring last in Atlantic water far off shore. He speculated then that the coastal bloom was stimulated by nutrients from land, and that polar meltwater (as Nansen had suggested) or the movement of

6. Gran 1902a, p. 116; Gran 1930, p. 71.
7. Nathansohn 1908, p. 50.

FIGURE 5.1 H. H. Gran in his office in Oslo, 1932. Photograph courtesy of Trygve Braarud, Institute of Marine Biology, University of Oslo; published with permission of the Department of Biology, Section for Marine Botany, University of Oslo.

coastal water off shore, could be responsible for the bloom in the open sea. Such regional blooms, according to Gran's early thinking, were likely to have the same causes as the differences Hensen had observed between polar and tropical seas. And before he had become aware of Nathansohn's ideas on vertical circulation—indeed, before they were formulated—Gran suggested, based on ideas of C. H. Ostenfeld, that mixing in coastal waters and at the edge of water masses could renew the nutrients available to phytoplankton cells.[8] His emphasis, in his early work, was on the mixing of rich coastal or polar water into depleted water masses. Thus the bloom did not really progress from shore toward the open ocean, it occurred seaward later because mixing and nutrient renewal were later there. In both cases, the nutrients necessary for blooms originated on the shore or near it, either on the Norwegian coast in the case of the coastal bloom, or on the distant arctic coasts in the cases of blooms in the open Atlantic.

If the seasonal cycle were analogous to the geographical differences in plankton stock observed by Hensen, studies of the annual cycle were likely to provide important general information about the control of plankton abundance. This insight came to Brandt and Gran at about the

8. Gran 1902a, pp. 116–118.

same time.[9] Brandt became involved in more and more detailed studies of the seas near Kiel, Gran in highly original studies of the phytoplankton of Oslo Fjord and the North Atlantic, the first of them in collaboration with Alexander Nathansohn.

Originally, Gran and Nathansohn had planned on a detailed description of the plankton cycle in the fjord, correlating it with changes of light, temperature, and nutrients. The work did not begin smoothly; Gran described their experience as follows:

> We soon found that our task was more difficult than we had at first imagined. The quantity of plankton fluctuated greatly in the course of short periods of time, yet the variations could not be ascribed directly to conditions of existence [light, temperature, nutrients], since these remained fairly constant. . . . A quite satisfactory explanation presented itself, however, for the variations turned out to be closely connected with the direction of the winds and currents. The outflowing current in the surface layers might reduce the quantity of plankton to a mere fraction of the normal amount in the course of a day or two, while the inflowing current might perhaps double the quantity in a few hours.[10]

Foiled by the hydrographic complexity of the fjord from doing a conventional study, they decided to assess the production of the phytoplankton by estimating the turnover rate of *Ceratium*, and by doing experimental studies on the effect of nutrient enrichment.[11] Thereafter, because of their early failures, Gran and his colleagues carried out a series of unorthodox and original investigations, many of them experimental, based on what they perceived to be inadequacies of the methods used by the Kiel school.

Gran, as professor of botany in Oslo (from 1905), worked extensively on the systematics and distribution of phytoplankton both in Scandinavian coastal seas and in the North Atlantic, notably during the lengthy cruise of *Michael Sars* under John Murray and Johan Hjort in 1910.[12] He stated his priorities to be 1. the biology of the phytoplankton—especially their adaptations for pelagic life, 2. the geographical distribution of

9. To Brandt about 1902 (see Chapter 4 near n. 20); to Gran by 1907; see Gran 1912a, p. 372.

10. Gran 1912a, pp. 372–373; see also Gran and Gaarder 1918 and Gaarder and Gran 1927, p. 4.

11. On the production of *Ceratium*, see Gran 1912a, pp. 373–376. Just as Brandt had concluded several years later, Gran and Nathansohn calculated that *Ceratium* divided most rapidly during the summer but reached its peak of population in autumn. They attributed the lag to losses due to currents during the summer; Brandt concluded that it was due to grazing pressure by copepods in the summer (see Chapter 4 near n. 75).

12. See his chapter in Murray and Hjort 1912, pp. 307–386. The cruise of *Michael Sars* from Scandinavia to Newfoundland and the southern North Atlantic in 1920 is outlined in Murray and Hjort 1912, 1914–1956.

phytoplankton species, and 3. "the laws of production in the sea."[13] Gran's work after 1910 shows that his training as a plant physiologist, his sympathy with Nathansohn's theory of vertical circulation, and the influence of Scandinavian physical oceanography (through Otto Pettersson, Gustaf Ekman, and Bjørn Helland-Hansen in particular) eventually outbalanced the systematic tradition in which he had been schooled. In my opinion, this was reinforced by Gran's remarkable intuitive understanding of the conditions under which phytoplankton lived. All his publications, beginning with his doctoral thesis of 1902, reveal a sophisticated appreciation of the dynamic interaction between cells and their environment that is strikingly absent in the publications of the Kiel school.[14]

According to Gran, a major failing of Hensen's methods, then traditional in the Kiel school's work, was their inability to resolve the vertical distribution of plankton organisms. Vertical plankton hauls, begun by Hensen, carried on by Apstein and Brandt, had become the standard on the quarterly cruises of the ICES nations. Yet as Lohmann had shown during his work at Laboe in 1905–1906, the vertical distribution of plankton organisms varied greatly in time. Gran was aware of Lohmann's findings and of the earlier work of Otto Pettersson and P. T. Cleve, which showed that the vertical structure of the Skagerrak and Kattegat was hydrographically and biologically complicated, each stratum carrying its own plankton flora with distinctive abundance and composition.[15] Each flora was likely to show a unique response to the environmental conditions, such as light, temperature, and nutrients, to which it was exposed. Thus it was important, according to Gran, to know how the phytoplankton varied with depth.

During the ICES meetings in April 1912, Gran proposed and undertook a study of the vertical distribution of small phytoplankton cells, using centrifugation and special preservatives.[16] His work on material from the cruise of *Michael Sars* in 1910 had shown that in the open Atlantic phytoplankton abundance was usually greatest at 50 meters rather than near the surface; just as in Scandinavian coastal waters the

13. Gran 1912a, p. 311.

14. The roots of Gran's basically ecological understanding of the water column and its occupants are not easy to determine. Somehow he was able to fuse creatively taxonomic, physiological, and physical oceanographic approaches to problems of the water column in a more sophisticated way than did any member of the Kiel school.

15. Gran 1912a, p. 359.

16. Discussion in 1912, *Conseil International pour Exploration de la Mer. Rapports et Procès-Verbaux des Réunions* 14: 90–94. Gran (1912b) proposed using Flemming's solution (a mixture of osmic, chromic, and glacial acetic acids) to preserve small phytoplankton cells. He stressed the importance of making detailed vertical sections; the resulting collections would be too numerous to be examined alive as Lohmann had done.

water column was apparently not uniform vertically. According to his view then,

> a vertical water column can never offer to the organisms quite equal conditions of life, even if water layers do not, as at most places in the Baltic, the Skager Rak and the North Sea, appear one above the other, and present widely different hydrographical characters. The intensity of the light will at any rate vary so much with the depth as to render it absolutely necessary to determine the character of the plankton, qualitatively and quantitatively, at stated depths, where all the conditions of life can be precisely determined.[17]

The results of Gran's study, reported in 1915, showed that at 61 stations off Norway and Scotland there were significant variations in the vertical distribution of the phytoplankton, as he had expected on the basis of their physiology and the hydrography of the water column. Frequently there were sharp decreases in abundance below the surface, apparently correlated directly with the decrease of light.[18] Often, though, cell numbers increased below the surface, suggesting that cells were sinking out of the well-lighted upper waters; occasionally cell numbers were equal from the surface to the bottom, indicating that vertical mixing was occurring.[19] And the thermocline, in unmixed areas, could act as a boundary at which sinking cells accumulated.[20]

In Gran's eyes: "Two main points are here of particular interest, viz: 1) at what depth the production of organic substance by photosynthesis in autotrophic plants occurs most richly, and 2) to what depth in the oceans the photosynthesis still takes place in a degree sufficient to maintain the predominance over the diminishing metabolism (respiration) of the assimilating algae, so that a true production becomes the result of the change of matter."[21]

To his surprise, Gran discovered that G. C. Whipple had considered the second point years before in the reservoirs around Boston, where he suspended diatoms in bottles at various depths to determine the effect of diminishing light on their metabolism.[22] Gran's contribution between 1912 and 1915 was to make an indirect estimate of the depth at which photosynthesis was balanced by respiration—later called the *compensa-*

17. Gran 1915, pp. 8–9.
18. Gran 1915, p. 116.
19. Gran 1915, pp. 118, 121.
20. Gran 1915, pp. 123–124.
21. Gran 1915, p. 113.
22. Gran saw a brief account in Whipple's *The Microscopy of Drinking Water* (1899, p. 97) perhaps as early as 1907 (Gran 1912a, pp. 372, 380). The original work was described by Whipple 1896.

tion point or *compensation depth.*[23] At a station in the Kattegat he noted that the bottom water and the water at 25 meters had virtually the same low oxygen tension, whereas at 20 meters the oxygen was higher because of photosynthesis by the phytoplankton. Thus the "balance depth of the photosynthesis" as Gran later called it, had to be at about 20 meters.[24]

Shortly afterwards, Gran and his colleagues at Oslo devised a technique that made it possible to measure the compensation depth directly. This, their second major contribution to understanding the dynamics of phytoplankton production in the water column, resulted from the chemist Torbjørn Gaarder's observation in 1915 that the oxygen content of the water in Oslo Fjord varied directly with the abundance of phytoplankton.[25] Based on this observation, their aim was to measure production by quick methods that did not involve the tedium and time required to count cells.[26]

> It has been found from earlier investigations . . . that there is often a very distinctly demonstrable agreement between the occurrence and extent of the phytoplankton and the changes in the quantity of oxygen in the upper-most layers of water in the fjord. Both the quantity of oxygen and the content of carbonic acid of sea water and thereby also its reaction or hydrogen ion concentration are influenced directly by the metabolic processes of the plankton . . . and must therefore be assumed through their changes to make it possible to estimate the production, independently of the determination of plankton.[27]

Beginning in March 1916, Gaarder and Gran hung one-liter glass bottles at various depths in the Oslo Fjord near Drøbak. Each was filled with surface seawater of known phytoplankton content, oxygen level, and salinity. The deepest bottle (at 40 meters in the first experiments) was wrapped in black cloth so that only respiration, not photosynthesis, could proceed (Figure 5.2A).[28] This technique became known later as the light and dark bottle method.

The first results showed, just as Gaarder had claimed, that oxygen was produced in proportion to the increase of the algal population, and that

---

23. *Compensation point* was used by Marshall and Orr in 1928. *Compensation depth* is modern usage.
24. Gran 1915, p. 126.
25. Gaarder and Gran 1927, p. 5. See also Gaarder 1915. Later, 1917, Gaarder showed that the pH of the water was also affected by phytoplankton photosynthesis.
26. Gran's study of small cells, carried out for ICES beginning in 1912, took three years; his earlier studies of the division rate of *Ceratium* (see Gran 1912a, pp. 373–376) was also very time-consuming.
27. Gaarder and Gran 1927, p. 5.
28. Gaarder and Gran 1927, p. 9.

FIGURE 5.2 *A*, The light and dark bottle method as first used by Torbjørn Gaarder and H. H. Gran in the Oslo Fjord, March 22–25, 1916. One-liter glass bottles were suspended at 0, 2, 5, 10, 20, 30, and 40 meters, each containing phytoplankton cells in seawater of known salinity and oxygen content. The deepest bottle was usually wrapped in black cloth to prevent photosynthesis. Redrawn from Gaarder and Gran 1927, p. 9, with permission of the Conseil International pour l'Exploration de la Mer. *B*, The light and dark bottle technique as modified by Marshall and Orr for their work on the light relations of photosynthesis in Loch Striven, Scotland, in March 1927. Each light bottle was accompanied by a dark (covered) one. Note the surface bottles at far right. Gaarder and Gran had used only one dark bottle, usually the deepest on the string. Redrawn from Marshall and Orr 1928, p. 323, with permission of Cambridge University Press.

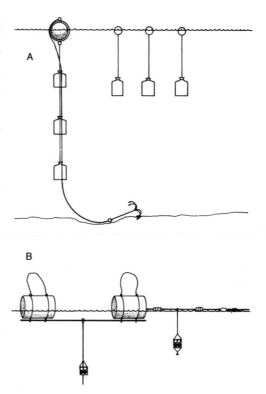

the pH of the water increased where production was high, indicating the uptake of carbon dioxide and release of oxygen by the photosynthesizing algae. Later they showed that bacterial respiration had a significant effect on the changes observed in the bottles.[29] Gaarder and Gran's technique, revolutionary in its implications for the measurement of production, was not taken up immediately, largely because Gran's first description of the results took place during the war, when Germany and Great Britain were absent from ICES meetings. In addition, his first published account of the work was in Norwegian.[30] An account of the work in English was not published until 1927.

Gaarder and Gran first used the light and dark bottle method to measure the effect of nutrient enrichment on the growth rate of phy-

29. Gaarder and Gran 1927, pp. 8–11; Gran 1918.

30. Gran described his work with Gaarder in 1916–1917 to ICES delegates from the "neutral nations" (Denmark, Sweden, Norway, and the Netherlands) at a meeting in Copenhagen on May 24, 1918. Germany formally withdrew from ICES in 1915 (until 1926), while Britain was inactive because of the war from 1914 until 1920. As a result, and because Gran's first publication (1918) was in Norwegian, Gaarder and Gran's work did not reach the audience most likely to make use of it until it was published in English in 1927.

toplankton, although there is no doubt they were also greatly interested in the way photosynthesis changed with increasing depth. Later, in 1922, Gran used the technique off the west coast of Norway north of Bergen, showing that the light and dark bottle technique gave higher values of production than those estimated using changes of oxygen in the open water column.[31] Perhaps because of the publication of Gaarder and Gran's first results in English in 1927, the Scottish planktologists S. M. Marshall and A. P. Orr became aware of the technique. They modified it to its modern form, accompanying the clear bottles at each depth with dark bottles, but using diatom cultures rather than natural phytoplankton (Figure 5.2B).[32] Their aim, working in Loch Striven in March 1927, was to assess quantitatively the effect of decreasing light with depth on the photosynthesis of diatoms, a factor that Gran had recognized to be significant some time between 1912 and 1915.

The most significant result of these early uses of the light and dark bottle method was that the compensation depth could be measured directly, rather than estimated, as Gran had done at first, from changes of oxygen in the water column, estimates that could be affected by mixing, currents, and losses to the atmosphere.[33] Gaarder and Gran's first experiments, in March 1916, showed that at 10 meters off Drøbak "there prevails equilibrity between the production and consumption of oxygen," though they recognized that this depth was likely to change seasonally, depending on the clarity of the water and other factors.[34] Marshall and Orr's work in 1927 showed just such a change: the compensation point in Loch Striven occurred near the surface in winter; it deepened to between 20 and 30 meters in summer as a result of increased light intensity.[35] A few years later, in 1931 and 1932, Gran and Trygve Braarud, who were assessing the potential effects of a tidal dam across Passamaquoddy Bay in North America, used the technique to assess the

31. First published by Gran in 1923. See esp. Gran 1927, pp. 34–35.

32. Marshall and Orr 1928, pp. 323–324. They stated explicitly for the first time the reason for using both light (uncovered) and dark (covered) bottles. In the light bottles the increase of oxygen represented photosynthesis minus respiration (now called *net production*); in the dark bottle oxygen decreased because of respiration of the cells. Adding the oxygen produced in the light bottles and that consumed in the dark bottles gave the total oxygen production of the algae (*gross production*).

33. Gran used indirect techniques later when it was not possible to use the light and dark bottle method, for example, during the cruise of *Michael Sars* between Norway, Iceland, and Greenland in 1924. By measuring the oxygen content of the water at various depths, Gran estimated that the "balance depth" of photosynthesis (compensation depth) lay between 25 and 50 meters over a large area of the North Atlantic (Gran 1929b, p. 4).

34. Gaarder and Gran 1927, pp. 10–11, 29.

35. Marshall and Orr 1928, pp. 335, 340–341.

effect of vertical mixing on phytoplankton production in the Bay of Fundy.[36]

Gran's increasing interest between 1912 and the early 1930s in the vertical distribution of phytoplankton and his realization that physiological processes—the balance of photosynthesis and respiration in the water column—must govern the production of phytoplankton slowly but effectively began to undermine the prevalent concept, based on Nathansohn's publications, that there was always a positive link between vertical circulation and phytoplankton production. It was the foundation of a significant reorientation of ideas about the factors controlling production which took place between 1929 and 1932. But before that occurred, the "classical theory" of vertical circulation—that production was promoted by mixing—finally undermined Brandt's denitrification hypothesis.

Brandt's complex theory of the control of the annual plankton cycle was based on the interplay of three essential nutrients, potential Liebigian limiting factors: nitrate, phosphate, and silicate. His early concentration on nitrate as the major limiting factor was superseded by a more complex reality, for diatoms clearly removed large amounts of silicate from the water in spring, while, according to Raben's results, phosphate rather than nitrate disappeared first and reached lower levels than nitrate during depletion after the blooms. During the 1920s, as a result of these observations and of new methods of analysis, phosphate came to be considered *the* nutrient ion limiting production.

About 1915, D. J. Matthews, working at Plymouth as hydrographer to the Marine Biological Association of the United Kingdom, happened upon a sensitive colorimetric test for phosphate that had been developed by the French chemists Pouget and Chouchak in 1909.[37] When he used it he found phosphate values in the English Channel off Plymouth that were roughly one-fourth those reported by Raben in the Baltic and in the North Sea. He also noted that phosphate was very low in the channel surface waters during spring.[38]

Matthews's results lay fallow for a few years until W. R. G. Atkins was appointed head (and only member) of the Department of General Physiology at the Plymouth Laboratory in 1921.[39] Atkins quickly began a thorough series of studies of chemical and physical variation of the water

36. Gran and Braarud 1935, esp. pp. 423–424.
37. On D. J. Matthews's (1873–1956) varied, distinguished career in oceanography, see Deacon 1984, p. 278ff., p. 293ff.; Carruthers 1956.
38. Matthews 1916, pp. 253–257.
39. W. R. G. Atkins's career is described by Cooper 1960 and Poole 1960. See also Chapter 8 near n. 5.

in the English Channel, confirming Matthews's observations using another new method for estimation of phosphate. His first results, published in 1923, were amplified in 1926 and 1927 when he showed that Raben's results had been too high because, at least in part, his technique had measured arsenate along with phosphate. Atkins's early work on phosphate resulted from technical necessity as well as scientific fashion; phosphate was the only major nutrient ion that could be measured rapidly using a sensitive colorimetric technique until H. W. Harvey's development of the strychnine sulphate method for nitrate in 1926.[40]

Atkins's work at sea in the Channel showed him that phosphate was depleted after the spring bloom, just when the surface layer became distinctly warmer than the underlying waters. Was this warming—and the change in phosphate—caused by local processes, or by displacement of the surface water by warmer, nutrient-depleted water from elsewhere? Atkins rapidly concluded that local processes, not advection of new surface water, caused the formation of the summer thermocline. He based this conclusion on vertical temperature profiles, which were slightly different even at nearby locations, and on analogies with lakes, where the development of the surface mixed layer, the epilimnion, had been described qualitatively and quantitatively by the American limnologists E. A. Birge and C. Juday.[41] Thus local heating gave rise to a stable surface layer that was highly resistant to mixing, for, as Atkins showed in 1925, the farther the water temperature is from the temperature of maximum density the greater is its decrease of density on warming. The strikingly nonlinear relationship between water temperature and decrease of density resulted in the summer surface layer being very difficult to mix into the deeper, cooler water below it.[42] In the stable surface layer phytoplankton would absorb phosphate and carbon dioxide, releasing oxygen to solution. The result was what Matthews and Atkins had observed: virtually phosphate-free surface water in which the pH was increased, both changes caused by the productive metabolism of the algae. For these reasons, it was obvious to Atkins that "the phenomena of thermal stratification are . . . of great importance in the study of phytoplankton."[43] More explicitly, he emphasized that "the phenomena of thermal stratification are of great importance for the development of the phytoplank-

40. See Harvey 1926. Until 1926, Raben's time-consuming, error-prone distillation method was the only one available for the estimation of nitrate.
41. Atkins 1924, pp. 320, 323. He referred to Birge and Juday's famous monographs of 1911 and 1922. Frey (1963, pp. 24–26) and Mortimer (1956, pp. 179–181) discuss Birge and Juday's work on thermal stratification of lakes.
42. Atkins 1925b, p. 699.
43. Atkins 1925b, p. 693.

ton, for . . . the supply of nutrient salts is regulated thereby."[44] Clearly, stratification played a negative role, overcoming the positive, nutrient-restoring effects of mixing, for his observations showed that the deep water of the Channel retained significant amounts of phosphate even during the peak of summer stratification and depletion at the surface.[45]

By 1926 Atkins was willing to place his ideas on a large stage. A few samples were becoming available which showed that surface phosphate values in the temperate and tropical oceans were low, while the deep water contained significant amounts of the nutrient. But the renewal of phosphate to the surface should be great at high latitudes, for

> as we proceed northwards the surface cooling becomes greater and the deep water temperatures are also lower, so the vertical circulation will proceed to progressively greater depths. Conversely near the equator the temperature changes throughout the year are small, so that vertical circulation there must be due in the main to wave motion—which cannot be effective to any very great depth—and not to density changes. Thus no considerable seasonal change in phosphate content is to be expected.[46]

The implications were great. Without mentioning Hensen or Brandt, Atkins described how nutrients and vertical mixing could account for the geographical patterns of plankton abundance that Brandt's hypothesis had been formulated to explain.

> Owing . . . to the fact that no great cooling occurs in winter, vertical mixing proceeds to lesser depths the further one moves towards the equator. . . . In fact as the range of the annual temperature variation becomes reduced, so also is that of the phosphate cycle. Furthermore, since dead organisms sink as they finally disintegrate and since the ocean waters are very deep, it is obvious that the bottom water—or the water sufficiently deep to become really rich in phosphate—will not enter into the vertical circulation at all. Thus a lesser phytoplankton might be expected as the temperate zones are left, which is precisely what is found. . . .
>
> The richness of northern waters may accordingly be considered as due to the better vertical circulation as regards the temperate zone; also to the fact, which becomes more marked in high latitudes, that since light is the limiting factor for plant growth for a considerable portion of the year, the water accumulates a greater store of phosphate. Thus when daylight returns, or lengthens its duration, a great outburst of the phytoplankton takes place.[47]

44. Atkins 1925b, p. 122.
45. Atkins 1926c, pp. 208–211, esp. figure 14.
46. Atkins 1926a, p. 463; Atkins 1926c, pp. 212–213.
47. Atkins 1926c, pp. 212–213.

Aiming directly at Brandt and the Kiel school, he concluded that "the data presented furnish reasons for the proved richness of northern and coastal waters and support Nathansohn's views."[48] Gran's and Brandt's early intuitive belief that local and global processes should be similar would be vindicated if Atkins's view were correct, but for reasons that they had not recognized. Now both the spring bloom and geographically varying patterns of phytoplankton abundance could be explained as functions of the same causal mechanisms: nutrient regeneration, nutrient uptake by phytoplankton, and the extent of vertical mixing.

By the late 1920s, Brandt was prepared for this attack. Familiar with Atkins's and Harvey's new techniques, he conceded that Raben's values probably had been too high.[49] In addition, he had seen the preliminary results of the German expedition on *Meteor*, in which Herman Wattenberg, the chemist on board, showed that uniformly low values of phosphate occurred in tropical surface waters, but high values were found below the permanent thermocline; as a result of the thermocline, convection could never restore nutrients to the surface except in special locations such as the West African coasts where conditions were suitable for upwelling.[50] Ernst Hentschel, on the same expedition, confirmed that the areas with very low nutrients at the surface also had very little plankton. Brandt now accepted the importance of vertical circulation (or its restriction) in controlling the plankton cycle, citing in addition to Whipple's work that of Birge and Juday on thermocline formation and mixing, conceding that "the variation of thermal layering and mixing in the sea is of even greater significance for plankton production than it was known to be in lakes."[51] He agreed too, on Wattenberg's authority, that in the tropics nutrients could never reach the surface from the great deep water reservoirs because the density contrast was too great for convection to occur. Atkins's suggestion, bolstered by Harvey's work on nitrate and Wattenberg's on the tropical Atlantic, that differences in vertical mixing at different latitudes could account for the patterns of abundance first

48. Atkins 1926c, p. 224.
49. Brandt 1929, pp. 10–17. He had defended Raben's work in 1927. Harvey's strichnine sulphate colorimetric method for the measurement of nitrate was published in 1926. It indicated that nitrate was much lower than previously suspected, thus that it could be a limiting factor (Harvey 1926, p. 77). Both phosphate and nitrate reached very low levels during the summer; often nitrate was lower than phosphate (Harvey 1928b, p. 186, figures 1, 2) unlike Raben's results, which had led Brandt to believe that phosphate was the major limiting nutrient.
50. Brandt (1929, p. 18) quotes Wattenberg 1927. On Herman Wattenberg (1901–1944) see Rakestraw 1956.
51. Brandt 1929, p. 24. He seems to have been aware of Birge and Juday's work about 1927; see Brandt 1927, p. 292.

noted by Hensen was so compelling that Brandt then made his final concession.

> This explanation is so evident that my explanation of 1899 that denitrity-ing bacteria are the cause of plankton deficiency in the tropical oceans is invalidated by it. However, I still maintain the view "that denitrifying bacteria break down an excess of nitrogen compounds and that it is they that maintain the existing equilibrium in nature."[52]

It is remarkable that Brandt was able to give up his long-held hypothesis so readily. He probably went through a long period of increasing doubt as a result of three factors: 1. the success of the vertical circulation hypothesis, 2. his difficulty in finding bacteria of the nitrogen cycle (especially nitrifiers), and 3. the prevailing climate of doubt about the ability of denitrifying bacteria to control the abundance of nitrate. Nonetheless, Brandt's rejection of his long-held brainchild in 1928 was a remarkably open-minded and objective act.[53] A year or two before, when reviewing the nitrogen cycle in the sea, he had held tenaciously to the denitrification hypothesis, even though it had not been the main object of his scientific attention for several years. But in 1928, Brandt's denitrification hypothesis finally died through his own actions. After thirty years of existence it was a casualty of the widely held belief that vertical circulation, a positive factor, and thermal statification, a negative one, were the main controlling influences on plankton production.

Despite the triumph of Nathansohn's ideas, vertical circulation and stability were not without their problems. Alexander Nathansohn, promoting the theory that vertical circulation could explain anomalies evident in the theory of plankton dynamics during the first decade of the twentieth century, had not been modest about its power. "Indeed one can say that if one does not pay attention to the details of hydrographic variations the variation of plankton appears to be totally incomprehensible and capricious, whereas on the basis of our theory the phenomena observed will be completely explicable."[54]

---

52. Brandt 1929, pp. 25, 32. The lecture was delivered in 1928. The English version accompanying Brandt's original does not fully convey his view that only the denitrification hypothesis, not marine denitrification in general, had to be abandoned. As Chapter 2 shows, modern research has vindicated Brandt's belief in the general significance of denitrification in the seas.

53. It was all the more remarkable because Brandt recanted before an international audience at a special meeting of ICES in Copenhagen on June 4, 1928. I suspect that he had doubted the validity of the denitrification hypothesis for several years and used the occasion to modernize his views. Although Brandt retired in 1921–1922, in 1927 he was still planning to write synoptic summaries and reassessments of the Kiel school's work during the preceding thirty years.

54. Nathansohn 1909b, p. 613.

He agreed that a residue of difficult problems remained—for example why at Monaco during rainy periods in the winter there were bursts of phytoplankton production—but in general he believed that enhanced vertical circulation and increased plankton production always went hand in hand.

This opinion was widely accepted during the first three decades of the twentieth century, despite the reluctance of Brandt, at least initially, to accept Nathansohn's ideas. Gran, an early convert, writing in 1931, believed that Nathansohn's ideas had "been splendidly proved" by a large number of investigations from all over the world, in regions as different as the west coast of Norway, the upwelling region off West Africa, and the tidally mixed straits in Puget Sound.[55] From their quite different viewpoints, though, both Brandt and Gran could see difficulties in accepting a uniformly positive role for vertical circulation. Commenting on G. C. Whipple's belief that vertical circulation in lakes was necessary both to renew nutrients and to keep the diatoms in the lighted surface waters, Brandt pointed out that vertical circulation must, inevitably, mix diatoms uniformly throughout the water column unless they had physiological mechanisms to keep themselves near the surface. He believed that phytoplankton had active means of modifying their specific gravity, allowing them to overcome the negative aspects of vertical circulation.[56]

Gran qualified his acceptance of the idea of vertical circulation in a different way. He was particularly impressed by the richness of Norwegian coastal waters compared with the open Atlantic. This could be accounted for in a general way by renewal of nutrients from the bottom by Nathansohn's mechanism, but the high level of production required an additional source of nutrients, Gran believed, which had to be sought on land. Then if the influence of coastal waters spread seaward in the spring—perhaps by the seaward extension of enriched waters from the Baltic Current (the source of the Norwegian Coastal Current)—the retarded development of the bloom in the open Atlantic could be explained.[57] But Nathansohn's mechanism could be invoked too, for at the edge of the continental shelf there should be vertical circulation, bringing nutrients to the surface.

> Where a deep-going current is pressed by the effect of the earth's rotation against the coast banks, as off the west coast of Scotland, or at the Nor-

55. Gran 1931, pp. 39–40.
56. Brandt 1920, pp. 413–418.
57. Gran 1915, pp. 130–133.

wegian coast banks off Romsdal and Lofoten, or at the Faeroe Bank, eddying movements, not only horizontal but also vertical, will necessarily arise, and water masses from the depths will be carried up to the surface.[58]

When he expressed these ideas in 1915 there was little evidence for either mechanism along the Norwegian coast. But Gran predicted that they should have directly observable results.

> The production centres for phytoplankton in our waters should . . . according to these suppositions, be the surface layers near the coasts and the water masses, which are pushed against the bold coast banks. At the seasons when one or other plant community is commencing its development, we may expect to find locally limited maxima in the neighbourhood of the production centres; later on, when the surface layer overflows greater areas, the conditions will be equalized.[59]

This prediction, combined with Gran's appreciation that there must be a depth in the water column at which photosynthesis was just balanced by respiration—that below that depth no net production could occur—provided the basis for a revision of Nathansohn's theory. Between 1922 and 1935, vertical circulation and stratification came to be viewed in a new, less contrasting way; by degrees, a theory of control evolved which used positive *and* negative aspects of vertical circulation, and, most strikingly, the positive role of stratification. This sophisticated modification of Nathansohn's original, simple theory came about largely as a result of practically oriented Norwegian work off the Romsdalsfjord and near the Lofoten Islands.

Scandinavian scientific interest in the relationships among climate, conditions in the sea, and the fluctuations of important fish stocks, especially cod and herring, dated from the mid-nineteenth century; it had been the underlying reason for the formation of ICES at the turn of the century (see Chapter 3). By the 1920s, plankton dynamics was sufficiently sophisticated to be applied to local problems of fish abundance. At the Lofoten Islands, Oscar Sund of the Norwegian Fishery Department began a study of the relationship between the abundance of cod larvae and fluctuations in the abundance of plankton, research that Gran had proposed in 1913. Gran's research assistant Birgithe Ruud (later Føyn) was responsible for the plankton investigations, which began in 1922 in the Vest Fjord. There, and on the seaward edge of the continental shelf,

58. Gran 1915, p. 133.
59. Gran 1915, p. 134.

plankton collections and temperature, salinity, and oxygen determinations were made (but not measurements of nutrients).[60] Gran himself was working farther south in 1922, off Bergen, also concentrating on the distribution of plankton (both horizontal and vertical) and on oxygen, temperature, and salinity.[61] A few years later, Gran joined Johan Hjort, then professor of marine biology at Oslo,[62] and Oscar Sund in a study of plankton off the Romsdals Fjord and the Lofoten Islands, aiming his work at the phytoplankton while others studied euphausiids and copepods. This detailed survey, intended to link fluctuations in the plankton with those of the fish stocks, involved sections across the continental shelf, beginning in the fjords, on cruises between March and June 1926 and in March 1927.[63]

Birgithe Ruud's first results from the Vest Fjord region off the Lofoten Islands confirmed Gran's prediction that two production systems existed in the early spring, one near the coast, the other at the seaward edge of the banks. As the season advanced the blooms spread toward midshelf, where they merged.[64] A similar pattern occurred off the Romsdals Fjord, as Gran discovered using data from his sections.[65] His prediction was verified, and the pattern could be explained as the result of nutrients from land (in the case of the coastal bloom) and from vertical mixing (in the case of the seaward bloom). Ruud's analyses of the results indicated that the coastal bloom increased as the salinity of the coastal water decreased, prima facie evidence for enrichment of nutrients by melting snow, as Gran suggested after seeing the first results.[66] Off shore, there was evidence that the water was well mixed before the bloom, presumably restoring nutrients to the surface, though Gran found it puzzling that by

60. Ruud 1926, pp. 1–20; see also discussion in 1913, *Conseil International pour l'Exploration de la Mer. Rapports et Procès-Verbaux* 19: 124–131. In 1913, while he was chairman of the plankton section of ICES, Gran proposed that several nations attempt to study the interrelations of plankton and fish abundance. The war and other complications prevented an immediate start; when it finally began, only Norway, not a group of nations, was involved in the study.

61. Gran 1927, pp. 5–7.

62. On Hjort's career see Andersson 1949, Maurice 1948, and Hardy 1950a.

63. Hjort 1927; Gran 1929a; Gran 1930, p. 5. Hjort, Gran, and Sund met in December 1925 to adopt a plan for 1926. They decided to concentrate on a small area and to sample frequently so that fluctuations in hydrography, plankton, and fish could be seen easily. Cruises every second week in the Vest Fjord and off Møre would sample from land to the edge of the continental slope ("the edge"). Hjort and Johan Ruud were to work off Møre, Sund at the Lofoten Islands, while Gran (with Birgithe Ruud) would study the phytoplankton samples. This program appears to have been a lineal descendant of the one proposed by Gran in 1913.

64. Ruud 1926, pp. 11, 15, 23, 29; Føyn 1929.

65. Gran 1929a, pp. 27–28, 48, 59; Gran 1930, pp. 5–6.

66. Ruud 1926, pp. 6, 22–25, 29; Føyn 1929, pp. 57–60. Gran's hypothesis was published in 1923. See also Gran 1928, p. 200; Gran 1930, pp. 6, 71. H. B. Bigelow (1926a) attributed the beginning of the spring bloom in the western Gulf of Maine to nutrients carried by melting snow.

the time the bloom developed, a 10-meter-thick stable surface layer was present. He speculated that it might be the explanation for the end of the bloom.[67]

Until 1927, the Norwegians had done no nutrient analyses to accompany their work on hydrography and plankton. But once Atkins's rapid colorimetric method for phosphate analysis became available in 1923, then Harvey's for nitrate analysis in 1926, it became possible to test Gran's ideas. Standard versions of the new analyses were established by ICES in 1928,[68] but Oscar Sund already had experimented with Atkins's and Harvey's methods at stations across the shelf at the Lofoten Islands.[69] His results, later corroborated by more phosphate and nitrate analyses, showed that there was something seriously wrong with Gran's hypothesis that nutrients were enriched at the coast and at the shelf edge. In fact, nutrients were high at the beginning of the bloom right across the shelf, from the coast to the edge of the banks, also in the open ocean, where the bloom began last. Later work confirmed and extended Sund's results.[70] Referring to his hypothesis that production near shore was promoted by meltwater, Gran was forced to concede that

> these conclusions cannot be maintained, if the supply of nitrates and phosphates is the only factor limiting the growth of the phytoplankton besides the physical factors, light, temperature and salinity. With the knowledge we now have of the distribution of nitrates and phosphates, we should expect that the diatoms would develop simultaneously over the whole area, and that the Atlantic waters outside the coastal Bank should have a production as rich or richer than the waters of the coastal current. As this is not the case, we are . . . forced to suppose that there is one or more factors limiting or stimulating the growth of the phytoplankton besides those which we have until yet taken into account.[71]

It became clear at this time that the coastal waters were not particularly rich in nutrients and that meltwater from snow did not contain high

67. Gran 1928, p. 200; Gran 1929a, p. 57.

68. At Hjort's suggestion, a meeting of chemists was held in Oslo in October 1928 to standardize the new colorimetric techniques for phosphate, nitrate, and silicate analysis. Present were Kurt Buch (Helsingfors), Torbjørn Gaarder (Bergen), H. H. Gran (Oslo), H. W. Harvey (Plymouth), Ernst Schreiber (Helgoland), and Herman Wattenberg (Kiel). Laboratory demonstrations and experiments accompanied the deliberations, carried out by Trygve Braarud, Birgithe Ruud Føyn, and Alf Klem. For the results see ICES (1928) *Rapports et Procès-Verbaux des Réunions* 56: 95–115.

69. Sund 1929a, pp. 80–84.

70. Sund 1929a, p. 85ff.; Sund 1929b, p. 77. See also Gran 1930, pp. 7, 71; Gran 1932, p. 352. Braarud and Klem (1931, pp. 67–68) noted that the offshore Atlantic water contained more nutrients than the shelf water.

71. Gran 1930, p. 71.

levels of nutrients, especially when compared with the store available in deep water.[72] What then were the "other factors" that accounted for the beginning of the bloom near shore?

Accepting Brandt's theory of nutrient limitation, but faced with the fact that the standard nutrients such as phosphate and nitrate could not account for the beginning and location of the coastal bloom, Gran suggested that

> if the productivity of the coastal waters is dependent on any factor of a chemical nature acting as a minimum factor, it must be an element which in its circulation does not follow the nitrates and phosphates accumulating in solution in the deep sea and reaching the surface again by vertical circulation of any kind. If such minimum stuffs exist, they must irreversibly go out of circulation in the sea, so that they can only be renewed from land.[73]

Iron was such a stuff. Gran considered that "iron-containing humus-compounds" from land might be the missing minimum substance; indeed iron compounds and soil extracts in a series of experiments did appear to enhance the growth of phytoplankton cultures.[74]

But hydrographic factors as well as nutrients were implicated in some way, as Gran sensed by 1931, for after discussing iron-humus compounds he recounted the puzzling results of an investigation in the Antarctic by his student Johan Ruud. Ruud went south to the Weddell Sea with Johan Hjort in the austral summer of 1929–1930 on the Norwegian whaler *Vikingen*. There, between September 7, 1929, and April 7, 1930, he collected plankton and water samples, analyzing the latter using Atkins's method for phosphate and Harvey's for nitrate.[75] At first not much occurred; the specific gravity and phosphate were uniform to at least 75 meters, and there was no evidence of a bloom. Vertical circulation was evident, nutrients were high, yet the cells did not respond. Then, in early January, the surface layer freshened, probably because of melting ice, a thermocline formed around 25 meters, and a dramatic increase of diatoms occurred. The bloom had begun.[76]

The absence of a bloom early in the season was easy to explain; according to Gran: "The diatoms also move with the vertical movements of the water, and . . . therefore no accumulation is found in the illumi-

---

72. Gran 1931, p. 40; Braarud and Klem 1931, pp. 65–69.
73. Gran 1931, p. 41.
74. Gran 1931, pp. 41–43.
75. J. Ruud (1930) describes the work and summarizes the results without any theoretical discussion.
76. Gran 1931, pp. 43–45.

nated zone, with the effect that the whole production is retarded, because too many of the diatoms sink below the balance depth of the photosynthesis." But once stabilization of the surface layer occurs, "it has the effect that the vertical circulation is stopped and the sinking of the diatoms retarded, and now it is clearly seen, how the phosphates (and nitrates) are consumed in the surface layers as far down as the photosynthesis is effective."[77]

Now nearly all the elements were in place for a reorientation of ideas about the role of vertical mixing and stabilization in the initiation of the bloom. Gran was aware by the summer of 1931 that vertical circulation might have to be limited before phytoplankton could profit from high levels of nutrients in the surface waters, but he did not recognize fully the implications of this idea. He was nudged into full awareness of its significance by a perceptive comment made by Atkins.

According to Gran's recollection, when the ICES delegates visited Plymouth in 1929 Atkins had commented that "the maximal growth [of phytoplankton] cannot take place until the vertical movements are completed, because the diatoms are prevented from accumulating in the surface layers where the light is sufficient for photosynthesis."[78] Earlier that year, the Norwegian work off the Romsdals Fjord had been extended, with the intention of doing detailed work along the section from the coast to the edge of the shelf before, during and after the spring bloom. Trygve Braarud and Alf Klem, students of Gran's, were responsible for the work on phytoplankton, the results of which they prepared for publication nearly two years after Atkins's comments.[79] Their work showed with great clarity how the distribution of production across the Norwegian continental shelf could be explained.

Braarud and Klem were aware that in early spring nutrients were more than abundant for phytoplankton production anywhere on the shelf. Despite this, the pattern was just as B. Ruud and Gran had observed, a coastal bloom, followed by an offshore one, then finally more or less continuous high production across the shelf from the fjord mouth to the

77. Gran 1931, p. 45.
78. Gran 1932, p. 351. After the regular meeting in London, April 8–15, 1929, some ICES delegates, including Gran, visited the Plymouth Laboratory and attended a meeting of the Challenger Society. According to a contemporary account, "at this meeting scientific papers were read and useful and stimulating discussions took place" (1930, *Journal of the Marine Biological Association of the United Kingdom* 16: 989).
79. Gran 1930, pp. 7–8; Braarud and Klem 1931, pp. 5–6. Braarud and Klem's manuscript was presented to the Norwegian Academy of Sciences in March 1931, nearly two years after the ICES meeting in Plymouth at which Atkins spoke. During 1929 and 1930, they made cross-shelf sections in March, April, and May.

shelf edge. They showed that this pattern of events coincided closely with the stratification of the surface waters. Using an index of stability, *E*, devised by H. U. Sverdrup and Theodor Hesselberg,[80]

$$E = 10^6 \frac{d\sigma_z}{dz}$$

where σ_z = a measure of the specific gravity of the water, and z = water depth, they showed that the surface water stabilized first near shore, then at the edge of the shelf. At the midshelf locations, where the bloom occurred latest, stability was low throughout the water column as a result of nearly uniform temperature and salinity from top to bottom, combined with the effects of wind (Figure 5.3). Inshore, they attributed the early beginning of the bloom to the effect of fresh water (which stabilized the surface layer) and to low wind speeds; offshore the bloom resulted from transport from the south, allowing cells advected from better lighted areas to grow readily under stable conditions with high levels of nutrients at the shelf edge.[81] It is curious that after demonstrating the stability of the water at the shelf edge, Braarud and Klem should have chosen such an ad hoc explanation of the bloom there. That they did suggests that they had not fully understood the implications of their own work. Even Gran found it hard to grasp the significance of their findings immediately, but enlightenment came in the next two years, when he had the opportunity to compare their results with those from other coastal areas in Europe and North America.

Once he had reflected on Braarud and Klem's results, Gran began to see the underlying causes of the initiation of blooms in many coastal regions as well as in the open sea. In the Bay of Fundy, which he and Braarud studied in 1931–1932, it was clear that the most highly mixed areas had the lowest phytoplankton content, just as would be expected if some degree of stability were necessary to keep the cells above their compensation depths. The situation Johan Ruud had observed in the Weddell Sea could be explained in the same way: stabilization of an intensely mixed water column, allowing the cells to stay near the surface long enough to absorb nutrients and out-balance their respiratory metabolism.[82] By 1932, both Braarud and Gran had seen the key to solving the

80. See Hesselberg and Sverdrup 1915, pp. 65–67; also Sverdrup 1929. In 1915, both were students of V. F. K. Bjerknes in Leipzig. Sverdrup's career is described by Fjeldstad 1958, Revelle and Munk 1948, and Spjeldnaes 1976. Hesselberg became the director of the Norwegian Meteorological Institute (see Eliassen 1982, pp. 5–6).

81. Braarud and Klem 1931, pp. 67–71, 77–78.

82. Gran 1932, pp. 351–352; Gran 1933a, p. 69; Gran and Braarud 1935, pp. 421–423.

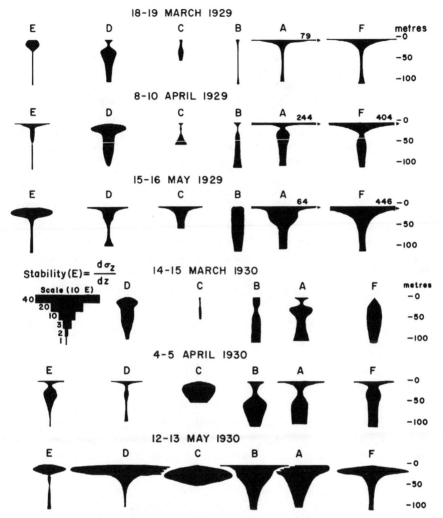

FIGURE 5.3 Diagrams devised by Trygve Braarud and Alf Klem showing the stability of the water column across the Norwegian continental shelf in 1929 and 1930. Stations E and D were near the shelf edge, stations C and B at midshelf, stations A and F close to the coast. The greatest stability is indicated by greatest width of the shaded areas. Note that inshore water stabilized first, followed by the water at the edge of the continental shelf. The midshelf stations B and C were the last; these were the locations where the bloom began last according to Birgithe Ruud's and H. H. Gran's observations. By May the spring bloom extended across the entire shelf (lower row in each series). From Braarud and Klem 1931, pp. 26–27, with permission of the Norwegian Research Council for Science and the Humanities.

problem of the bloom: *both vertical mixing and stability had positive and negative aspects.* The bloom resulted when the proper balance was struck. Gran concluded that

> These examples show the importance of the stabilization of the surface layers. A marked stratification excludes circulation of the nutrient salts with the result that the surface layers become depleted of plant nourishment and the diatoms with their high requirements sink and disappear. On the other hand, continuous vertical mixing prevents them from accumulating in the lighted zone and from utilizing the nutrient salts present to such an extent as might have been expected.[83]

In 1932 he could not quite understand, despite the revelation above, how the bloom spread to the open ocean from the continental shelf. Was it a question of some essential nutrient carried to the open ocean by coastal or polar waters, or perhaps the spores of phytoplankton species adapted to growth under oceanic conditions? But the stabilization hypothesis was insidious, for he felt compelled to add to these speculations that "it may perhaps be taken into consideration that the thin layers of coastal waters covering the Atlantic current during the spring and summer may have a stabilizing effect on the surface waters and thus promote the growth as mentioned above."[84]

Thus by 1935 a new synthesis of theory had come into being to explain the preconditions, origin, maintenance, and eventual decline of plankton blooms. The basic elements were the Kiel school's application of Liebigian limiting nutrients; Nathansohn's theory of vertical circulation made plausible by limnologists, especially Whipple; Gran's concept of compensation depth; and the recognition (mainly by Gran between 1931 and 1935) that the bloom resulted from a dynamic balance between vertical circulation and stability of the water column. This recognizably modern theory of plankton dynamics came into being piecemeal during the early years of the 1930s from German, Norwegian, American, and English sources. The solution may have been recognized first by Braarud, but it was synthesized by Gran, a plant physiologist with the ability to reconstruct in his imagination the dynamic nature of the marine water column. The formulation of the new theory was neither particularly rapid nor particularly clear-cut. It was several years before even Gran, the

---

83. Gran 1932, p. 352. Braarud's doctoral thesis, based on his work in the Arctic with Johan Ruud in 1929, explored all the factors governing phytoplankton production, including stability, without placing great emphasis on its role in initiating blooms (Braarud 1935, esp. pp. 18–21).

84. Gran 1932, p. 354. Stability in the open ocean actually results mostly from heating of the surface layer, except in highly productive frontal regions where water of lower specific gravity lies over nutrient-rich deeper water.

dominant figure in its development, could reorient his thoughts to fully accept the complexities of vertical circulation and stabilization of the surface waters. Nonetheless, by 1935 vertical circulation was well established as a complex factor governing the bloom. The ideas of the Kiel school became of merely historical interest.

6 The End of an Era: The Demise of the Kiel School

> The idea of decline is important. It has serious meaning and
> implication not only for the scientific enterprise, but also for
> the larger society to which the enterprise belongs.
>
> MARY JO NYE 1984

Trygve Braarud, writing in 1935, expressed the then prevailing opinion of the Kiel school's theory of plankton dynamics when he said that "Brandt's view concerning the production factors, as expressed for instance, in his basic papers 'Über den Stoffwechsel im Meere' has been generally accepted by the students of plankton and applied in the interpretation of quantitative observations from many parts of the ocean."[1] Without doubt, by 1935 the Kiel school's theory was accepted virtually dogmatically as the foundation of quantitative plankton biology; in being so accepted it receded into history as far as working marine biologists were concerned.

The preceding chapters have shown how the Kiel school's work originated in Victor Hensen's interest, during the 1870s and 1880s, in applying quantitative methods analogous to those used in physiology to the determination of production in the seas. Hensen, stymied by the complexity of nature, particularly the difficulty of sampling the oceans adequately, did not achieve his goal. But his aims were espoused by the zoologist Karl Brandt, whose students and junior colleagues provided the personnel and scientific skill that made the Kiel school's work noted and emulated during the first decades of the twentieth century.

It was the need to explain the results of the Plankton Expedition that diverted Brandt, subtly but surely, from Hensen's aim of measuring production in the oceans. Using the newest late-nineteenth-century approaches in microbiology and scientific agriculture, Brandt sought on the

1. Braarud 1935, p. 7.

local scale, beginning in Kiel Bight, general factors that could explain the paradoxical geographical patterns of plankton abundance observed during the expedition. In so doing, he shifted attention from the geographical pattern of plankton abundance to its seasonal changes. In this way, Brandt and the Kiel school were quickly caught up, late in the 1890s, in the problem of explaining the causes of seasonal changes in the plankton content of temperate seas.[2]

The "discovery" of the spring bloom by the Kiel school (and by the Norwegian H. H. Gran) was actually a change of viewpoint, at the turn of the century, from a mainly taxonomic approach to the phytoplankton to one in which chemistry played the greatest role. Later as the importance of nutrient salts in controlling plankton abundance was elucidated, puzzling problems in explaining different patterns of the bloom in apparently similar regions were resolved by bringing Nathansohn's physical oceanography to plankton dynamics, followed by a sophisticated fusion of biology, physics, and chemistry resulting from Norwegian work on the plankton cycle during the late 1920s and early 1930s.

Brandt's work could never have developed in such influential fashion without the aid of state financing, particularly Prussian support of the Internationale Meereslaboratorium at Kiel, which carried out Germany's share of the ICES program between 1902 and 1917. In particular, the work of Emil Raben, Brandt's chemist in the Internationale Meereslaboratorium, proved to be absolutely essential in revealing the chemical factors at work as the plankton of the Kiel Bight and North Sea went through its seasonal changes. Indispensable too were the laboratory assistants and junior colleagues such as Carl Apstein and Hans Lohmann who actually did the fieldwork, sorted the collections, and tabulated the results of the analyses. Their work, modified and molded (often unconsciously) to meet the criticisms of outsiders such as Gebbing and Nathansohn, led to the Kiel school's greatest triumph, its theory of the control of plankton abundance by nutrient salts and physical circulation, centered most notably around the marine nitrogen cycle.

But missing from this summary of the achievements of the Kiel school between the time of Victor Hensen and the last works of Karl Brandt is an analysis of the school itself. What gave it its individual character, its creative nature? How long did the creative period last? And, most significant, why did the Kiel school's creative period not last longer? When Brandt died in 1931, the school's synthesis had been achieved, but its most creative members (apart from Brandt himself) had long since left

2. The pattern is one that Mulkay (1979, p. 73) and I independently have called "differentiation" of a new area of research. This is what Mulkay describes as "the creation and exploration of a new area of ignorance."

Kiel and were working in other fields. The most creative years of the Kiel school had been twenty years before, between 1899 and 1912. Brandt's great monographs *Ueber den Stoffwechsel im Meere* (1920) and *Stick-stoffverbindungen im Meere* (1927) summarized work that, in the main, had been done before World War I by the members of the Kiel school. Yet it was not the disastrous effects of the war on German academia that reduced the creativity of the Kiel school, for there is solid evidence that its best days ended before the war.

Figure 6.1 shows the cumulative number of publications by members of the Kiel school (defined as Hensen, Brandt, and their associates who contributed directly to the theory of plankton dynamics) between 1887 and 1927. Beginning with Hensen's *Über die Bestimmung des Planktons* in 1887 and ending with Brandt's last monograph in 1927, the Kiel school's publications totaled 103, averaging two or three a year throughout the forty-year period. Between 1887 and 1901 the output of the Kiel group was steady; it included a number of significant reports on the Plankton Expedition and the first publications on the chemical analysis of plankton, in addition to Brandt's seminal 1899 monograph *Ueber den Stoffwechsel im Meere*, in which he outlined the denitrification hypothesis. After 1901 the rate of publication increased until 1912. During these years the majority of the school's work on microbiology, nutrient analyses, and the biota of the open ocean was first published. During this period members of the school published eight reviews of their work, 17% of their total output. The large number of publications in 1912 consists mainly of Lohmann's reports on the cruise of *Deutschland* to the South Atlantic in 1911 and his review of plankton studies, *Die Probleme der modernen Planktonforschung*; only two publications dealt directly with new information on plankton dynamics.

After 1912 the scientific output of the Kiel school consisted more and more of reviews (a few of them, like Brandt's monographs, with original syntheses): 44% of the total between 1912 and 1927, and 90% between 1919 and 1927. Clearly there was less and less original work to present after 1912. And what was original was mainly the use of information or ideas dating back several years, even a decade or more, such as Brandt's incorporation of vertical circulation into his theory in 1920, or his suggestion (in 1927) that the problem of nitrification could be solved if organic nitrogen, not ammonia, were the substance oxidized by nitrifying bacteria. Thus the cumulative curve of publications represents, at least roughly, the output of new data, new ideas, and original synthesis of their research by the members of the Kiel school. Without doubt, the school had lost its impetus after 1912. How can the life cycle and creativity of the Kiel school be explained?

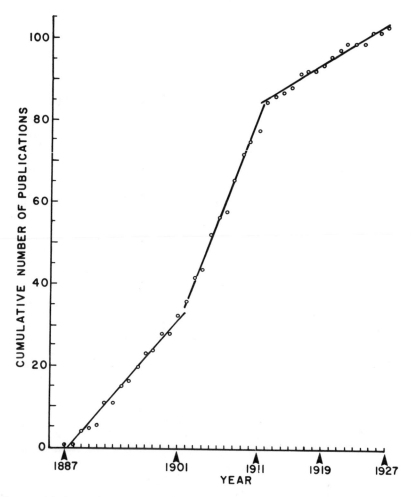

FIGURE 6.1 A cumulative plot of publications on plankton dynamics by the Kiel school (as defined in the text), 1887–1927. Note the rapid rate of publication between 1901 and 1912, when the majority of original work was done at Kiel, and that the rate began to fall off well before the beginning of World War I.

Nothing so abstract as loss of scientific motivation or the mining out of the field accounts for the decline of the Kiel school before World War I. Adolf Remane, one of Brandt's successors at Kiel, diagnosed the problem correctly as loss of creative personnel, noting that "bald nach 1910 begann aber die Kieler Meeresforschung viele ihrer Träger zu verlieren."[3] The Kiel school began to falter when it lost its most creative members,

3. Remane 1968, p. 172.

among them Carl Apstein and Hans Lohmann. What kinds of scientists were lost, and what did they have in common?

Franz Schütt, the first important young scientist to leave the Kiel school, responded to factors that affected most of Brandt's most productive, creative junior colleagues who followed him.[4] As a marine botanist, newly arrived at Kiel, he accompanied Hensen, Möbius, and Heincke to the North Atlantic in 1885. On his return he supervised Hensen's regular plankton-collecting cruises which had begun in 1883. These cruises, which were extended into 1888, showed the seasonal succession of species and regular changes of volume on which the Kiel school's theory of plankton dynamics eventually was based.

During the winter of 1888–1889, Schütt made quantitative plankton collections at Naples for Hensen and Brandt. Then, having proved his worth, he accompanied them on the Plankton Expedition. Using the results of the expedition and his early experiences in the North Atlantic and the Baltic, Schütt wrote two major monographs, both published in 1892, *Analytische Plankton-Studien* and *Das Pflanzenleben der Hochsee*. These established his reputation and are still quoted as founding works on the ecology of marine phytoplankton. Schütt's career at Kiel ended in 1895 when he became Professor ordinarius of botany and director of the botanical garden and museum at Greifswald. Thereafter, disabled by illness, he planned a great work on diatoms which was never finished. His work was continued mainly by students, who worked on the algae around Greifswald.

Brandt's early plankton collections, especially those on cruises between 1888 and 1893, following Schütt's lead, depended on the assiduous work of Carl Apstein (1862–1950).[5] Apstein, born in Stettin, then living in Tilsit, Coburg, and Naumburg, appears to have spent the *Wanderjahre* of a typical German student. He was schooled in Halle, then went to university in Leipzig, Freiburg, and Kiel. His doctorate was awarded at Kiel in 1889, where nine years later he habilitated, entitling him to teach in a German university. Apstein served the Kiel Commission and the Zoological Institute as Brandt's right-hand man, an appointment that ended when he joined Carl Chun on the German Deep-Sea Expedition of 1898–1899. Independently of his work on marine plankton, Apstein carried out a distinguished series of studies on the lakes of Holstein near Plön. Between 1902 and 1911, as plankton assistant in the Internationale Meereslaboratorium, Apstein was responsible for collecting and working

---

4. On Schütt's career see Volbehr and Weyl 1956, p. 212; Borriss 1956, pp. 531–532; other details have been compiled indirectly from publications of the Kiel school. See also Chapter 4, n. 9.

5. *Neue deutsche Biographie* 1:33; Remane 1968, p. 170; Volbehr and Weyl 1956, p. 212.

up the plankton during Germany's ICES investigations. Nicknamed "der Liebe Gott because of his white beard and serene eyes," according to Remane's memories, Apstein was capable of directing and doing copious amounts of work in quite separate fields of aquatic science. He gave up this hectic pace in 1911 to become a paid official of the Akademie der Wissenschaften in Berlin, where he spent the rest of his working life.

A close contemporary of Apstein, the contributor I call the *enfant terrible* of the Kiel school, Hans Lohmann (see Chapter 4 near n. 41), was undoubtedly the most original and productive of its young members.[6] Only the contributions of the chemist Emil Raben, who was responsible for nutrient analyses, had nearly equal significance during the years between 1901 and 1912. When Lohmann came to Kiel in 1886, he was captured by Hensen's spell; he then remained at Kiel, or worked for Brandt and Hensen in Italy or at sea, for twenty seven years. After habilitating in 1893, he was appointed Privatdozent in zoology and comparative anatomy in 1895, then Professor titular in 1904, modest recognition of the unusual volume and quality of his work.

Lohmann's first postdoctoral study, on the collections of the Plankton Expedition, resulted in his recognition of the importance of very small phytoplankton cells in the sea. This revelation led to his well-known studies on nanoplankton and his undeservedly less-known monumental work on the plankton cycle in Kiel Fjord during 1905–1906. His quantitative plankton collections, from Messina (1896–1897) and Syracuse (winter of 1900–1901) showed, as Brandt had expected, the low abundance of plankton in the Mediterranean. His development of pumping and filtration methods demonstrated the prevalence of very small cells nearly everywhere, including the open Atlantic. In 1913 Lohmann left Kiel for appointments at the Natural History Museum in Hamburg which were understood to be steps toward a chair in zoology in the University of Hamburg.[7] Because the university's founding was delayed by the war, Lohmann became first Professor ordinarius of zoology in the new university only in 1919. Thereafter he held the chair as well as directorships of the state zoological institute and the zoological museum until his death.

There are many similarities among the careers of Schütt, Apstein, and Lohmann. Their doctoral degrees were awarded between 1885 and 1889. Each habilitated and was appointed a Privatdozent; later each was given the moderately prestigious but unpaid position of Professor titular.

---

6. Hentschel 1933; Klatt 1935; Remane 1968, pp. 169–170; Volbehr and Weyl 1956, p. 213.

7. Hjalmar Thiel, personal communication in Hamburg, March 1982.

Each became involved, directly or indirectly, in the work of the Plankton Expedition; Apstein and Lohmann continued their work on plankton in association with Brandt for many years. Their research work was of high quality and voluminous, resulting in international reputations for all three, certainly adequate qualifications for senior academic positions at Kiel. Despite this, each found it necessary to leave Kiel to maintain or advance his career. Schütt, Apstein, and Lohmann left Kiel, Schütt early, the others later, because of the deficiencies of the German university system.

The Christian-Albrechts-Universität zu Kiel, dating from 1665, was typical of many other "reformed" German universities in the period between the 1870s and World War I.[8] The major academic subjects were taught by full professors, the Ordinarien, who received salaries supplemented by student fees for lectures and who wielded most of the academic power within the university. Privatdozenten, usually younger men who had received the doctorate, supplemented it by additional research to achieve Habilitation, a requirement for the state-issued license to teach in universities. Privatdozenten were not usually paid but received fees for their lectures, which were frequently on specialties peripheral to the teaching of the Ordinarius, or on newly developing topics.[9] A few Privatdozenten were given paying appointments as Professor extraordinarius (ausserordentliche Professor) usually as assistants to Ordinarien, or as the directors of research institutes and seminars. Others, like Schütt, Apstein, and Lohmann, whose qualifications merited reward, were granted the title Professor titular, which brought some prestige but no salary; to live they still had to teach as Privatdozenten or work as assistants to Ordinarien like Brandt. The distinction in ranks was tangible indeed, for, in about 1900, Ordinarien in German universities received an average salary of 10,000 to 12,000 marks (roughly U.S. $45,000 to $54,000), but it ranged from 4000 to 40,000 marks, depending on the popularity of the professor's lectures, the importance of the field, and the ability of the professor to campaign on his own behalf with the Kultusministerium (ministry of education) of his state. By contrast, Extraordinarien could expect a salary of 3000 to 5000 marks and Privatdozenten about 1500 marks from lecture fees.[10] It was no accident that, in the words of

8. The voluminous literature on German universities in the nineteenth and early twentieth centuries may be approached through Ben-David 1971, chapter 7; Ben-David and Zloczower 1972, pp. 47–62; Busch 1962; Farrar 1976; Jarausch 1982, 1983a, b; Lundgreen 1980; McClelland 1980, chapters 3, 4, 7, 8; McCormmach 1974; Ringer 1969, chapter 1; Ringer 1979; and Turner 1971, 1981.
    9. On the difficulties faced by Privatdozenten, see especially Busch 1962.
    10. From Ringer 1969, pp. 37–38, and McClelland 1980, pp. 270–271.

Alexander Busch, "the poverty of the Privatdozenten became an almost unquestioned tradition."[11] Nor was the position of the Extraordinarius necessarily an enviable one, for it was often the "consolation prize" awarded to the second-rate or unlucky intellectual who never achieved the position of Ordinarius.[12] At least he was paid to stifle his disappointment, unlike the struggling Privatdozent, whose desperation and poverty increased as he grew older, especially if he had a family and was not promoted or offered a chair.

Many German universities after about 1870, Kiel included, were characterized by an increasing number of research institutes. In Prussia, under the official of the Kultusministerium responsible for universities between 1882 and 1907, Friedrich Althoff, universities added research institutes in subjects as varied as law, theology, medicine, and zoology.[13] Typically an institute (like Brandt's at Kiel, founded under Möbius in 1868) was the charge of an Ordinarius who negotiated its financing directly with the Kultusministerium. Often a Professor extraordinarius or another senior academic directed the actual work, which was carried out by paid assistants who might be Privatdozenten (e.g., Apstein and Lohmann), doctoral students, or just employees (e.g., Raben). The role of Privatdozent had been created early in the nineteenth century to bring vitality to the universities, through independent scholarship. But late in the century, as more and more Privatdozenten worked in institutes to assure an income, their independence decreased and they became increasingly dependent on the research ideas, opportunities, and good will that issued, or did not issue, from the director, whether Ordinarius or Extraordinarius. Increasingly too, as the institutes developed, more and more science was done by the lower ranks, for the Ordinarien became caught up in heavy loads of administration and teaching.[14]

The facilities of many institutes in the sciences, including Brandt's at Kiel (united with the Kiel Commission's laboratory), often matched E. Cittadino's description of botanical institutes in the late nineteenth century.

> The larger institutes had a separate lecture room, a classroom for undergraduates, a laboratory for advanced students, an office and separate laboratory for the professor, and space for a herbarium and small library. Beginning students heard lectures, saw demonstrations, and performed a

11. Busch 1962, p. 326.
12. The evaluation is Zloczower's (1981, p. 48).
13. McClelland 1980, pp. 279–283; Jarausch 1982, pp. 28–31.
14. On the role of assistants see Busch 1962, pp. 328–330.

minimum of laboratory work. Advanced students had their own space in the laboratory, where they were expected to work eight or ten hours a day. In addition to the professor there were one or two assistants and, depending on the institution, one or two assistant professors and a varying number of privatdozents, all of whom also regularly worked at the institute.[15]

The cost of these nearly autonomous academic research factories was high. At Berlin in 1890, which was by no means atypical, the annual running expenses of the institutes were twice the cost of professorial salaries.[16] Brandt's good fortune at Kiel lay in the fact that he controlled what amounted to a superinstitute, his own Zoological Institute, for which he negotiated the funding, combined with the Internationale Meereslaboratorium, which had its funding assured by the German government for at least the initial five years requested by the ICES nations. And, from the point of view of the junior researchers, the facilities for work on the sea provided at Kiel far exceeded anything they could hope to find in another university. At least for a time—more than two decades in the cases of Apstein and Lohmann—they were prepared to accept the low status of non-Ordinarien for the distinct advantages Kiel gave to their careers as independent or nearly independent researchers. The results are obvious in the creativity of the Kiel school between 1901 and 1912, even using an imperfect indicator, such as their record of publications, or a more satisfactory assessment of their work, acceptance of the group's theory of plankton dynamics.

In an ideal world, Schütt, Apstein, and Lohmann would have proceeded along the accepted route from doctorate to Habilitation to Privatdozentur to assistant to ausserordentliche Professor, culminating in a chair at Kiel which would have allowed each to carry on marine scientific research and teaching. But the ideal career was achieved by few academics in Germany after 1870 because of the huge increase in university enrollment that began then and was still in progress in 1914. Total German university enrollment in 1865–1866 was just under 14,000. By 1910 enrollment had increased four times, to nearly 60,000.[17] Much of the increase took place, beginning in the 1880s, in the philosophical faculties of the universities, which housed both humanities and sciences. The results were profound for all academics from Ordinarien to Privatdozenten.

15. Cittadino 1981, p. 23.

16. McClelland 1980, p. 280. As early as 1870 costs of institutes and of salaries were about equal (pp. 204–205).

17. Jarausch 1982, pp. 28–29, 136; Jarausch 1983a, pp. 12–18; McClelland 1980, p. 247; Titze 1983.

The greatly increased number of students in German universities was met by increasing the number of teachers—but Privatdozenten, not Ordinarien, carried the increasing load. Admittedly the number of new chairs increased—by a factor of 1.7 between 1864 and 1910—but the number of Extraordinarien and Privatdozenten increased about threefold during the same period.[18] Ordinarien continued to give the financially profitable general courses, which also grew, occupying more professorial time. The Ordinarius found himself with less and less time for research, particularly if he was also responsible, as many were, for a growing research institute. Despite this, for obvious reasons a host of younger men were hungry for his position.

As the universities and the lower academic staff expanded, professorial resistance to the creation of new chairs grew. Rather than creating competition by increasing the number of chairs, many universities chose to assign special topics to Extraordinarien, or to institutionalize new fields in research institutes, both subject to the Ordinarien.[19]

> From a financial point of view, it began to make a very great difference to every academic whether his field of specialization was popular or not, whether students were attracted to his lectures, whether or not his courses were basic for a certain state examination, whether he himself gave many of these examinations, whether he had to share his area of competence with a colleague, and even whether or not a new academic chair was likely to take students away from his classes.[20]

The result, at Kiel as elsewhere, was that a rapidly increasing number of young, ambitious Privatdozenten sought a much more slowly expanding number of professorial chairs. The delays between Habilitation and attainment of a chair increased inexorably, worsening a difficult situation for many Privatdozenten. There were a few advantages—the long wait for a chair gave many ambitious young men, such as Schütt and Lohmann, the opportunity to travel widely and to undertake lengthy research projects[21]—but the price was financial privation and emotional insecurity.

> The old *Privatdozent* who somehow managed to keep his head above water by such activities as keeping boarding houses for foreign students, by

18. Lundgreen 1980, p. 217; Ringer 1979, p. 419. Similar figures for a different period are given by Busch 1962, p. 319.

19. Zloczower (1981) discusses the significance of these changes for physiology in Germany during the period of university expansion. This is the most detailed study available.

20. Ringer 1969, p. 52.

21. Cittadino (1981, chapter 6) describes the advantages gained by botany and young botanists by the ability to travel widely during these years.

giving private tuition or aiding in the revision of lectures was a failure in his academic career, simply by virtue of not being appointed to a professorship. Such men became typical figures in the universities. Their lives were miserable. Many of them were married and had families for whom they could not provide an appropriate standard of living.[22]

Undoubtedly it was this fate that Schütt, Apstein, Lohmann, and a host of other junior academics in German universities sought to avoid.

The problems of academic advancement were severe for all Hensen's and Brandt's junior colleagues in Kiel. Botany and zoology had gone through a period of expansion in Germany much earlier, in the mid-nineteenth century. By the 1880s the chairs were static, and what is more, they were filled by young men who could hold them for years. In Schütt's case, he arrived in Kiel in 1885, the same year that Johannes Reinke moved from Göttingen to become Professor ordinarius of botany.[23] Schütt's move to Greifswald, decidedly a less prestigious university than Kiel, reflected his judgment that to be an Ordinarien there was superior to the uncertainties of a lengthy Privatdozentur at Kiel. His move mirrored the judgment of many young academics that less-well-regarded universities were preferable, as "first class waiting rooms" to no chair at all.[24] Schütt's decision was wise, for Reinke held the chair of botany at Kiel until 1921.

Apstein and Lohmann were in similar positions. Both received their doctorates in 1889, the year following Karl Brandt's appointment as Professor ordinarius of zoology and director of the Zoological Institute at Kiel, positions he held until 1922. No new chairs were likely to be founded at Kiel, thus both must have realized early that they would have to leave eventually. It is remarkable that they stayed as long as they did.

When Schütt, Apstein, and Lohmann considered leaving Kiel, each faced the general problem of Ordinarien during the era of university expansion, an overload of teaching and administration. In addition they had to reconcile themselves to leaving behind the superb research facilities at Kiel. Apstein's decision was to become a scientific administrator with the Akademie der Wissenschaften. Schütt abandoned plankton dynamics, concentrating his work on the taxonomy and morphology of diatoms. Lohmann founded a major school of zoology in Hamburg, which, although it became well known for marine research, never produced work as original as his own at Kiel. Although he might have inspired students to expand his early work on marine production, Loh-

22. Busch 1962, p. 334.
23. Volbehr and Weyl 1956, p. 150.
24. Quoted from Zloczower 1981, p. 47.

mann's students at Hamburg were assigned safe taxonomic or morphological studies that would eventually assure them jobs.[25] Thus Lohmann made his concessions to the pressures of the German university system.

At Kiel, Brandt found himself increasingly isolated, even before the war. The departure of Apstein and Lohmann was a serious loss, even though there was no lack of doctoral students to do routine studies.[26] Through the prewar and war-time years the Kiel school's work was maintained mainly by Emil Raben, working on nutrients in the Internationale Meereslaboratorium. With the disbanding of the laboratory in 1917 and Raben's death in 1919 this came to an end.[27] The war had its direct effects as well, for a number of Brandt's students were killed, including Karl Müller, who had been given the special task of duplicating Lohmann's famous study of the yearly cycle in Kiel Fjord. It would be a mistake however to equate Karl Brandt with Viktor Jacob, the semifictional Professor extraordinarius in Russell McCormmach's 1982 novel *Night Thoughts of a Classical Physicist*. The aged Jacob realized his personal failure as a scientist and his lack of achievement as an academic; Brandt, by contrast, had been eminently successful as both. But, burdened with administration and teaching, he could not maintain the originality of the Kiel school without creative younger men. Unlike many Ordinarien, Brandt was able to keep up specialized research even after Apstein and Lohmann left, but it was based less and less on original data and new ideas after 1912. The Kiel school ran well on assistants and increasing numbers of doctoral students between the 1880s and about 1912; inevitably it ran out of energy when these men had to find professorial chairs during the first two decades of the twentieth century.

Superficially, the rise, most creative period, and demise of the Kiel school fits accepted models of the development of research groups, especially the one outlined by Michael Mulkay and his co-workers.

> The onset of growth in a new area typically follows the perception, by scientists already at work in one or more existing areas, of unresolved problems, unexpected observations or unusual technical advances, the pursuit of which lies outside their present field. . . . An early lead in the competition for results is usually taken by those with best access to such resources as graduate students, research funds, suitable techniques, publication outlets and the legitimacy conferred by the support of eminent scientists. . . . Research networks develop typically in response to basic contributions which appear early in the growth sequence. Subsequent

25. Professor Dr. Adolf Bückmann, personal communication in Hamburg, March 1982. Bückmann was the last of Lohmann's students.

26. See Brandt 1921a, 1925a, on the succession of students and their work.

27. Germany did not rejoin ICES until 1926, after Brandt's retirement.

work tends to consist primarily of elaborations upon these contributions. . . . This means that opportunities for making a notable scientific
advance and the chances of receiving an unusual amount of scientific
recognition decline very quickly after the earliest period. . . . As the decline
of interesting and/or solvable problems continues, accompanied by a
growing scarcity of professional recognition and career opportunities,
recruitment falls away and established members of the network move
elsewhere into problem areas in process of formation.[28]

Such a development takes a sigmoid form, represented in my analysis by
scientific output (the number of papers) (see Figure 6.1). Similarly, according to Mulkay et al. a research group begins slowly, achieves a period
of logarithmic growth, then stabilizes or declines as the research area
becomes well-worked conventional science with fewer and fewer major
rewards.[29] During the period of decline, according to their model, the
school or research group has the following characteristics, quoted (with
some paraphrasing) from their description:[30]

1. A firmly established intellectual framework
2. Cognitive and technical standards keeping innovations within well-
   defined limits
3. Institutionalized recruitment and funding
4. A stable rate of recruitment
5. Research teams or groups of collaborators with special areas of
   competence and interest
6. Productive, influential leaders who are highly cited in papers and
   who are at the center of informal communication networks
7. Declining opportunities to make significant findings
8. Declining opportunities for professional recognition and advancement.

Most of these points apply well to the Kiel school, but the list parts
most strikingly from the situation of the Kiel school in item 7. To the
biologists of the Kiel school a considerable number of significant problems of high scientific merit still remained in 1912. Despite Lohmann's
work during 1905–1906, an acceptable estimate of marine production
still did not exist. Nitrification remained a puzzling and important problem. Vertical circulation, though an integral part of the theory of plankton dynamics, could not readily be related to regions of high production.

28. Mulkay 1975, pp. 520–522. The model is explained in greater detail by Mulkay et al.
1975.
   29. Mulkay et al. 1975, figure 2.
   30. Mulkay et al. 1975, p. 198.

And many problems related to nutrients still existed, such as which nutrients were limiting under given conditions and at specific times, not to mention the uncertainties and inaccuracies of nutrient analysis. Each of these problems was taken up in the 1920s and 1930s by research groups outside Germany; this is powerful evidence that plankton dynamics, as perceived when the Kiel school declined, was not exhausted of important problems.[31]

The potential for scientific reward was still great after 1912; yet the Kiel school declined. At Kiel the problems faced by Schütt, Apstein, and Lohman were not the result of scientific exhaustion of the research area. Instead, there was a lack of opportunities for new research in plankton dynamics caused by purely institutional factors, especially the inability of the German university system to make places with high professional status for even its most distinguished junior members.

Avraham Zloczower's well-established hypothesis that German physiology in the nineteenth century became less creative when the number of new professorial positions decreased[32] applies in general to the situation at Kiel, for there seems little doubt that plankton dynamics would have continued to advance at Kiel long after 1912 if, for example, Lohmann had been given a chair there. But without major institutional changes it is unlikely that the Kiel school could have continued indefinitely, for the structure of German academia, especially during the years of expansion, eventually stifled protracted creative work. Plankton dynamics, which developed most creatively at Kiel between the 1880s and about 1912, fell into an institutional gap; botany and zoology had expanded in German universities before 1880; oceanography became a recognized field and began to grow in the 1920s.[33] Thus, although the volume of work at Kiel between 1899 and 1912 allowed Brandt to work for twenty years more and established the careers of Schütt, Apstein, and Lohmann, the Kiel school itself could not survive without a way to keep its most creative members at home. Once the Kiel school disbanded, its members at Greifswald, in Berlin, or at Hamburg, found no facilities for work on plankton or had to subordinate their science to establish new institutions of their own.

In general, research groups have short lives, because scientific fashion, the availability of resources, and the relevance of scientific questions

31. For a contemporary account of the problems remaining in 1912 see Lohmann's *Die Probleme der modernen Planktonforschung.*

32. Zloczower 1981. Turner et al. (1984) confirm and modify Zloczower's conclusions.

33. The first university chair in oceanography was created in Paris in 1906, the second in Liverpool in 1919. Also in that year the Deutsche wissenschaftliche Kommission für Meeresforschung was founded, replacing the earlier commission that had participated in ICES.

change quickly in tangible and not-so-tangible ways.[34] The decline and demise of the Kiel School, however, may be accounted for nearly entirely by institutional factors, particularly the inflexibility of the German academic system. The Kiel school's short life was due to the lack of career opportunities in Kiel, not to exhaustion of ideas, nor directly to the effects of war and social upheaval.

The remaining chapters will be concerned with the development of plankton dynamics outside Germany, mainly in the United Kingdom and the United States. Writing of German science early in the twentieth century, Joseph Ben-David has said that "because of the internal tensions between the ranks and the difficulty of obtaining recognition for new fields, the center of scientific activity . . . started to shift to Britain and the United States."[35] But the shift was a slow one, and once established proved to be as episodic and contingent on social and scientific forces as the achievements of the Kiel school had been at the turn of the century.

34. See, for example, the comments of Collins (1981, pp. 12–13) and Geison (1981). The dynamics of change are discussed by Mulkay (1975) and Mulkay et al. (1975).
35. Ben-David 1971, p. 137.

PART 2

Biological Oceanography in Britain and the United States, 1921–1960

Turning to a very fundamental matter of purely scientific investigation, we do not know with any certainty what causes the great and all-important seasonal variations in the plankton (or floating minute life of the sea) as seen, for example, in our own home seas, where there is a sudden awakening of microscopic life, the Diatoms, in early spring when the water is at its coldest. In the course of a few days the upper layers of the sea may become so filled with organisms that a small silk net towed for a few minutes may capture hundreds of millions of individuals. And these myriads of microscopic forms, after persisting for a few weeks, may disappear as suddenly as they came, to be followed by swarms of Copepoda, and many other kinds of minute animals, and these again may give place in the autumn to a second maximum of diatoms or of the closely related Peridiniales.

W. A. HERDMAN 1920

7 Food from the Sea: The Origin of British Biological Oceanography

> Now the two pressing problems which present themselves, from the practical point of view are: Do we at present make the best possible use of the harvest of the fishing grounds? And how can the yield of human food in the form of fish be increased?
>
> E. J. ALLEN 1917

Present-day biological oceanographers are scarcely aware that the Kiel school made the first significant contributions to the theory of plankton dynamics. The early demise of the group at Kiel, the effects of World War I, and anti-German sentiments that persisted long after the war were partly responsible for its disappearance, as far as scientists are concerned, into the mists of history. More important than those factors, real though their effect may have been, was the development of a new group of biological oceanographers in Britain, working mainly in the Plymouth Laboratory of the Marine Biological Association (MBA) of the United Kingdom. Their work during the 1920s and 1930s provided new, highly sensitive techniques for the determination of nutrient salts in seawater, extended and made more sophisticated the causal analysis of the plankton cycle that had originated at Kiel, and demonstrated the role of biological factors as well as chemical ones in controlling the cycle. Publishing sophisticated work in English at a time when German science had lost its international influence as a result of the war, they produced the outline of what present-day biological oceanographers consider a truly modern theory of plankton dynamics.

This perception of the foundations of biological oceanography, unfair as it may be to the Kiel school, contains a significant grain of truth. Plymouth was important in providing a recognizably modern conception of the plankton cycle, a viewpoint that has become a cornerstone of the modern science, even if not the full edifice. But while the work of Plymouth scientists such as E. J. Allen, W. R. G. Atkins, H. W. Harvey, F. S. Russell, and others now has a mythic quality as "the origin of it all," it has never been clear why the Plymouth group was so successful, nor why its major triumphs occurred over such a short period, the decade and a half beginning in 1921. The laboratory, after all, opened in 1888, yet significant work, it appears, did not get under way until the 1920s and ended before 1940. What conditions in British science and at Plymouth existed between the 1880s and the early 1920s to result in such a short episode of truly creative work? Why did a significant change in plankton research occur there and not elsewhere? And why did the Plymouth group's work take the course it did, culminating in the 1930s but declining significantly thereafter? The answers lie in the history and organization of the Plymouth Laboratory as well as the work of the Plymouth scientists.

After the success of the International Fisheries Exhibition in London in 1883, it occurred to E. Ray Lankester, professor of zoology at University College, London, that a laboratory for fisheries research could be established in England, capitalizing on the enthusiasm engendered by the exhibition and on its profits. His colleague Albert Günther, keeper of zoology at the Natural History Museum, London, hearing Lankester's ideas, suggested that the scheme could be advanced if an association were set up to found the laboratory.[1] After gaining support from many British scientists and the patronage of the Royal Society, the Marine Biological Association of the United Kingdom was established at a meeting chaired by T. H. Huxley in London on March 31, 1884. Its avowed aims were, in the words of the Duke of Argyll, "accurate researches . . . leading to the improvement of zoological and botanical science, and to an increase of our knowledge as regards the food, life, conditions and habits of British food fishes, and molluscs in particular, and the animal and vegetable resources of the sea in general."[2]

The association's laboratory was begun in 1886, on the strength of financial pledges from individual scientists, members of the aristocracy, and the Company of Fishmongers which by 1887 amounted to £10,000.

---

1. Anonymous 1888a, p. 137. Lankester was professor of zoology at University College, London 1874–1891, Linacre Professor of Zoology at Oxford 1891–1898, and director of the Natural History Museum 1898–1908. Aspects of his career are outlined by Austerfield 1979, Bourne 1930, DeBeer 1973, and Goodrich 1930. The funds from the exhibition went elsewhere, to support the gathering of fisheries statistics by the Board of Trade.

2. Anonymous 1884d, pp. 5–6. More details are given in Anonymous 1884a, b, c; Anonymous 1887; and Lankester 1884. A short but detailed history of the Marine Biological Association and the laboratory is provided by Southward and Roberts (1984, 1987).

It was also supported by a grant from the British treasury of £5000 and the promise of £500 per year for five years, subject to the conditions that the association have an annual audit, submit annual reports, work on "procuring practical results with regard to the breeding and management of food-fishes," provide space to fisheries investigators from other agencies, and collaborate with quasi-governmental agencies elsewhere, such as the Fishery Board for Scotland.[3] On Saturday, June 30, 1888, the new laboratory was opened on land obtained from the War Office beside the Citadel in Plymouth, "remarkable as being the first institution in this country designed purely for scientific research which has been originated and firmly established by the efforts of scientific men appealing to the generosity and confidence of wealthy individuals and corporations who desire the progress of knowledge for practical ends and the general good of the community," according to a panegyric in *Nature*.[4]

The Plymouth Laboratory was organized and opened while state support of science was being actively debated in Britain. To Lankester, who had visited the well-known Stazione Zoologica established at Naples in 1872 by Anton Dohrn, it was evident that Britain lay far behind many European nations in supporting marine science.[5] Naples, the Kiel Commission, and the U.S. Fish Commission, dating from the early 1870s, were models of the pattern science should follow.[6] An anonymous author in *Nature*, supporting the foundation of the Marine Biological Association, believed that the national interest lay in "national encouragement and Government support" of zoology as "an important handmaiden to national wealth." He concluded that "patriotism, as well as the desire for the advancement of human knowledge, would therefore urge with all possible earnestness the establishment of a national Marine Zoological and Physical Survey, whereby the fauna and conditions of every portion of our coast should be carefully investigated."[7] Although this ambitious scheme did not come about, the sentiment that science and national aims were at least complementary if not synonymous was felt by an increasing number of scientists during and after the 1880s, just at the time the

---

3. Anonymous 1886, p. 177; Committee on Fishery Investigations 1908, p. ix. The treasury grant of £500 was increased to £1000 after 1892, an amount that was maintained until World War I, when it was reduced to £500. The grant of £1000 was restored in 1919. Throughout this long period, the treasury grant was the mainstay of the Plymouth Laboratory.

4. G.C.B. (G. C. Bourne) 1888, p. 198. On the opening and the laboratory see also Anonymous 1888a, b; Heape 1887.

5. Marine laboratories were widespread in continental Europe by 1888. Oppenheimer (1980) discusses the formation of the Stazione Zoologica; further details of the founding of biological stations are given by Kofoid 1910 and M'Intosh 1896.

6. Anonymous 1884d, pp. 10–14.

7. Anonymous 1884a, p. 25.

Plymouth Laboratory was conceived and realized. Moreover, many of them believed that British science was being overtaken or had been left behind by state-supported research, particularly in Germany.[8]

The covert social program rooted in professional aspirations and nationalism that lay behind the foundation of the Marine Biological Association was reinforced by a lengthy debate, beginning no later than the 1860s, about the state of British fisheries. Beam trawling by sailing smacks from British ports began early in the nineteenth century; about 1878 steam trawling began, a technological change that accelerated during the 1880s and was virtually complete by the 1890s. As the intensity of fishing increased so too did the fishermen's complaints, especially inshore, about the decline of their catches. The Royal Commission of 1863, including T. H. Huxley, which was appointed to look into these claims, concluded that the value of the fishery was still increasing, and that in its current state of development beam trawling did not harm the supply of young fish. In sympathy with the political thought of the times, the commissioners concluded that the fisheries should be regulated as little as possible, and that many laws regulating local sea fisheries should be repealed.[9]

But doubts persisted as the intensity of fishing increased. A Royal Commission on Trawling was appointed in 1883 under the earl of Dalhousie (again including T. H. Huxley; the zoologist W. C. M'Intosh of St. Andrews was scientific adviser) to examine the possibility that trawlers harmed the conventional line and drift-net fisheries. The main complaints centered around flatfish, which even the trawler fishermen believed might be decreasing. The commissioner's report, which was published in 1885, while the Marine Biological Association was in its early days, concluded that trawling did not damage fish eggs, the majority of which were pelagic, nor did it affect the benthic animals eaten by flatfish such as plaice and sole or the immature fish present on the banks. But there was little gainsaying that a decrease of flatfish had occurred on inshore trawling grounds, due perhaps to fishing itself, perhaps to other causes. For this reason, the commission recommended that the Fishery Board for Scotland, the only effective regulatory agency in the United Kingdom at that time, be allowed to close inshore areas to trawling, a course of action the board followed in 1886.[10] But the biological questions remained unresolved about what governed the abundance of commercial marine fishes.

8. See especially Cardwell 1972, chapter 6; Alter 1982, 1987; Haines 1958; Haines 1969, chapters I, III; MacLeod 1971, 1976; F. M. Turner 1978, 1980.

9. Johnstone 1905, chapters II, XIV; Jenkins 1920, p. 145–158.

10. Johnstone 1905, pp. 29–41; Jenkins 1920, p. 150.

During the 1880s there was no authority in England empowered to regulate marine fisheries, nor to carry out research on fishes the way the Fishery Board for Scotland (given modern form in 1882) could do.[11] English and Welsh fisheries were governed only by local regulations. Although a fisheries department of the Board of Trade was established in 1886 its duties were mainly to gather statistics on marine fisheries and to administer a few locally significant fisheries, such as those for salmon. Thus the Marine Biological Association was in a potentially powerful position to fill at least the scientific vacuum while politicians and politically active scientists attempted to develop effectively a way of administering the fisheries of England and Wales. In doing so, it could help to halt the "decline" of science that some decried in Britain while achieving practical ends that could be espoused even by the many who saw no role for the state in the support of science.

This mixture of motivations, of opportunism and altruism, is evident in the sentiments expressed by the founders of the Marine Biological Association and their friends. The founders themselves, constituted as the Governors and Council of the Association, were a cross-section of the British scientific establishment who regarded government support as at least helpful if not absolutely necessary to restore and maintain English science. Lankester was to be secretary, joined by T. H. Huxley (also president of the Royal Society), John Lubbock (president of the Linnean Society) and W. H. Flower (president of the Zoological Society and director of the Natural History Museum) in addition to the duke of Argyll, the earl of Dalhousie, and the prime warden of the Company of Fishmongers along with other scientific and political worthies. Although their day was nearly past, three council members, Huxley, Lubbock, and Joseph Dalton Hooker, were members of the influential X Club which, as a powerful, only barely invisible college of British science, had met regularly since 1864, favoring state support and a more prominent place for science in British life.[12]

A leader writer in the *Times* (it was Lankester), writing of the inaugural meeting of the association in March 1884, was emphatic in his support of science, for "the utility in its highest, and even in its lowest, sense of encouraging scientific research may now be taken as recognised in all civilized countries. All the most valuable "practical" discoveries have been made by men who were not seeking for them, but whose sole aim

11. On the history of the Fishery Board for Scotland, see Fulton 1889; Johnstone 1905, chapter VI; Lucas 1956, pp. 223–224. Freshwater fisheries did get some attention in England; see Bartrip 1985 and MacLeod 1968.

12. Jensen 1970; MacLeod 1970. F. M. Turner (1978) explores the context in which the X Club functioned.

was to satisfy a noble acquisitiveness."[13] This was a sentiment shared by
W. H. Flower, speaking at the opening of the laboratory. He claimed that
"it is scarcely possible to name one of the marvellous improvements
which have taken place in late years, that have added so much to the
convenience, the comfort, the capabilities of human life, that has not
been, when traced back to its source, the outcome of scientific research
undertaken originally for its own sake." But this was not all, for Glad-
stone's Liberal government "recognised the practical importance of our
work, as calculated to benefit not only the interests of the fishing industry, but those of the community at large."[14] The practical side of science
was taken further into the realm of politics by the earl of Morley, who as
undersecretary of state for war had helped to obtain the laboratory's site.
He asked rhetorically, "How was it that we, who more than any other
nation in the globe reaped the richest harvest from the sea, had never yet
endeavoured scientifically to inquire into the sources of this great industry and article of food? How was it that we had lagged behind other nations?" To him "it seemed extraordinary . . . that so many years should
elapse without scientific efforts being made; for we must recognise more
and more that the wealth of nations and individuals depended on the
economical and ample use of the powers of nature."[15]

Lankester agreed. He would concede no points to the critics of the
association, already visible on the horizon. Science and the practical
could not be separated.

> Such a distinction could not be drawn. All purely scientific research had a
> practical end. They might not be able to tell what the practical end might
> be; but they pursued scientific knowledge with the conviction that the
> progress of knowledge must lead to practical benefits. On the other hand,
> they also know that any attempt to make inquiry with a practical end in
> view which should ignore scientific methods and aim too directly at the
> practical end was fraught with danger and almost certain failure. The only
> way to attain success was to cultivate the tree of science first, and then
> gather the fruit; they could not grow the fruit without attending to the
> tree.[16]

By the summer of 1888 the tree of marine science had at least taken
root in Plymouth. But despite the vigor of its origin, the early years of the
Plymouth Laboratory were difficult. Although glowing sentiments were

13. *The Times*, March 31, 1884, quoted in Anonymous 1887, p. 20. Bourne (1930, p. 367)
claims that Lankester was the author.
    14. Anonymous 1888a, pp. 128–129.
    15. Anonymous 1888a, pp. 132, 134.
    16. Anonymous 1888a, pp. 136–137.

expressed at its opening, the laboratory and its tiny staff faced serious difficulties. Not the least of these was the chronic poverty of the association, a state not aided by the skepticism of some treasury officials about the usefulness of the laboratory. As one first-class clerk there wrote in 1885, before the laboratory was built:

> I do not believe that any addition to food supply will be secured by anything that is done at the Biological Laboratory. I think that when the time comes, asking the Association to show cause for a continuation of their grant, they will find this very difficult to do. The grant will be spent on scientific research and will be very difficult to withdraw from the people who will get the handling of it, but I don't believe that a single herring or sole will be caught the more for all the Association's efforts.[17]

Although the facilities for research were excellent,[18] and the building itself (Figure 7.1) was worthy of pride, its early directors found the dire state of day-to-day financing and the administrative inconveniences of satisfying themselves, the secretary of the association, and the council onerous, if not impossible. The first "resident superintendent," Walter Heape, lasted only a few months. He was succeeded before the opening by the Oxford biologist G. C. Bourne,[19] who passed on the directorship to W. L. Calderwood in 1890. He in turn was succeeded in 1893 by a Cambridge zoologist, E. J. Bles, who, after a year was replaced by E. J. Allen of University College, London. Administrative changes then made life (though not finances) easier for the director, and a long period of stability in poverty set in for the laboratory.[20]

Despite these teething troubles, the early work at the laboratory was prolific and significant. Even before the building opened Heape had compiled a list of the local marine plants and animals. And Bourne, although his was also a short stay, made a good start on assessing the plankton off Plymouth. Only four years after his appointment Allen published a lengthy monograph on benthic animals and their environments.[21] As the laboratory's first naturalist, J. T. Cunningham, during ten years in Plymouth, carried out an impressive amount of work on the distribution, reproduction, and development of commercially important

17. F. A. A'Court Bergne in November 1885, quoted by MacLeod 1976, p. 148.
18. Described by Heape 1887 and G.C.B. (Bourne) 1888.
19. Heape's varied career is described by F.H.A.M. (F. H. A. Marshall) 1930. On Bourne, a significant figure in British zoology, see Baker and Pitman 1949, *Who Was Who* (1929–1940), p. 142; and Harmer 1933.
20. Best described by Bidder 1943, pp. 671–677. On Calderwood, see Clark 1952. Bles's career is described by Hopkins 1926.
21. Allen 1899. Council reports in each early volume of the Journal of the Marine Biological Association of the United Kingdom (begun in 1887) summarize the work.

FIGURE 7.1 The Plymouth Laboratory of the Marine Biological Association in June 1888 just before its official opening. The Citadel lies to the right of and behind the laboratory. From *Nature* 38: 199 (June 28, 1888).

marine fishes.[22] He also began drift-bottle studies, modeled on work by the Fishery Board for Scotland and by W. A. Herdman of Liverpool in the Irish Sea,[23] intended to show the planktonic movement of fish eggs and larvae. The laboratory itself proved attractive; in 1889 the first independent investigators from English universities appeared during the summer vacation, and in 1895 Walter Garstang brought the first of a long series of university classes there for special instruction in marine biology.

Controversy over the state of the North Sea trawl fisheries was still increasing in 1890. As a result, late in 1891 the association appointed E. W. L. Holt to examine "the actual condition of the North Sea trawling grounds."[24] He began work at Grimsby early in 1892, remaining there through 1894 to examine the size structure of the catches brought in by the trawlers. His analysis, based mainly on knowledge of the size at which plaice and sole became mature and on clear evidence that large numbers of immature fish were taken by the trawlers, led to the conclusion that trawling could harm the fishery and that the damage could be limited by imposing a size limit on fish landed.[25] Holt's work, imperfect and limited in scope but widely read, placed Plymouth squarely in the middle of a furious controversy over the alleged depletion of the fishing grounds by trawlers and the means by which the apparent depletion could be mitigated. The controversy was fueled by more work from Plymouth, Walter Garstang's *The Impoverishment of the Sea*, published in 1900.

Garstang (1868–1949) (Figure 7.2A), who had been associated with the Laboratory since 1888, succeeded Cunningham as naturalist in 1897.[26] Although part of his work involved an extensive hydrographic survey of the English Channel in 1899 and 1900, his main duties were to

22. J. T. Cunningham (1859–1935), fellow of the Royal Society of Edinburgh and fellow of University College, Oxford, while on the staff of the University of Edinburgh, had been superintendent of John Murray's Scottish Marine Station in Granton. He left Plymouth in 1897 to work for the Cornwall County Council. Later he became a lecturer in the University of London (Anonymous 1941, Bidder 1935).

23. Herdman's work began in 1894 off Port Erin (Herdman 1895, pp. 63–66). Cunningham's hydrographic work was also influenced by Otto Pettersson and his colleague at Plymouth (from 1891 through 1897), H. N. Dickson; see Cunningham 1896.

24. Directors Report, 1891, *Journal of the Marine Biological Association of the United Kingdom, New Series* 2: 88. Southward and Roberts 1984, pp. 164–165.

25. Holt 1895, pp. 339–344, 444–446.

26. Walter Garstang (1868–1949) was an assistant naturalist at Plymouth from 1889 until 1890, when he left for Manchester. He returned in 1892 as naturalist, leaving in 1894 when he was appointed a lecturer and fellow of Lincoln College, Oxford. In 1895, Garstang brought the first university classes to Plymouth, an activity considered so important by Allen that a special room was provided for them beginning in 1896. Garstang was appointed naturalist once again in 1897, an appointment he held until 1907, when he became professor of zoology at Leeds. Until 1907 he also played an important role in ICES (Hardy 1950b, c; Hardy 1951; Baker and Bayliss 1984).

FIGURE 7.2 *A*, Walter Garstang, who was associated with the Plymouth Laboratory between 1888 and 1907. He first worked on faunal studies and hydrography, later on fisheries of the North Sea at Lowestoft. Garstang was the main link between Plymouth and ICES from 1902 until 1907, when he became professor of zoology at Leeds. From Hardy 1951, with permission of Michael Hardy, Reading, England. *B*, E. J. Allen, director of the Plymouth Laboratory and secretary of the Marine Biological Association of the United Kingdom from 1894 to 1936. Allen's work on the laboratory culture of phytoplankton, reinforced by his holistic, nonvitalist philosophy of science, prepared the ground for Atkins's and Harvey's work on plankton dynamics during the 1920s and 1930s. From Kemp and Hill 1943, with permission of the Royal Society of London.

carry on the work begun by Cunningham and Holt. Taking aim at those such as W. C. M'Intosh who claimed that fishing had not affected the trawl fisheries,[27] Garstang showed with great originality that between 1889 and 1898 the total catch had increased only 30% although the catching power of trawlers had increased by 300%. Moreover, when corrections were made to compare vessels of different catching power, the catch per standard vessel had been halved during that ten-year period; even in M'Intosh's home area the catch of plaice and lemon sole was decreasing.

Garstang's contentious conclusions were only part of a confusing, changeable situation in English fisheries research and administration at the turn of the century. By default, the Plymouth Laboratory was *the* English fisheries laboratory, for, unlike Scotland, England had no other comparable institution, nor did it have a basis in law for one. A number of minor acts regulating sea fisheries were administered by the English Board of Trade (an arm of government), while broader fisheries regulations and research were delegated to district sea fisheries committees appointed by county or borough councils under the Sea Fisheries Regulation Act of 1888.[28] In 1902, when the Committee on Ichthyological Research was appointed by the Board of Trade to examine the state of fisheries research and the possibility of creating a central administration governing all of Britain, only one local committee was carrying out an effective program of research. This was the Lancashire and Western Sea Fisheries Committee, centered in Liverpool. Its secret of success was W. A. Herdman, Derby Professor of Zoology at Liverpool since 1881, who first brought the resources of his Liverpool Marine Biology Committee (founded in 1885) before the Lancashire committee, then persuaded them to support a Lancashire Sea Fisheries Laboratory, which began in 1892 as a wooden shed on the roof of University College, Liverpool. In the same year Herdman inaugurated a biological station at Port Erin (succeeding one on Puffin Island built in 1887) devoted to marine biology and fisheries research; later he and his colleagues, supported by the Lancashire committee, established a fish hatchery at Piel near Barrow in Furness.[29]

27. Best expressed by M'Intosh 1899, 1907, 1919. The controversy was a long-standing one; see Garstang 1919 and M'Intosh 1921.

28. Committee on Ichthyological Research 1902, pp. v–vi. The original bill was titled *Sea Fisheries Regulation 1888. A bill for the regulation of the sea fisheries of England.* [51 and 52 VICT] [Bill 322]. British Parliamentary Papers 1888 VI, 5 pp. (plus 3 pp. of Lords amendments). See also Johnstone 1905, pp. 57, 97–113, 337.

29. Johnstone 1905, pp. 57, 337; Jenkins 1920, p. 167. On Herdman see Hickson 1925, Johnstone 1925. The background of Herdman's work is described in Herdman 1893a, b; Herdman 1903.

Herdman's group patterned its work on the success of the Kiel Commission, investigating, in addition to fish biology, the distribution and abundance of plankton, hydrography, and invertebrate morphology.[30] Much of his personal scientific effort went into studies of the plankton in the Irish Sea, beginning in 1906–1907, but he was also a fundraiser supreme and promoted the continued decentralization of British fisheries research, which he believed was best carried on by groups like his own, responsive to local control, a view shared in Scotland by W. C. M'Intosh.[31] Although Herdman was a long-time member of the Council of the Marine Biological Association, his views differed strikingly from the opinions of Lankester and Garstang, both of whom favored some sort of central administrative authority in which the Plymouth Laboratory and the association would play a large part.[32]

When the committee (which included Herdman) reported in 1902, its recommendations favored both a central authority in England, which would control the funds given to institutions such as the MBA, and the autonomy of local organizations such as the Lancashire committee. The Scottish and Irish fisheries boards were to be left intact, but an English fishery council would be established to oversee the local committees, control funding, and ensure that grants (such as those to the MBA) were being properly used. The work itself would be divided up: the MBA would study the south coast, the Lancashire committee the west coast, and a yet-to-be-established laboratory would work on the northeast coast. In short, the MBA, certainly at Herdman's insistence, was being maneuvered into a position of equality with groups like Herdman's and would be forced to give up its special call on treasury funds for fisheries research.

Although these attempts to dethrone the MBA were eventually successful, British entry into ICES in 1902 delayed them for several years and markedly improved the financial status of the Plymouth Laboratory. Britain's role in ICES (see Chapter 3 near n. 33) was to gain control of the North Sea fisheries from its European neighbors. Because of Lankester's lobbying, the MBA, joined by the Scottish fisheries board, was designated by the treasury to carry out the British work beginning on January 31, 1902, with a grant of £18,500 for three years. Allen viewed the situation

30. Most of the work is described in detail in the *Proceedings and Transactions of the Liverpool Biological Society* between 1888 and 1920. Herdman, who created the first chair of oceanography at Liverpool in 1919, is an under-appreciated and under-studied figure in early-twentieth-century marine science.

31. Committee on Ichthyological Research 1902, pp. 29–33, 48–51; Herdman 1907; M'Intosh 1907.

32. Committee on Ichthyological Research 1902, pp. 22–28, 78–84.

as a chance for "English fishery science" to "take its proper place as compared with that of the other countries of Europe."[33] More concretely, the staff of the laboratory, which had been three in 1900, was increased to eight in 1903 (and to 12 in 1910), with the establishment of a fisheries laboratory at Lowestoft directed until 1907 by Garstang.

Using its newly acquired steamer *Huxley*, leased by the laboratory from G. P. Bidder, the Lowestoft group under Garstang set out on a threefold program: 1. to do trawl surveys of the fishery banks in the southern North Sea, 2. to work out the life histories and food of the commercial fishes (in addition they tagged and transplanted fish), and 3. to study the hydrography of the sea in relation to the fisheries. Plymouth, for its part, was responsible for the English Channel, where, on the ICES quarterly cruises, using the steamer *Oithona*, the Plymouth-based staff studied plankton and hydrography. During these years, in addition, the first edition of *Plymouth Marine Fauna* (1904) was published, and Allen began laboratory work on the rearing of marine larvae.[34]

These years were only a respite for the Plymouth Laboratory. The ICES investigations were expensive, using funds that made no appreciable difference to the research that the MBA considered its proper work. And by linking itself to ICES, which was widely criticized for its inadequate sampling and lack of results, the MBA remained the focus of criticism, especially from those who, like Herdman, had not been able to milk ICES's financial cow. He complained, quite justifiably, that ICES completely ignored the west coast of Britain, concentrating only on the North Sea and the English Channel. Local committees were not involved in the ICES work (only his committee could have made any significant contribution in 1902), which, at any rate, was unduly expensive and subject to "grave doubts . . . as to the adequacy of the methods to solve the problems and as to the sufficiency of the observations to justify the economic conclusions." If a "central authority" was set up to oversee English fisheries research, the MBA laboratory at Lowestoft could well be taken over by the government "in order to furnish the Government Department with the laboratories, experimental tanks and scientific assistants, without which the officials cannot be expected to carry on original investigations."[35]

Government institutions now existed to make Herdman's suggestion a reality. In 1903 the fisheries department of the Board of Trade was transferred to the agriculture department to form the Board (later Minis-

33. Allen 1903, p. 656.
34. Marine Biological Association 1903; Allen 1905. Bidder played an important role during these early days; see F. S. Russell 1955.
35. Herdman 1907, pp. 114, 120–121.

try) of Agriculture and Fisheries.[36] Officials of the newly placed fisheries department were annoyed at being left out of the treasury's grant for ICES work. Perhaps worse, their professional integrity had been questioned by the outspoken Lankester, who claimed that the work of the MBA had been "hindered and threatened by the jealousy of highly salaried Government servants who were unable to do the work achieved by the Association, or to understand its importance." W. E. Archer, an assistant secretary and head of the fisheries department, counterattacked, emphasizing "the danger of the present system of entrusting the National Fishery Research to men [the MBA] who had themselves no special knowledge of the subject and have not the necessary time to test the statements put before them by others." He gave his testimony to the Committee on Fishery Investigations of 1908 "with a view to showing how dangerous it was to have a system . . . of what I may call private individuals being entrusted with large sums of money when it was certain questions which the government desired to solve which had to be investigated."[37] The committee in its report attempted to steer a neutral course between the warring parties, recommending that the Board of Agriculture and Fisheries be strengthened to coordinate fisheries research in England, that the British contributions to ICES be continued, and that the MBA be supported as before to carry on unrestricted marine research.[38] But the situation was only truly resolved, with some bitterness, when the Lowestoft Laboratory was transferred to the Board of Agriculture and Fisheries, which took over its staff with the exception of Garstang. He, scenting the direction of events, left in 1907 for the chair of zoology at Leeds. The transfer of Lowestoft to the board was accompanied by a treasury grant of £8240 enabling it to carry on the ICES work in the North Sea previously carried out by the MBA, which now found itself in 1910 where it had been before 1902.[39]

Although the MBA had lost more than half its scientific staff and had sold the steamer *Huxley*, essential to all-weather work off shore, the situation was not entirely bleak in 1910. A small staff remained in Plymouth, consisting of Allen as director, two naturalists, and a hydrographer. And the government grant of £1000 per year remained, supplemented by donations and a few other limited sources of income, including funds to continue ICES surveys in the English Channel. Allen's ingenuity and enthusiasm are evident in the aims the council stated in 1911, despite its dire financial straits.

36. Lucas 1956, pp. 225–226; Johnstone 1905, pp. 99–100, 338.
37. Committee on Fishery Investigations 1908, pp. 292, 357.
38. Committee on Fishery Investigations 1908, pp. xv–xix.
39. Jenkins (1920, pp. 244–246) summarizes these events.

In future it is proposed to confine the economic work of the Association to special scientific problems of a fundamental character, which bear directly upon fishery investigations. At the same time it must be pointed out that the Plymouth Laboratory will still afford precisely such training as required by men who may afterwards be employed in scientific investigation in the service of the government, and that the general scientific work of the Association, though it may have no immediate economic value, is of such a character as to form an important part of the necessary foundation upon which the applied science of fisheries must be built.[40]

The new direction had actually been set (in ways unforeseen by the council) in 1905, when Allen began work on the laboratory culture of marine larvae and phytoplankton. Increasingly after 1910 both Allen's work and his philosophy of research dominated the way scientists at Plymouth approached marine biological problems.

E. J. Allen (1866–1942) (Figure 7.2B) studied zoology, chemistry, and physics at Leeds and University College, London. After teaching school for a few years he studied further in Berlin, in London (under W. F. R. Wheldon, who was a member of the MBA's council), and at Plymouth. Wheldon appears to have been responsible for Allen's appointment as director of the Plymouth Laboratory in 1894 (he actually began in early 1895), succeeding E. J. Bles. He too may have been responsible for Allen's early success, suggesting that the posts of director and secretary of the council be united, thus removing the incubus borne by the previous directors.[41] Allen, who never married, began a life of devotion to the laboratory and its well-being, scrupulously managing the limited funds, keeping together a staff on ridiculously low salaries. His own salary in the early years (until 1919) was only £200 to £300, while the naturalists had to do with about half as much. In 1895 the laboratory's total income was £1950, of which the treasury provided £1000; in 1914 the total income of only £3500, supported vessels, the building, and a staff of six.

Under these conditions Allen did as everyone else on the staff had to do, he worked on virtually everything. His early research on the morphology and histology of Crustacea was succeeded by benthic studies, including polychaete taxonomy. Later he studied amphipod genetics, fish biology, and regeneration in invertebrates. In March 1905 he began an attempt to produce pure food supplies so that larval stages of invertebrates and fish could be raised under well-controlled conditions. In 1907 the technical side of the work was taken over by E. W. Nelson, who co-authored with Allen in 1910 the paper that began plankton dynamics at

40. Marine Biological Association 1911, p. 250.
41. Bidder 1943, pp. 671–672; see also Kemp and Hill 1943, p. 358. The change occurred in 1902.

Plymouth, "On the Artificial Culture of Marine Plankton Organisms." In it, laboratory culture of larvae took second place to the control of phytoplankton growth. When it appeared, Allen was completely familiar with Brandt's and Raben's work on the chemical control of phytoplankton growth as well as Whipple's studies on the growth of phytoplankton in reservoirs. Using defined media supplemented with natural seawater he noted that pure phytoplankton cultures maintained for long periods died back, but began to grow again after a few months, probably because of nutrient regeneration. Although he lacked sensitive methods for analyzing nutrients, Allen's work, among the earliest on phytoplankton culture, showed that diatom growth varied in proportion to the amount of nitrate in solution. Phosphate too appeared to be essential, but it seldom enhanced growth in the way nitrate did. And natural seawater was necessary in small amounts to add some unknown growth-promoting factor.[42]

Following up on the role of the unknown factor in seawater after 1910, Allen found that adding as little as 1% natural seawater would increase the growth of the diatom *Thalassiosira* in culture. Some trace "catalytic agent" was present, analogous to, or perhaps identical with, the "vitamines" that Frederick Gowland Hopkins of Cambridge and others were at work on then in human physiology. In an oceanographic rather than a physiological framework, as Allen suggested, adding natural seawater to cultures appeared to have the same effect as mixing coastal water into sterile oceanic water—phytoplankton growth was stimulated. He concluded, "There is reason to hope therefore that culture experiments may in time throw additional light upon the general questions relating to the production of animal life in the sea, questions which are of immediate importance to a study of the productivity of the fisheries."[43]

This sentiment was typical of Allen, whose interests were always broadly biological. He believed that the branches of biology had to proceed on a broad front, for "none can succeed without the others." Allen decried vitalism, saying that "it is brought into the story at the point where knowledge based on observation and experiment ceases, at the point where it seems to many of us more frankly satisfactory to say, I do not know."[44] Instead, he relied on a broad knowledge of physics, chemistry, and physiology, applied to the whole organism in its environment. In his view, "The successful research worker is one who, whilst carrying to

42. Allen and Nelson 1910, p. 455. The background of work on artificial culture media is given by Provasoli et al. 1957.

43. Allen 1914, pp. 438–439. The significance of the discovery was not lost on the MBA's Council; see Marine Biological Association 1915, p. 649.

44. Allen 1930, pp. 134, 138. By "vitalism" Allen meant special properties of organisms not based on their physical and chemical attributes.

the utmost limit he can achieve his search into detail, maintains as by instinct a true sense of proportion and holds firmly to the idea of the organism as a whole."[45] That organism was closely tied to its environment, for "the idea of adaptation is a recognition of the fact that organism and environment are interlocked, are fitted closely to form that harmony which is nature and life."[46]

Allen applied his general philosophy of biology to the sea and to the role of the laboratory.

> Until the natural fluctuations in fish populations are adequately understood, their limits determined, and the causes which give rise to them discovered, a reliable verdict as to the effect of fishing is difficult to obtain. If such problems as these are to be solved the investigation of the sea must proceed on broadly conceived lines, and a comprehensive knowledge must be built up, not only of the natural history of the fishes, but also of the many and varied conditions which influence their lives. The life of the sea must be studied as a whole.[47]

His work on the laboratory culture of marine diatoms was an important part of such a scheme. Their growth under defined conditions could

> throw considerable light upon the causes which bring about the variations in the quality of minute vegetable life which takes place in the sea itself. Since it is the minute vegetable life which forms the fundamental food supply of all marine animals, an exact knowledge of the conditions under which it can best flourish is of importance from both a theoretical and a practical point of view.[48]

Allen's ideas were expressed concretely in 1909, in a paper with the deceptively simple title "Mackerel and Sunshine." Postulating that copepod abundance should be controlled by the amount of phytoplankton, itself governed by the sunlight available for photosynthesis, he concluded that mackerel, which feed on copepods or on copepod-feeding fish, should be abundant when spring sunshine was above average. As expected, the mackerel caught per boat off Plymouth in May showed a close link with high levels of sunshine in February and March but no correlation with sea temperature and salinity, confirming the importance of oceanic production to the fisheries.

Very early, certainly no later than 1905, Allen had established a firmly

45. Allen 1922, p. 79.
46. Allen 1930, p. 119.
47. Allen 1922, p. 93.
48. Marine Biological Association 1912, p. 580. The style is not Allen's, but the idea is his. See also his essay "Food from the Sea" (1917).

held set of views about the nature of science and its aims. He also held beliefs about the administration of research that greatly affected the work done at Plymouth for three decades thereafter. According to those who knew him, the Plymouth Laboratory was Allen's Laboratory, run with considerable skill as a collection of individuals, not as a unified group.[49] Allen was full of ideas, but "he would never himself suggest a subject of research to a student: . . . he would encourage him to form his own ideas and would assist him in deciding how they could best be put into practice."[50] This was true of his relation with the staff as well: "Allen never directed. He just planted seeds."[51] He was opposed to team research and picked his staff for their ability rather than for their training. The staff were then given great freedom, but never completely outside Allen's influence, for he subtly directed them by asking questions that often evoked new research. Allen's importance as a director lay first in his experimental approach to the early ideas of Brandt and Raben; the work he began in 1905 was soon directed to experimental examination of factors governing the plankton cycle. His second great strength was his choice of colleagues, especially when the financial difficulties of the laboratory eased after 1920.

It was very difficult for Allen and his staff to increase research at Plymouth Laboratory until nearly 1920. When the expansion began, its origins lay in 1909 and 1910 when Parliament passed the Development and Road Improvements Funds Act, which allowed grants or loans to be made to or through government departments for forestry, agriculture, scientific research, and other purposes.[52] A development fund of £500,000 was set up, and five commissioners to oversee its use by recommendations to the treasury were appointed in 1910. A special committee to examine grant requests for fisheries work was soon established, consisting of G. C. Bourne, Arthur Shipley, and D'Arcy W. Thompson. Although applications were few in the early years, the Lancashire and Western Sea Fisheries Committee was given £1240 in 1912 (£100 of which went to Herdman's laboratory at Liverpool), and the MBA was granted £500 "in aid of their research work." Both continued to receive smaller grants, channeled through the Board of Agriculture and Fisheries, thereafter.[53]

49. Kemp and Hill 1943; L. H. N. Cooper, interview at Plymouth, May 25, 1982.

50. Kemp and Hill 1943, p. 361.

51. L. H. N. Cooper, interview at Plymouth, May 25, 1982.

52. *Development and Road Improvements Funds Act, 1909 and 1910.* [9EDW7] [Bill 312] British Parliamentary Papers 1909 I; Jenkins 1920, pp. 246–247.

53. Herdman et al. 1912, pp. 76–77; Herdman 1913, p. 178. Administrative complexities led to the amount being quoted differently in various places.

Although the MBA's grant through the board increased to £400 in 1918, preparation to expand the laboratory had been in the wind for some time. In 1919 the expansion began, based on private donations from G. P. Bidder and E. T. Browne, an enlarged treasury grant of £1000, and £300 from the board's development fund.[54] The old building was fully electrified, a new laboratory for physiological research was added (named the Allen Building), and the library was enlarged. But even more significant changes occurred that year when an Advisory Committee on Fishery Research was appointed by the development commission. A subcommittee under W. B. Hardy, consisting of G. C. Bourne, Walter Garstang, Henry Dale, E. W. MacBride, and J. Graham Kerr received an application from the MBA for increased salaries, upkeep, a new vessel, and staff and equipment for a Department of General Physiology.[55] On the condition that it raise matching funds, the MBA received the first installment of a major grant in 1920, enabling it to buy the steamer *Salpa*, improve the laboratory, hire new scientific staff, and increase salaries. In 1922 the increased funding allowed a building for teaching to be added; in 1926 another new wing was built, greatly increasing the space for research. Other laboratories too benefited from the increase of development commission funding; Liverpool's Port Erin Laboratory was expanded by means of a grant in 1922, and the Scottish Marine Biological Association added a biologist, Sheina Marshall, and a chemist, A. P. Orr, to its staff at Millport in 1922 and 1923.[56]

The grants from the development commission beginning in 1920 brought the first real financial security to the Plymouth Laboratory. But their most important effect was that they allowed the staff to expand. Allen chose as head of his new Department of General Physiology W. R. G. Atkins of Trinity College, Dublin, and as administrative and hydrographic assistant H. W. Harvey, a chemist from Cambridge.[57] Both dominated the work of the Plymouth Laboratory on plankton dynamics after their appointments in 1921. Their work, which contributed to the eclipse of Kiel, made Plymouth the center of research on the plankton cycle for nearly twenty years.

54. Bidder gave £850, Browne £500. Both had been associated with the laboratory for many years and were close friends with Allen. The treasury grant had dropped to £500 during the war. On Bidder, see Russell 1955; Browne's contribution is outlined in Allen 1938.
55. Kerr 1949, pp. 86–96; Marine Biological Association 1921, p. 562.
56. Johnstone 1925, p. 24. On Millport see Chumley 1918, pp. 1–7; Elmhirst 1939; Yonge 1946; Kerr 1949; Watson 1969; and Marshall 1987.
57. Marine Biological Association 1922, p. 836.

8 Surveying the Blue Pasture: Plankton Dynamics at Plymouth, 1921–1933

> In the best interests of biological progress the day of the naturalist who merely collects, the day of the anatomist and histologist who merely describe, is over, and the future is with the observer and the experimenter animated by a divine curiosity to enter into the life of the organism and understand how it lives and moves and has its being. "Happy indeed is he who has been able to discover the cause of things."
>
> W. A. HERDMAN 1920

Lucretius's dictum might well have been applied by Herdman to Brandt and the Kiel school, who provided a clear theory of the plankton cycle far more sophisticated than mere observation of the natural world and collection of data. Even in Britain, the day of the naturalist was over in marine biology by the early 1920s, when several currents of scientific thought about the environment and organisms converged. Herdman at Liverpool and Allen at Plymouth worked in a sophisticated way that owed a great deal to Germany, but just as much to the less definable development of ecological thought in the early twentieth century.

At Plymouth, early in the 1920s, the ocean was "regarded as a pasture." More concretely, specifying the issue that interested E. J. Allen, "from an economic point of view the sea may be regarded as a blue pasture for the raising of marketable fishes. Farmers know that a certain acreage will support or fatten a definite number of cattle, but similar precise information regarding the sea is lacking."[1] It became an article of faith at Plymouth that the life of the sea was governed by a mosaic of

---

1. See Atkins 1925a; Marine Biological Association 1923, p. 291. Allen was probably the author of the "blue pasture" analogy.

physical, chemical, and biological factors, but that no special weight could be given to any one a priori. This approach to the sea was certainly not *organismic*, in the sense that the views of the American ecologist F. E. Clements have been characterized, nor was it vitalist. Instead, judging by Allen's writings, in which he castigated those who invoked unknown forces, it was a moderate reductionism, a view of life as dependent on, though not subordinate to, the physical and chemical surroundings in which it occurred. Such an approach was not unique to Plymouth; it developed independently more than once according to the British ecologist Charles Elton, who in writing later of his book *Animal Ecology* (1927) says he attempted in it to "enter upon a new mental world of populations, inter-relations, movements and communities" and that "even by the nineteen-twenties considerable advances in these directions had been made in marine ecology and in the study of lakes, of which I was not then fully aware."[2]

By 1920 a number of leading British marine biologists, among them W. A. Herdman and J. Stanley Gardiner, expressed clearly their views that the problem of variable and declining fisheries production had its roots in basic biological relations. As Gardiner put it,

> The economic species is the final link in a chain of metabolism from the ultimate foodstuffs and energy of sunlight and each link has to be considered under two heads: 1) the relation of the organism to sea water, and 2) the relation of the organism to other organisms.[3]

Herdman put the problem more clearly to the British Association for the Advancement of Science in Cardiff in 1920:

> The prospects of a year's fishery may . . . depend primarily upon the rate of spawning of the fish, affected no doubt by hydrographic and environmental conditions, secondarily upon the presence of a sufficient supply of phytoplankton in the surface layers of the sea at the time when the fish larvae are hatched, and that in turn depends upon photosynthesis and physico-chemical changes in the water, and finally upon the reproduction of the stock of molluscs or worms at the bottom which constitute the fish food at later stages of growth and development.

Herdman saw the solution clearly: "Our need . . . is Research, more Research, and *still more Research*."[4]

But what kind of research was it to be, and how was it to be directed?

2. Elton 1927, pp. vii, viii of the author's preface to a reissue.
3. Gardiner 1920, p. 72.
4. Herdman 1920, pp. 24, 25.

At Liverpool, Herdman, benignly autocratic, attempted to direct a chronically poorly funded group of younger academics, assistants, and students in a variety of work centered around plankton abundance, fish reproduction, and fish rearing. The better funded laboratory at Plymouth, under Allen, might have been firmly directed toward fisheries work, but it was not. Rather, as an expression of Allen's belief that "the investigation of the sea must proceed on broadly conceived lines," it developed, beginning in 1921, through the efforts of a few scientists whose training was mainly chemical or physical, and whose work was broadly overlapping or complementary rather than closely integrated. The first of these, W. R. G. Atkins (1884–1959) and H. W. Harvey (1887–1970), began their work that year, nominally as physiologist and hydrographer, respectively, but with a great deal of freedom to follow their investigations of Allen's "blue pasture" in independent ways. The result, between 1921 and the 1940s, whether by luck or Allen's good management, was the establishment of a distinctively "Plymouth view" of the marine production system. This had its origins in Kiel, but it was increasingly refined by close attention to the effects of the physical and chemical environment on marine organisms using techniques that were not available to the Kiel school.

The appointment of Atkins (Figure 8.1A) to the Plymouth staff in February 1921 was a crucial one, for as the first of the staff hired on the development commission funds he was expected to set up new lines of inquiry, broadly called "physiology," at the laboratory. By 1921 Atkins's career was already distinguished, beginning with a First in physics, and a distinguished record in chemistry and biology at Trinity College, Dublin. After graduation in 1906 he worked in the university chemical laboratory, mainly on the chemical properties of milk, urine, and blood, later on the osmotic properties of fluids. After 1911, when he began work in the botanical laboratory of Trinity College, he continued work on the chemical and electrical properties of plant fluids, adding to this research on enzymes.

During World War I Atkins worked for a time as a chemist at the Woolwich Arsenal and at the National Physical Laboratory. Eventually he joined the Royal Flying Corps, establishing a laboratory in Egypt which did important work on fabric dopes and lubricating oils for aircraft. He returned to plan physiology at Trinity College in 1919, doing research on pH in plants, followed by a year in India as a research botanist working on indigo.[5]

Atkins's experience and training became manifest quickly in the prob-

5. Biographical details in Cooper 1960 and Poole 1960.

FIGURE 8.1 *A*, W. R. G. Atkins, the first physiologist appointed to the staff of the Plymouth Laboratory (in 1921) during its expansion under Development Commission funding. Atkins made the first production estimates for the English Channel, applied new techniques of nutrient analysis, and was a pioneer of light measurement in the sea. From Poole 1960, with permission of Godfrey Argent Studios, London. *B*, H. W. Harvey, appointed in 1921 as hydrographer to the Plymouth Laboratory. Harvey soon took up nutrient analysis, the culture of phytoplankton, and studies of the annual cycle of plankton production, culminating in a collaborative study with L. H. N. Cooper, Marie V. Lebour, and F. S. Russell in 1934–1935. From Cooper 1972a, with permission of Mrs. Marjorie Harvey, Penzance, England.

lems he attacked at Plymouth. Fifteen years earlier D. J. Matthews, hydrographer at the Plymouth Laboratory, following the seasonal change of phosphate in the English Channel, had noted that the high levels found in winter decreased in the spring, a phenomenon that was at least partially due to "the removal of phosphates from solution, at first by the fixed algae, and later in the spring by the diatoms and for a short time by *Phaeocystis*."[6] If diatom production was responsible for the changes, how might that production be measured? The answer lay in pH change, which Allen and Atkins believed indicated the shifting "balance between plant and animal life in the sea,"[7] for it responded rapidly to photosynthetic activity and could be determined quickly using colorimetry.

The link between pH and photosynthesis was not unique to Atkins's work, for in the spring of 1912 Benjamin Moore, E. S. Edie, and E. Whitley had noted that the growth of algae in a fish pond at Port Erin removed carbon dioxide, increasing the pH and resulting in the death of the fish. As a result of their observations they decided to follow changes of pH in the sea throughout the year. The result was clear: pH decreased in winter and increased in spring as a result of the changing photosynthetic activity of the plankton. Because the relationship between the increase of pH and the uptake of carbon dioxide by photosynthesizing plants could be calculated, Moore, E. B. R. Prideaux, and G. A. Herdman recognized in 1915 that the production of the sea, in terms of carbon, could be calculated using pH change. They estimated that seasonal plankton production to a depth of 100 meters off Port Erin was about 300,000 kg of carbohydrate per square kilometer (in modern terms about 125 g of carbon per square meter), assuming that most of the annual production took place during the spring season.[8]

There is little doubt that the work of Moore and his colleagues at Port Erin inspired Atkins to estimate the production of plankton off Plymouth using the pH changes he observed at sea on hydrographic cruises with H. W. Harvey during 1921 and 1922. Although he found that pH varied from place to place and with depth, the change of pH in late spring allowed Atkins to calculate that the spring production throughout the 83-meter water column at one location in the English Channel amounted to 250 g of dextrose per square meter (roughly 100 g of carbon per square

6. Matthews 1916, p. 257.
7. Marine Biological Association 1922, p. 840.
8. Moore et al. 1915, pp. 247–249; Herdman 1920, p. 25–26; Moore et al. 1921. Benjamin Moore (1867–1922) was professor of biochemistry at Liverpool from 1902 to 1914, later a staff member of the Medical Research Council, and eventually the first professor of biochemistry at Oxford, 1918–1922. It appears to have been his work on the buffering capacity of blood plasma that led to the insight that pH change, CO_2 uptake, and production in the sea could be linked quantitatively. On Moore see Johnstone 1923 and Kohler 1982, pp. 55–57, 347 (n. 62).

meter), a result strikingly similar to that off Port Erin.[9] Most of the carbon fixed, he believed, did not accumulate, but there was a residue of dissolved carbon in the water, resulting from a slight imbalance of photosynthesis and respiration. This could be estimated, for the oxygen that disappeared from seawater stored in the dark in closed bottles represented the oxidation of this dissolved material. It proved to be a significant amount, perhaps twice the amount of carbon fixed in photosynthesis annually.[10]

Atkins's careful measurement of pH resulted in the first estimate of primary production in the English Channel, but it showed too that pH was highly variable from location to location as a result of differences in metabolism and the mixing of surface with deep waters. An independent method was needed to verify the production figure based on pH change. Here phosphate entered, once it was possible to measure it routinely and quickly with a sensitive method. Matthews's early work had been revolutionary in using a sensitive colorimetric method, rather than Raben's awkward gravimetric technique. But Matthews's technique, derived from I. Pouget and D. Chouchak, was slow; five hours were required for duplicate estimates, moreover careful filtration of the sample to avoid contamination took up to sixteen hours. By late 1922 Atkins had discovered and modified Denigès's colorimetric method, which shortened the analysis to about ten minutes and increased its sensitivity threefold.[11] Now the seasonal changes of phosphate that Matthews had observed could be used to estimate production, provided the phosphorus content of the plankton was known.

Atkins's first analyses of phosphate in the English Channel in 1922 readily confirmed that a seasonal drop occurred; laboratory experiments showed him that phosphate could be virtually completely removed from solution by diatoms characteristic of the Channel plankton. But there were difficulties in knowing how much phosphate was in a typical diatom, for the phosphate absorbed in experimental cultures could not be fully recovered nor could the weight of a diatom be easily determined using the techniques Atkins had available in 1922. As a result, he chose an estimate of the phosphate content of leaves (later of macrophytic marine algae such as *Fucus*) as reasonable for diatoms. Once this assumption was made, the production of diatoms could be calculated knowing the amount of phosphate that had disappeared from the sea off the Eddystone Light between March and July. This simple method gave the

9. Atkins 1922a, p. 763; Atkins 1923a.
10. Atkins 1922b, 1923b; Marine Biological Association 1923, p. 291.
11. Atkins 1923c, pp. 146–147.

result that 1.4 kg of diatoms per square meter were produced in the 70-meter water column off the Eddystone Light, exactly the same figure as the production calculated from pH change when the appropriate conversions were made. As Atkins pointed out, "The exact agreement is, of course, fortuitous in view of the assumptions; but it shows that the methods must have a certain degree of reliability, or rather it confirms the alkalinity result, for the phosphate method involves only one assumption, that the percentage of phosphate in the algal plankton is close to that of the larger brown algae."[12] Then, rather uncharacteristically stretching his data to estimate the fish production, he concluded that fish production in the same area was likely to be about 0.2 to 2 g per square meter, on the assumption that much of the phytoplankton was eaten by invertebrates. Thus within two years of his arrival at Plymouth he had taken a significant step toward the goal of finding a definable relationship between plankton production and the abundance of fish.

Early in 1921, about the time of Atkins's arrival, the Plymouth Laboratory was asked by the Ministry of Agriculture and Fisheries to begin a hydrographic survey of the western English Channel, coordinating its work with French and Irish hydrographers and submitting the results to ICES. This program was possible only because the Plymouth Laboratory was about to acquire a new ship, *Salpa*, a steam-drifter modified for trawling. With its arrival in June 1921 a lengthy series of hydrographic surveys became part of the annual regime at Plymouth, requiring regular sampling at a network of stations stretching from just off Plymouth as far as Brittany.[13] E. W. Nelson, Allen's collaborator and naturalist at the laboratory, made the first collections. When he left Plymouth to become scientific superintendent of the Fishery Board for Scotland early in 1921, the work was taken over by Atkins, then in October 1921 by H. W. Harvey upon his appointment as hydrographer. For eighteen years the regular sampling cruises of *Salpa* provided abundant basic data on which the Plymouth group's distinctive contributions to the theory of plankton dynamics were based. Without the ship their work could not have taken the form it did—when *Salpa* was requisitioned by the Royal Navy in 1939 the scientists remaining in Plymouth during World War II had to discontinue their work at sea entirely.

H. W. Harvey (Figure 8.1B), appointed as Allen's assistant and as hydrographer only shortly after Atkins's arrival, was responsible for studying the circulation and dissolved gases of the English Channel region, an assignment that resulted in major increases in knowledge of

12. Atkins 1923c, pp. 140–142.
13. Marine Biological Association 1922, pp. 836, 839.

the physical circulation and heat budget of the western Channel during the eleven years that he remained responsible for physical measurements.[14] But Harvey's appointment was an unusual one, perhaps because Allen believed that he had a broader field of view than hydrography alone. In this Allen was correct, for Harvey's approach to plankton production became the most distinctive product of the Plymouth Laboratory's biological work during the late 1920s and 1930s. It developed from an approach to nature that was simultaneously intuitive and experimental, based more on chemistry than on physics.

At Cambridge, Harvey took a degree in chemistry in 1910. As an undergraduate he had worked with F. F. Blackman on the effect of poisons on unicellular algae, and during a year of research after graduation he worked with W. B. Hardy on problems spanning the boundaries of chemistry, physics, and biology.[15] After four years spent in his family's paint firm in Surrey, he went to sea as a navigating lieutenant in the Royal Naval Volunteer Reserve, commanding a small squadron of minesweepers in the Barents Sea. There, one of his colleagues believes, while he spent long watches on duty, he acquired his belief that the sea was a complex entity, not a phenomenon made up of isolated parts.[16]

During Harvey's early years at Plymouth he stated clearly that the physical oceanographic changes observed in the Channel could have profound biological effects on the animals. Basing his ideas on J. H. Orton's observations that sea temperature controlled the breeding cycle of marine animals, Harvey pointed out that variations of temperature from year to year could cause the reproductive cycles of animals to lose their synchrony with plankton production, with the result that even the fisheries might be affected.[17] But, characteristically, he was slow to begin work on the biological processes of production, for Harvey never proceeded without deep and lengthy thought, which some of his colleagues, notably the energetic Atkins, took for laziness. Harvey liked to spend long periods of time talking, exchanging ideas, or exploring the implications of other people's ideas. Only when this had led to clear hypotheses did he proceed to experimental testing. A colleague describes seeing Harvey come into a person's office and, observing him with his feet up, saying, quite seriously, "That's what I like to see, someone working." Thought was important to him and preceded action.[18] During the early

14. See especially Harvey 1923; 1925a, b, c; 1929; 1930; 1934b.
15. Cooper 1972a, pp. 332–333. See also Cooper 1972b.
16. E. I. Butler, interview at Plymouth Laboratory, June 14, 1983.
17. Orton 1920; Harvey 1925b, p. 61; Harvey 1925c, 1930.
18. E. I. Butler, interview at Plymouth Laboratory, June 14, 1983; also Cooper 1972a, pp. 344–345; Cooper 1972b, p. 775.

years, while Harvey compiled hydrographic data and thought about processes in the sea, the water samples from the *Salpa* cruises gave Atkins abundant information on pH, phosphate, silicate, organic matter, and physical stability, while Marie Lebour, who had joined the laboratory staff in 1915, surveyed the samples for phytoplankton and larval fish and invertebrates. Harvey's major contribution, to unify the work of the Plymouth group, did not begin to emerge until early in the 1930s.

An immediate consequence of Harvey's early work with Atkins was their observation that the English Channel was stratified for most, if not all, of the summer production season.[19] Atkins soon established that this was a local phenomenon, due to solar heating, and that as the surface water warmed it became increasingly resistant to mixing, being thus cut off from the supply of nutrients in deep water.[20] The implications of this for Brandt's hypothesis were great, as described in Chapter 5 (at n. 41), especially when it became clear, as Brandt himself had suspected, that deep ocean water was rich in nutrients, and that it could move upward to support phytoplankton at high latitudes only when and where seasonal mixing occurred.[21] The early *Salpa* cruises provided the information that made Brandt's hypothesis untenable; they also yielded information on the physics, chemistry, and biology of the water which was far more copious, abundant, and precise than the limited data that the Kiel group had been able to compile from occasional cruises and laborious chemical techniques.

In particular, the results of the first three or four years' cruises confirmed Matthews's early observations that the growth of phytoplankton and the disappearance of phosphate from the surface waters were indeed closely, causally, linked. After the spring bloom began each year, phosphate dropped, but with intriguing variations from year to year that could be attributed to light, for in years with a great deal of early sunshine (e.g., 1924) phosphate decreased early. This could be attributed to an early bloom, triggered by early spring sunshine.[22] Once the phosphate level decreased, production also declined until vertical mixing began in the late summer or autumn; then a new bloom ensued, limited eventually by the declining autumn light. On an annual basis, phosphate was efficiently cycled, for Atkins's estimates, based on changes in the amount of phosphate from month to month, showed that there was a close balance

19. Harvey 1923, p. 227.
20. Atkins 1924, 1925b.
21. Atkins and Harvey 1925; Atkins 1926a; Atkins 1926c, pp. 212–213. The British *Discovery* investigations, at almost exactly the same time as the German *Meteor* expedition, showed the extent of the deep-water reservoir.
22. Atkins 1925c, pp. 703–705; Atkins 1926a, pp. 455–457; Atkins 1926c, pp. 208–209.

nearly every year between phosphate regenerated from dead organisms and the amount taken up in growth by the photoplankton.[23] Atkins concluded "that the phosphate cycle is essentially a closed one," but he and Harvey were fully aware by 1926 that the details of the way the balance was achieved—and the timing of the cycle—were the result of varying physical processes, especially the amount of sunshine and the amount of vertical mixing that took place during the year. Brandt's intuitions were given quantitative expression, or, at the least, powerful inductive support.

A further result of the early *Salpa* cruises was the information they provided on the role of nitrate and ammonia in phytoplankton production. Of necessity, Matthews had concentrated on phosphate, for he had a good sensitive method that gave reliable results; there was no comparable method for nitrate. When he discovered that phosphate reached very low levels after the bloom, a result confirmed by Atkins's careful work in 1924, 1925, and 1926 based on a new method, the evidence seemed overwhelming that phosphate alone was the controlling nutrient salt. This conclusion was reinforced by the common observation that phosphate disappeared before nitrate decreased during the bloom.[24] With the application of Atkins's and Harvey's new methods for determining nitrate in 1925, the picture began to change subtly but profoundly, for it became clear that in surface waters of the English Channel in late summer some phosphate was present (that is, had been regenerated) but that nitrate was very low and could be limiting.[25] Harvey concluded that "during the summer months, plant life is at times limited by lack of nitrate and at times by lack of phosphate,"[26] a situation that could be explained, according to Atkins, by more rapid regeneration of phosphate than of nitrate, which, unlike phosphate, resulted from a lengthy series of chemical transformations.[27]

Like Raben and Brandt before him, Harvey found ammonia difficult to deal with. In principle it should have been the precursor of nitrate in the marine nitrogen cycle, yet its variation in the sea was slight and could not be related to changes in the nitrate content of the water.[28] Nitrate increased in the water of the English Channel in the autumn, just as predicted by theory, but its relation to the unpredictably fluctuating amounts of ammonia was a serious problem.[29] Atkins's suggestion in

23. Atkins 1926a, pp. 452–455; Atkins 1926c, pp. 211–212.
24. Atkins 1932b, pp. 168–169.
25. Atkins and Harvey 1925, p. 784; Harvey 1926, pp. 81–84; Harvey 1928b, pp. 183–186.
26. Harvey 1928b, p. 185.
27. Atkins 1926c, pp. 220–221; Harvey 1928a, pp. 179–180.
28. Harvey 1928a, p. 180.
29. Atkins 1930a.

1926 that ammonia could be used by marine plants was soon borne out by the work of Ernst Schreiber, Trygve Braarud, and Birgithe Ruud Føyn, among others and by Harvey himself, all of whom showed that algal cells could absorb ammonia, sometimes in preference to nitrate.[30] It was 1933 before L. H. N. Cooper's careful analytical work on a long series of samples showed that at times ammonia, nitrite, and nitrate changed in sequence in the sea, evidence that indicated complete nitrification in the water column.[31]

By the early 1930s, as a result of the *Salpa* cruises, the development of new techniques for nutrient salts, and the Plymouth group's attention to the processes of stratification, mixing, and nutrient regeneration, the group had evolved a sophisticated model of the annual plankton cycle, recognizably based on the Kiel school's ideas, but with far greater support, and with a more complex role for the influence of nitrogen and phosphorus. As a result of the bloom

> water is depleted of nitrate to about the same extent as of phosphate. It appears, however, that the phosphate is first used up, leaving a small quantity of nitrate. The gradual regeneration of phosphate, which appears to be set free as such or in labile organic compounds, results in a further depletion of nitrate up to the point of complete exhaustion in the upper layers. Phosphate then begins to accumulate in the deeper and in the surface water, accompanied by ammonium salts, by nitrite, and finally by nitrate in the deeper waters. The mixing of the water consequent upon its autumnal cooling brings phosphate and nitrate together again and a fresh outburst of diatom activity ensues, limited however by the lessening power of the sun and by the fact that the continuance of a vertical circulation of the water tends to plunge the algal cells into the darker depths.[32]

It is clear that Harvey had carefully thought through the implications of these changes by 1926, when his paper "Nitrate in the Sea" was published. In it he described clearly the closed cycle of nitrogen in the sea, well known for years, but he added an original idea, that the amounts of nutrient salts were less important in governing the intensity of marine production than was the *rate* at which they were supplied to the organisms.

> The fertility of any area of the open oceans . . . depends upon three main factors. (a) The length of time protein formed by phytoplankton remains

30. Atkins 1926c, p. 223; Schreiber 1927; Braarud and Føyn 1931; Harvey 1933, pp. 254, 264–265.
31. Cooper 1933a, pp. 713–714; Cooper 1937a, pp. 189–190.
32. Atkins 1932a, pp. 309–310.

part of the plant or nourished animal's body. (b) The time which elapses during the decay and formation of ammonium salts and phosphate from corpses and excreta. To this must be added the time taken for nitrate forming bacteria to convert the ammonium into nitrates. (c) The time which elapses before the reformed nitrate and phosphate again reaches the upper layers where there is sufficient light for photosynthesis.

Of these processes, he concluded "there is every reason to suppose that this third factor [mixing of nutrients from deep water into the lighted surface layers] regulates the speed at which nitrogen and phosphorus pass through the complete cycle in the sea *as a whole*, being the slowest in the series of changes."[33]

Processes on this grand scale underlay the marine production system, according to the viewpoint developing at Plymouth by the early 1930s. But the varying details of the annual cycle were also important, for they, resulting from fluctuations in light and perhaps temperature, could control the strength of fish populations through their need for a reliable seasonal food supply. In the short term, even from day to day, production occurred in seemingly random pulses—as Atkins said "productivity proceeds in a series of bounds"[34]—some of which could be attributed to winds, vertical mixing and nutrients, others to less clearly determinable factors. Overriding all was the effect of light, which was self-evidently important, but still problematic. Its role could be inferred, but it was hard to determine definitively without direct measurements.

The link between light and the bloom had been known since at least 1908 when Lohmann had shown that it was possible to account for the onset and decline of the bloom using the product of daylength and temperature (see Chapter 4 at n. 48). Furthermore, during their work in 1916–1917 in the Oslo Fjord, Gaarder and Gran had demonstrated experimentally that the intensity of light in the water column controlled the amount of photosynthesis that occurred (see Chapter 5 following n. 18). Their work, which became available to the English-speaking scientific world in March 1927, made it clear that net photosynthesis was limited to the top 10 meters or so, at least in near-shore Norwegian waters.

Atkins concluded very early that a two-pronged attack was necessary on the problem of how light affected the bloom. Physical measurement of light was important to understand how the bloom began and why it varied from year to year. But indirect methods, based on simple estimates of sunshine and daylength, were essential too, because the quantitative

33. Harvey 1926, pp. 85–86.
34. Atkins 1930b, p. 835.

measurement of illumination was difficult, time-consuming, and uncertain. These approaches were not united in a fully quantitative way until the mid-1930s, when the instruments used to measure light became reliable and easy to use, allowing photosynthesis and light to be studied together.

During the 1880s, the study of light in the sea and in lakes began with the use of photographic plates to indicate the amount of light reaching great depths. Bjørn Helland-Hansen used a similar method to measure the penetration of light to deep North Atlantic waters in 1910, and by 1911 it was known that the spectral composition of light changed with depth in the sea. But quick and reasonably reliable methods for measuring the amount of light in the sea did not exist until photocells became available in the early 1920s.[35]

Late in 1922, Victor Shelford, professor of zoology at the University of Illinois, and F. W. Gail, professor of botany at the University of Idaho, published a study of light in the sea near Friday Harbor, Washington, based on the use of a gas-filled potassium hydride photocell.[36] Within a few weeks, Atkins had seen the paper and written to H. H. Poole, his friend and chief scientific officer (later registrar) of the Royal Dublin Society, suggesting that they do similar work at Plymouth.[37] Their work began in 1924, using a photocell similar to Shelford and Gail's, soon replaced by a more stable one developed by the English General Electric Company.[38] A year later they took the photocell to sea on *Salpa*, using special circuitry to indicate the current produced by the cell rather than the unseaworthy galvanometry employed by Shelford and Gail.[39]

The first photocells were difficult to use, partly because they had to be standardized individually and because their characteristics changed with time. In addition, the early photocells, based on alkali metals such as sodium or potassium, had a peak of sensitivity in the 400-nm range but were insensitive to the green and red wave lengths above about 500 nm. Despite this, with care and careful calibration, Atkins and Poole soon showed that light decreased rapidly in the surface waters, permitting photosynthesis, according to an early estimate, to depths of 10 to 20

35. Brief historical summaries are given by Atkins 1926b, pp. 103–120; Atkins 1932c, pp. 186–202; and R. W. Holmes 1957, pp. 109–112. On the early use of photocells see Shelford and Gail 1929.

36. Shelford and Gail 1922.

37. Poole 1960, pp. 7–8.

38. Poole 1925, pp. 99–109.

39. Poole and Atkins 1926, pp. 177–180. More information is given on methods in Poole 1925; Shelford and Kunz 1926; Poole and Atkins 1926, 1928, 1929; Atkins and Poole 1930a; Atkins 1932c; Atkins and Poole 1933a; Poole and Atkins 1935; Atkins and Poole 1936; and Poole 1936.

meters in summer but restricting it to the top 5 meters at less well-lighted times of year.[40] They also showed that there was a predictable relationship between a simple optical method of estimating the attenuation of light in water (use of the Secchi disk) and attenuation measured with a photocell, and that a chemical method of estimating light intensity (the decomposition of oxalic acid in uranyl acetate) was more sensitive to short wavelengths than was the photoelectric cell, but that it was less sensitive overall.[41]

By 1931, when D. C. Gall and Atkins built a marine photometer for the new U.S. ketch *Atlantis*, Atkins and Poole were the recognized authorities on the measurement of light in the sea.[42] But the complexities of photometry and the difficulty of getting reliable results made the routine use of their photocells in studies of planktonic photosynthesis difficult. The turning point came with the development of selenium rectifier photocells in the early 1930s. After experimenting with a cuprous oxide cell in 1932, they first used a selenium cell in 1933. Thereafter, benefiting from the larger currents produced by the selenium cells, their great stability, and the simpler circuitry needed to use them at sea, Atkins and Poole began to use them exclusively. By 1938, selenium cells were recognized as standard by a special ICES committee on light in the sea which included Atkins and Poole.[43] The advent of the new cells allowed the measurement of light at sea to be applied to quantitative studies of photosynthesis, linking Atkins's observations of phosphate uptake and the role of light to the controlled studies of photosynthesis begun by Gaarder and Gran two decades before.

From the beginning of his work on the depletion of phosphate during the spring bloom, Atkins believed that a simple relationship existed between light and the onset, timing, and extent of the bloom. Observing that phosphate had disappeared earlier in 1924 than in 1923, an indication that the bloom had begun earlier, he showed that the mean daily sunshine had been nearly twice as great early in 1924 as in 1923.[44] Later,

40. Atkins 1926b, pp. 118–120.
41. Poole and Atkins 1929, pp. 309–311. The relationship they gave, that (using modern terminology) $k = 1.7/Z_{SD}$ (where k = vertical extinction coefficient and Z_{SD} = Secchi disk depth), was calculated incorrectly; the true relation is $k = 1.45/Z_{SD}$ (see Walker 1980). On the oxalic acid method see Atkins and Poole 1930b.
42. Gall and Atkins 1931. The new photometer, built at the request of G. L. Clarke of Harvard, was picked up in Plymouth by Clarke, H. B. Bigelow, C. O'D. Iselin, F. Zorell, and R. B. Montgomery during the inaugural cruise of *Atlantis* from Copenhagen to Woods Hole.
43. Atkins and Poole 1933b; Poole and Atkins 1934; Poole 1936. Selenium cells were described by the German physicist Bergmann in 1931. For the recommendations of the ICES committee see Atkins et al. 1938.
44. Atkins 1925c, pp. 702–708.

when data were available on the bloom in 1925 (when early sunlight had been intermediate), he found that the bloom had occurred later than in 1924 but earlier than in 1923. He concluded that "the vernal outbursts occur in the three years in the same order as do the amounts of spring sunshine," that the correlation was lost in the summer when low phosphate levels controlled production, and that "in the autumn as the light becomes reduced and regeneration preponderates, the quantity of light will again determine how far the photoplankton die out."[45]

Atkins's conclusions were contested by Sheina Marshall and A. P. Orr (Figure 8.2) who, from their base at Millport on the Firth of Clyde, were studying the spring bloom to provide biological information on the Scottish herring fishery.[46] Marshall and Orr began their work with cruises in the Clyde Sea in 1924 and 1925. A year later they undertook a detailed study in a pristine, isolated sea loch, Loch Striven. They intended to determine the chemical and physical factors that controlled the bloom, contending that only "general relationships" had been inferred previously. Marshall and Orr's sampling (weekly in 1926 and at two-day intervals during March and April in 1927 and 1928) included counts of phytoplankton cells, so that unlike Atkins (who complained of the lack of help to count cells) they could correlate changes in nutrients with the growth of phytoplankton populations rather than having to infer growth from nutrient depletion. Their sampling showed that the bloom in Loch Striven began toward the end of March, that it resulted in depletion of phosphate (but not nitrate or silicate), that the bloom sank as time went on. and that the chemical changes attributable to the bloom did not occur below about 20 meters, presumably because there was not enough light below that depth.[47] Because the bloom began near Millport at almost exactly the same time each year between 1924 and 1926 even though the early sunshine was different from year to year, they suggested that "the amount of sunshine is not the factor which starts off the spring increase. A measurement of the total amount of incident light and not only sunshine would be a more reliable guide."[48]

Atkins was inclined to disagree at first, attributing, rather tentatively, Marshall and Orr's results to some local factors such as wind or tidal mixing that might not have been taken into account.[49] Only when the

45. Atkins 1926a, pp. 455–457.
46. On Marshall see H. Barnes 1966; F. S. Russell 1978. Orr's life and career are outlined in Barnes 1966, Marshall 1963, and Yonge 1962.
47. Marshall and Orr 1927, 1930. The data on which these papers were based still exist in the University Marine Biological Station, Millport, Scotland.
48. Marshall and Orr 1927, pp. 862–863.
49. Atkins 1928, pp. 191–195. Wind and other factors seemed to have no part in affecting the onset of the bloom.

FIGURE 8.2 *A*, Sheina Marshall (1896–1977) in 1923, the year after her appointment as naturalist at the Scottish Marine Biological Association's laboratory at Millport, Scotland. *B*, Andrew Picken Orr (1898–1962) about 1929, perhaps at the time of the Great Barrier Reef Expedition. He was appointed chemist at the Millport Laboratory in 1923. Marshall and Orr's early work on phytoplankton in the Clyde region, especially in Loch Striven, had a profound effect on Atkins's and Harvey's work at Plymouth by showing that experiments on photosynthesis could be carried out in the sea, relating oxygen production, nutrient uptake, light intensity, and vertical circulation. Later, after the Great Barrier Reef Expedition of 1929, in which they both took part, they began a forty-year study of copepod biology for which they are best known. Photographs from the Marshall and Orr papers, University Marine Biological Station, Millport, Scotland; with permission of J. A. Allen.

neat correlation between sunshine and the date of the bloom off Plymouth broke down between 1926 and 1929 did Atkins begin to search for factors more complex than sunlight alone. As Marshall and Orr had predicted, it was a combination of sunlight and diffuse daylight—what he called the "effective length of day"—that restored order, allowing him to account for an anomalously early bloom in 1929.[50] Atkins was forced to agree with Marshall and Orr's belief, based on all the information they could muster on blooms near Millport from 1924 to 1929, that

> the date of the spring increase is decided chiefly by the total light which depends both on length of day and brightness. Only such a comparatively constant external factor could account for the narrow limits of time within which the increase begins. Vertical mixing may alter this date a little, but the increasing light gradually overcomes this factor and allows the spring increase to begin.[51]

Atkins's analysis soon showed that total light could account for the anomalies as well as the regularities of the early bloom.

The first steps in relating quantitative measurements of light to photosynthetic properties of bloom cells may be traced to Marshall and Orr's work in Loch Striven during March–April 1927 and 1928 when they attempted to account for the absence of changes caused by the bloom below about 30 meters in the water column. To test their early idea that light was too low for effective photosynthesis below that depth, they moored light and dark bottles containing laboratory cultures of the diatom *Coscinosira* at 0, 5, 10, 20, and 40 meters for 24 hours (see Figure 5.2B). In general, photosynthesis decreased below the surface, except when the illumination became intense later in the season; at those times, photosynthesis was highest at 5 meters, decreasing both above and below that depth. And, in general, what they called the *compensation point* (the depth at which oxygen produced in photosynthesis was equaled by the oxygen consumed in respiration) never went deeper than 20 to 30 meters.[52] These results were the foundation of their conclusion in 1930 that

50. Atkins 1930b, pp. 842–846.
51. Marshall and Orr 1930, pp. 871–872. Their paper, in the same issue of the *Journal of the Marine Biological Association* as Atkins's (1930b), does not refer to his.
52. Marshall and Orr 1928. The data used in this paper were selected from 47 experiments done between March 10 and December 7, 1927, in Loch Striven and at Farland Point, Millport. Marshall and Orr did about 50 more, unreported, light and dark bottle experiments in 1928, including some under controlled conditions in the laboratory, to work out the effect of temperature, light intensity and quality, different kinds of bottles, and alternating sun and shade. The results are still available in Marshall's papers at the University Marine Biological Station, Millport.
It is not clear how Marshall and Orr developed their version of the light and dark bottle method. Gaarder and Gran's paper describing the technique and its results was published in

the bloom developed when spring light penetrated farther and farther into the vertically mixed water column; furthermore they showed how photosynthesis and light might be related quantitatively once photometry could be used in conjunction with the light and dark bottle method.

Just such a step was in Atkins's mind in 1932 when Penelope Jenkin of Newnham College, Cambridge, was appointed Ray Lankester Investigator at the Plymouth Laboratory to work with Atkins on the "correlation of photosynthesis of diatom cultures in the sea with photoelectric measurements of light penetration."[53] She continued work at the laboratory even after moving to Bristol in 1935, relating light measurements by Atkins and Poole to light and dark bottle experiments in the Channel, using a modification of Marshall and Orr's method in which six bottles with cultures of the diatom *Coscinodiscus* were placed at each of several depths on long buoy strings in 55 to 65 meters of water. The novelty of Jenkin's approach was both practical and conceptual. If light in the sea could be quantitatively related to photosynthesis, it might be possible to estimate production using light alone, greatly reducing the time required for the estimate. And because Atkins and Poole had carefully standardized their photometer, the light energy available for photosynthesis at each depth could be compared with the amount of photosynthesis, measured as oxygen produced. Jenkin's results, published in 1937, showed unequivocally that light energy and photosynthesis in the water column were closely linked (Figure 8.3).[54] At low and moderate light intensities oxygen production and light energy available per hour were linearly related; only at high light intensities near the surface was photosynthesis inhibited, just as Marshall and Orr had observed in Loch Striven.[55] The compensation point too could be precisely determined. A lengthy experiment on August 9–10, 1934, in which L. H. N. Cooper, L. E. Bayliss, and G. I. Crawford of the Plymouth staff helped Jenkin at sea, enabled her to calculate that photosynthesis was just balanced by respiration during 24 hours when light fell to 0.23 g cal $\cdot$ cm^{-2} $\cdot$ h^{-1}, a value found as deep as 45 meters in the Channel during summer.

English in March 1927, the same month that Marshall and Orr began using their version in Loch Striven. They had probably learned of the technique from Gran earlier. If so, the influence was mutual, for Gran used Marshall and Orr's version (with paired light and dark bottles) during his work with T. G. Thompson around the San Juan Islands, Washington, in the summer of 1928 (Gran and Thompson 1930). Marshall and Orr's papers in Millport do not indicate how these transfers of ideas occurred.

53. Marine Biological Association 1933, p. 449. Jenkin left Plymouth in 1935 to become an assistant lecturer (later lecturer and senior lecturer) in zoology at Bristol, where she taught comparative physiology.

54. Jenkin 1937, pp. 323–331, figures 3–8.

55. Jenkin 1937, pp. 332–336 and figure 9.

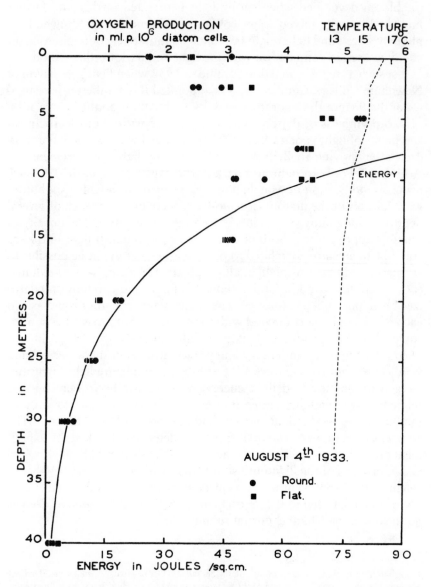

FIGURE 8.3 The relationship between photosynthesis and light energy available in the water column of the English Channel off Plymouth. Penelope Jenkin's work between 1932 and 1935 with W. R. G. Atkins, who provided the measurements of light, showed that light and photosynthesis in the sea were directly related except near the surface, where strong light inhibited the process. Redrawn from Jenkin 1937, figure 4, with permission of Cambridge University Press.

Penelope Jenkin's work showed quite clearly that there was a precise relationship between photosynthesis and light energy in the sea. But it remained to be established what quantitative relations existed between light, nutrient uptake, and the growth of cells. Both Atkins and Harvey began to examine these problems in the laboratory about 1930, basing their work on Allen's long-established cultures of the diatom *Nitzschia closterium* (now known to be *Phaeodactylum tricornutum*).

Laboratory uptake of nutrients by unicellular algae had been known since at least 1905, when Brandt mentioned his unpublished culture experiments (see Chapter 4, n. 39). Gran too had investigated nutrient uptake in the laboratory or in light and dark bottles as early as 1916. Later, his assistant Birgithe Ruud Føyn worked with the German phytoplankton specialist Ernst Schreiber (Figure 8.4A), who had shown that the growth of algal cultures could be used as a bioassay of phosphorus and nitrogen in solution in seawater.[56] Schreiber's careful techniques were passed on to a blue-ribbon panel of experts from ICES during a special session to standardize methods held in Oslo in October 1928 (Figure 8.4B). Harvey was present, representing the Plymouth Laboratory, where the tradition of laboratory work on nutrient uptake dated to the work of Allen and E. W. Nelson between 1905 and 1910.[57] But while the work of Allen and Nelson had been mainly to develop culture solutions, the pressing problem by 1930 was to determine quantitatively the effect of light and nutrient solutions on the growth of the phytoplankton.

Atkins's first work with phosphate in the sea in 1923 showed that it declined rapidly as the bloom began, and that the decline was due to uptake by the cells. Using Allen's cultures of *"Nitzschia"* he demonstrated that as the number of cells increased in a phosphate-containing culture medium the amount of phosphate decreased to very low levels, indicating its role as a limiting nutrient.[58] Later, when his photometer was available he worked with F. A. Stanbury of the Plymouth Technical College on the quantitative effect of light on the growth of diatom cultures. During 1930, Stanbury placed light filters over some of Allen's cultures, which were then grown at ambient light intensities determined using a photometer on the roof of the laboratory. She found that in the relatively narrow range of 400 to 720 nm growth (determined by counting cells) was relatively independent of wavelength; it depended mainly on the total illumination the cells received, provided it was not so high that growth was inhibited.[59]

56. Schreiber 1927, pp. 20–31; Schreiber 1929.
57. Allen and Nelson 1910.
58. Atkins 1923c, pp. 122–123.
59. Marine Biological Association 1931, pp. 598–599; Stanbury 1931, pp. 634–641. Stanbury's results showed the effect of absorption by accessory pigments in addition to chlorophyll, a phenomenon not fully understood in 1931 (see Jenkin 1937, pp. 331–332).

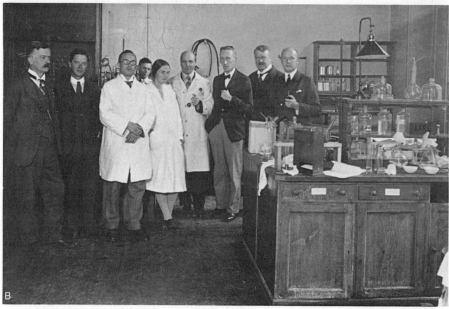

FIGURE 8.4 *A*, Birgithe Ruud Føyn (H. H. Gran's assistant) and Ernst Schreiber with phytoplankton cultures, in Oslo in October 1928. Schreiber, from Helgoland, was among a group (B) who met at Johan Hjort's suggestion to standardize methods of nutrient analysis. *B*, The ICES "meeting of experts" in Oslo, October 1928, during which standard methods were tested and recommended for the analysis of the major plant nutrients. From left to right, H. H. Gran, H. W. Harvey, Ernst Schreiber, Alf Klem, Birgithe Ruud Føyn, Kurt Buch, Hermann Wattenberg, Johan Hjort, and Torbjørn Gaarder. Photographs courtesy of Trygve Braarud, Institute of Marine Biology University of Oslo; published with permission of the Department of Biology, Section for Marine Botany, University of Oslo.

Harvey's work on phytoplankton cultures, beginning in 1929, provided evidence that both nutrients and light governed diatom growth in a fully quantitative way, but that the role of nutrients was complex. He had set out to investigate the role of ammonia, since, as he said, "it is not clear how the plants get a sufficiency of nitrogen salts during the summer months, at least in such places as the upper layers of the English Channel unless they utilize ammonium as it is being formed."[60]

Ammonia was indeed absorbed by phytoplankton cells in culture, but this proved to be only one thread of a "skein of possible factors which regulate the growth of diatoms in various places," for phosphate, nitrate, iron, and soil extract, as well as ammonia, all promoted the growth of diatoms in laboratory culture.[61] Moreover, using nutrient-depleted cultures, he found that phosphate uptake began as soon as the cultures were illuminated and that the growth of the cells (measured as oxygen produced) was directly proportional to the amount of phosphate provided to them.[62] Thus it became possible for the first time to calculate the release of oxygen in photosynthesis per unit phosphate absorbed and from this to determine the amount of carbon fixed in photosynthesis.

This dramatic achievement resulted from three decades of attention to phytoplankton growth at Plymouth, first in the hands of Allen and Nelson, and then redirected during the 1920s and early 1930s by Atkins's work at sea and in the laboratory using new techniques of nutrient analysis. Harvey's careful analyses and laboratory experiments complemented Atkins's work, so that by the early 1930s a still sketchy but increasingly quantitative model of the control of processes in the water column was being fitted together, mainly in Harvey's thoughts. But as the structure grew, so too did its complexity, taxing the resources and time of the two men.

Early in the 1930s, Atkins and Harvey had a conception of the relationships among organisms in the sea not greatly different from Lohmann's two decades before. According to this view, the phytoplankton provided a food base for a variety of invertebrates and fish, which in turn were preyed upon by marine carnivores. All these animals provided energy to decomposers, the bacteria and protozoans, which in turn were eaten by heterotrophic organisms (Figure 8.5). Their contribution in the early years at Plymouth was to examine this scheme in greater and greater detail, determining the factors that control the rate of production, especially of the phytoplankton, and clarifying the mechanisms by which nutrients were regenerated. Harvey's insight in 1926, reemphasized in his

60. Harvey 1933, p. 254.
61. Harvey 1933, pp. 253, 265, 272.
62. Harvey 1933, pp. 259–261, figure 2.

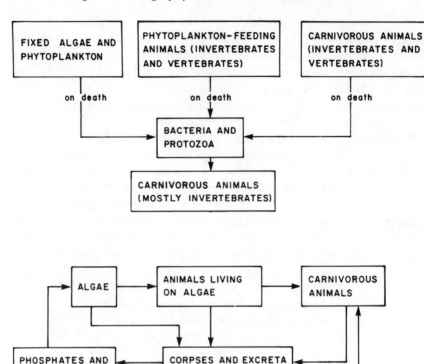

FIGURE 8.5 Harvey's formulations in 1928 of the relationships among planktonic organisms (upper) and of the marine food chain and its nutrient supply, expressed as a closed system (lower). Harvey's "carnivorous animals" today would be more realistically called "heterotrophic organisms." Harvey emphasized that the rates of conversions governed the overall fertility of the oceans. Much of Harvey's work in the 1930s demonstrated the relationships or processes implied in these deceptively simple models. From Harvey 1928a, pp. 167, 168, with permission of Cambridge University Press.

book *Biological Chemistry and Physics of Sea Water* in 1928, that "the fertility of an ocean will depend for the most part upon two factors, namely the length of time taken by the corpses of marine organisms and excreta to decay, and the length of time taken by the phosphates and nitrates so formed to come again within the range of algal growth,"[63] provided a powerful and subtle new way of looking at production that emphasized the whole system rather than its components.

63. Harvey 1928a, p. 168.

By 1928 Harvey had taken another conceptual step, recognizing that the production system in the open ocean was likely to be a closed one in which decomposing organisms provided nutrients for the growth of phytoplankton, they, in turn, being nearly completely consumed by grazing animals. As he said, considering the biological links in the marine food chain (shown in Figure 8.5) "leads directly to the next step showing that the cycle in the sea forms a closed system."[64] To justify this step required a great deal of new information about nutrients, the growth and decomposition of the organisms, and the activities of grazing animals. Their deceptively simple pattern of relationships (Figure 8.5) was far from proved: in particular, details of nutrient regeneration and of the fate of the phytoplankton remained to be clarified. Nor was it clear what the ultimate limits of production were, so that the efficiency of agricultural and marine production could be compared.[65]

The work involved in building a scientifically documented scheme of this complexity was beginning to overtax Atkins late in the 1920s, especially when his work with Poole on light became more difficult and time-consuming. He was forced to give up the analyses of all but the main nutrients, nitrate and phosphate, in the spring of 1929 or face the prospect of being overwhelmed by the samples collected several times a year on the regular hydrographic cruises. Help was needed. During 1930 it came, in the person of L. H. N. Cooper (1905–1985), a chemist from the University College of North Wales.

Leslie Cooper, who had worked for a time in a rubber company, later with Imperial Chemical Industries, took over the hydrographic sampling from Harvey and the analyses of "the minor constituents of sea water" (nitrate, ammonia, silicate), pH, oxygen, and alkalinity, from Atkins on a cruise of *Salpa* in August 1930.[66] He quickly added a large amount of new data to the lengthy series, dating back to Matthews's time, of nutrient analyses, bringing formidable analytical skill and a quantitatively oriented mind to the job. His first two papers, both published in 1933, which were based on sampling through January 1932, showed that examining the array of nutrients, major and "minor," as a group could be fruitful. By comparing the patterns of abundance of nitrate, nitrite, and ammonia, he could discern the pattern of change to be expected in the nitrogen cycle—from ammonia to nitrite to nitrate—during the production season. And both ammonia and nitrite appeared to be used by

64. Harvey 1928a, p. 167.
65. Atkins 1932a, pp. 304–306.
66. Marine Biological Association 1931, pp. 588, 599. While being interviewed for the job as assistant chemist, Cooper was sent to sea for a day on *Salpa*. He was sick, but apparently less sick than the other applicants, a situation that he believes got him the position (L. H. N. Cooper, interview, Plymouth, May 25, 1982).

phytoplankton in nature, just as Harvey's experimental work had indicated.[67]

Another significant pattern emerged during Cooper's examination of nutrient changes between 1926 and 1932. Although phosphate clearly limited the bloom initially, the roles of phosphate and nitrate were complex because of their different rates of uptake and regeneration. The outcome was that

> once the spring outburst is complete, it would seem that nitrate is much more likely to prove a limiting factor hindering further growth than is phosphate and that a marked increase in phosphate at midsummer may not necessarily be followed by a second phytoplankton outbreak. Thus during this period of the year nitrate is likely to prove the more reliable guide to the subsequent economy of the sea.[68]

Harvey's belief that nitrate had a limiting role was supported by Cooper's new evidence; he also provided evidence that silicate was regenerated from the tests of decomposing diatoms far more rapidly than had been suspected.[69]

The abundance of new data on nutrients, pH, oxygen, and carbon dioxide permitted Cooper to check Atkins's early estimates of production in the English Channel, which had been based on pH change (from which Atkins inferred carbon dioxide uptake) and phosphate uptake. New information on the carbon dioxide system, which had just become available as a result of Harvey's work in 1931 in Helsinki with Kurt Buch, Herman Wattenberg, and Stina Gripenberg, ensured the reliability of calculations based on pH.[70] The nutrient and oxygen data too were more detailed and accurate than in 1923, when Atkins had made his pioneering estimates. Cooper's calculations, based on changes of carbon dioxide, oxygen, phosphate, and nitrate, vindicated Atkins—the estimates ranged from 1000 to 1600 metric tons of phytoplankton per square kilometer. Only the silicate changes yielded a much lower figure than Atkins had obtained—a result consistent with Cooper's conclusion earlier that silicate was recycled many times in the course of a production season.[71]

Once it was clear that a reasonable minimum figure could be calculated representing production in the channel, Cooper proceeded to relate fish

67. Cooper 1933a, pp. 701, 714.
68. Cooper 1933a, p. 707.
69. Cooper 1933a, pp. 695–698.
70. Buch et al. 1932.
71. Cooper 1933b, pp. 741–745; Marine Biological Association 1933, p. 452. Cooper (1938a) corrected his production values when it became known that the early phosphate values had been in error because of the presence of salt.

production (landings of fish taken in the English Channel) to phytoplankton production. His conclusion that "only 0.06% of the phytoplankton produced each year is harvested as fish for the use of man and this in a region where the fishing has become so intense that many large English stream trawlers now find it unprofitable"[72] was the first use of a transfer efficiency (in modern terms) in marine ecology.

The sophisticated use of his data was a hallmark of Cooper's approach. Recognizing that plants used nutrients as atoms in their syntheses of new material, he introduced the use of milligram-atoms as a way of making easy, physiologically significant comparisons between nutrient salts and gases.[73] When he detected small changes in the alkalinity of the water he tested the reliability of his data, comparing the mean values with their standard errors. Once he was convinced that the changes were real, he computed the production of calcareous algae permitted by the tiny changes in calcium throughout the water column.[74] Even information on the wind could be used quantitatively to show that when the windstress (defined as the cube of the Beaufort number) dropped in winter, brief phytoplankton blooms resulted; when strong winds occurred in summer the breakdown of the summer thermocline occurred.[75]

Cooper's contribution to the Plymouth group's theory of plankton dynamics was rapid and important, but his main job was to provide chemical analyses and chemical advice to his colleagues. Though seemingly prosaic, Cooper's analyses—determination of the phosphorus and nitrogen content of phytoplankton—provided Harvey with the basic currency, phosphorus content, needed to allow the nutritional values of plankton samples to be compared.[76] Next, at Harvey's request, he provided information about the rate at which phosphorus was released from decomposing plankton cells. These innovations, Cooper's contributions to a revolution in methods, led to a synthesis of the Plymouth group's work, based on Harvey's conceptual model of a closed planktonic system, within a year of their introduction.

72. Cooper 1933b, p. 745.

73. Cooper 1933a, pp. 714–715; Cooper 1933d.

74. Cooper 1934b, pp. 752, 753–754. With a sure instinct, Cooper went for advice to the great statistician of the Rothamsted Experiment Station R. A. Fisher (1890–1962). His use of statistics was, and remained, unique at Plymouth in the 1930s.

75. Cooper 1933c.

76. Cooper 1934a; Marine Biological Association 1934, p. 956. Harvey (1934c) needed a standard measurement of the abundance of plankton which could be related quantitatively to their pigment content. Cooper provided this information in 1933 (see Chapter 9, paragraph 1). During 1934, at Harvey's urging, he determined the rate at which phosphorus was released after the death of phytoplankton; see Marine Biological Association 1935, p. 468; Cooper 1935b.

9 Plankton Production and Its Control: The Marine Ecosystem at Plymouth, 1934–1958

The most profitable fields of study frequently demand a combination of knowledge and techniques rarely within the competence of one man. Yet if fundamental research is to be fertile the individual worker needs great freedom to follow his own bent. The necessary informal team spirit for the fusion of these opposed requirements has been happily achieved by the laboratory of the Marine Biological Association at Plymouth, with the result that many biological, chemical and physical investigations, pursued apparently independently, integrate into a cohent central theme—the productivity of the sea.

L. H. N. COOPER 1948

In the early 1930s, H. W. Harvey's work at Plymouth was handicapped by the difficulty of estimating quantitatively the abundance of phytoplankton cells. Counting cells, the standard technique, was tedious and slow. Even when the counts had been made, a great deal of extra effort was required to estimate the size of the cells and their organic content. Rapid assessment of the bloom—when it started and in what abundance—was achieved in 1933 when Harvey designed and built a small, metered plankton net and invented a rapid colorimetric method of determining phytoplankton abundance (Figure 9.1). The net was simple—a conical device 60 cm long and 13 cm in diameter, into which a calibrated flow meter was fitted.[1] Its catch of phytoplankton cells was transferred to acetone, extracting the pigments, which Harvey then compared with

---

1. Harvey 1934c, p. 761; Harvey 1935; Marine Biological Association 1934, p. 957.

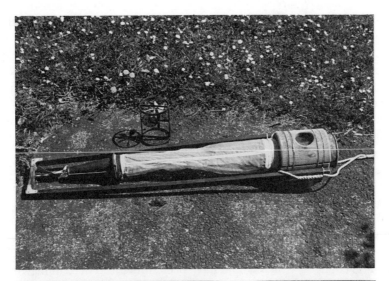

FIGURE 9.1 Harvey's metered plankton net (top) and phytoplankton color standards (bottom), used to quickly estimate the size of phytoplankton populations. Missing on the right is a tapered canvas cone to which the towing bridle was attached. Phytoplankton caught in the net were transferred to 80% acetone, which dissolved the photosynthetic pigments. Standard volumes of the pigment in solution were compared with the graded series of green color standards, indicating the amount of phytoplankton as relative units of pigment color (Harvey plant pigment units). Photographed at the Plymouth Laboratory, June 1983, courtesy of E. I. Butler and A. D. Mattacola.

green color standards, yielding a relative estimate of phytoplankton abundance which came to be called the Harvey plant pigment unit (HPPU).[2] These could be converted to absolute units by counting the cells in duplicate samples, or better, by using values Cooper had obtained for the relationship between pigment units and the phosphorus content of phytoplankton.[3]

Harvey took his new net to sea early in 1933 to follow the course of the production cycle between Plymouth and the Eddystone Light. The spring bloom was dramatically evident as an increase in plant pigment units at the end of March, but within a week the phytoplankton decreased "to near winter values" despite the presence of abundant phosphate and nitrate; at the same time zooplankton increased dramatically and their fecal pellets, green with plant material, appeared in the net collections. Throughout the summer, zooplankton proved to be abundant, but when they decreased in the autumn another diatom bloom occurred.[4] These dramatic changes, which would have been difficult to determine without Harvey's net and method of estimating phytoplankton abundance, indicated that the zooplankton appeared to have a strong influence on the spring and summer phytoplankton population.

Because the plant pigment units could be converted to phosphate values, Harvey could compare the increase of phytoplankton observed during the bloom with its production estimated from the decrease of phosphate at the same time. The difference was striking—the production of diatoms during six weeks up to March 28 was nearly eight times the stock present on that date. Cells had disappeared in large numbers, despite the presence of abundant phosphate and nitrate (Figure 9.2B). No evidence of dead or settling cells could be found, but larval zooplankton and a pulse of their green fecal pellets late in March indicated the solution:

> The evidence as a whole indicates that with the lengthening days and increasing sunshine the stock of diatoms increased from February to March 28th, producing more daily than was eaten by the herbivores. Then a sudden outburst of larvae rapidly ate down the stock of diatoms to a low level and kept it eaten down closely for the following five weeks, during

2. Harvey 1934c, pp. 770–771; Marine Biological Association 1934, p. 957. This technique was based on the work of Kreps and Verjbinskaya (1930), who had extracted the plant pigments in alcohol and determined their abundance spectrophotometrically. Harvey did not have a spectrophotometer at Plymouth and could not afford one (L. H. N. Cooper, interview at Plymouth, May 25, 1982), so he used a simple, though relative, optical method.

3. Harvey 1934a, p. 782.

4. Marine Biological Association 1934, pp. 957–958; Harvey 1934a, pp. 778–780, esp. figure 5.

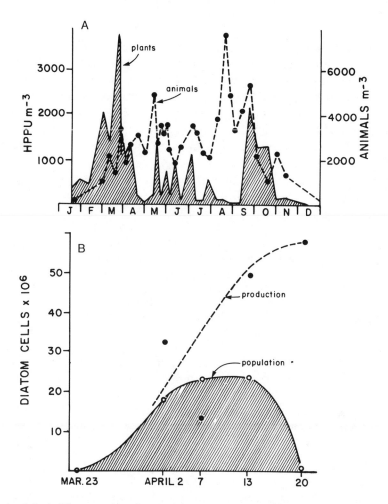

FIGURE 9.2 *A*, The annual cycle of plankton in the English Channel off Plymouth, observed by Harvey and his co-workers in 1934. Note the spring bloom (in March) and the increase of animals during the bloom. *B*, The estimated production of diatoms in Loch Striven, Scotland, in 1926 (as explained in text) compared with the population observed by Marshall and Orr. According to Harvey's calculations a similar relationship occurred in the English Channel in 1934. Harvey's comparison of production with the phytoplankton population that year led directly to his conclusion that grazing by zooplankton controlled the bloom. Redrawn from figures in Harvey et al. 1935, pp. 408, 432, with permission of Cambridge University Press.

which time the small stock was producing vegetable food for the herbivores.[5]

Rapidly dividing phytoplankton cells were held in check, even reduced,
by zooplankton grazing, indicating that at least in spring "the sea was a
closely-grazed pasture"[6] in which the daily production was high but its
removal by grazing could be even higher. The grazers themselves, according to new evidence at the same time from Sheina Marshall, A. G.
Nicholls, and A. P. Orr at Millport, had breeding cycles that were at least
temporally, and perhaps causally, linked to the spring and early summer
phytoplankton blooms.[7]

Harvey's remarks at the time show that grazing was rapidly being
recognized as an important process involved in the closed or nearly closed
cycle of materials in the sea. His work in 1933 led him to conclude that

> in the first place . . . several times more vegetation is produced during the
> few weeks' proliferation than is found at the spring maximum. It is next
> concluded that this excess of production over increase in population was
> eaten and did not die. It is then concluded that at the end of March the
> population was rather suddenly eaten down to a low level. For this to
> happen would only require a moderate increase in the intensity of grazing
> by herbivores.[8]

The result was that illumination, nutrients, and grazing interacted to
shape the form and determine the course of the annual plankton cycle.

> During the winter months at the beginning of the year the amount of
> sunshine and transparency of the water . . . keeps the daily production of
> diatoms from greatly exceeding the herbivores' requirements, [whereas] in
> March the lengthening days and greater illumination allow an increase in
> vegetation. From then onwards the diatom population is regulated by the
> intensity of grazing, until it has used up the nutrient salts in the water. The
> utilisation of the nutrient salts and onset of barren conditions would
> presumably occur in the spring, if the herbivores did not keep the breeding
> stock of plants eaten down.[9]

5. Harvey 1934a, p. 784.
6. Marine Biological Association 1934, p. 958. Harvey calculated division rates from the
change of phosphate; the bloom diatoms were dividing at least once every three and a half days
(Harvey 1934a, p. 783).
7. Marshall et al. 1934, pp. 805–808. No comparable work was being done at Plymouth.
F. S. Russell had been working on vertical migration of zooplankton and on indicator species in
the Channel.
8. Harvey 1934a, p. 789.
9. Harvey 1934a, p. 790.

By the beginning of 1934 a distinctively Plymouth view of the marine production system had come into existence. Harvey summarized this outlook by stating that "an intimate relation or balance exists between the ever-varying populations of carnivores, herbivores, and vegetable food, which in turn is sometimes controlled and always affected by the available nutrient salts and illumination."[10] Carefully executed work during the preceeding thirteen years led to this brief verbal model, with its background of knowledge about illumination, nutrient levels, the uptake of salts, regeneration, phytoplankton production, and grazing by zooplankton.

In 1934 Harvey took steps to verify and amplify this model of marine production by doing a synoptic, year-long study of production in the English Channel off Plymouth. Its foundations were laid late in 1933, when, "somewhat against our wills," according to L. H. N. Cooper,[11] Harvey persuaded Cooper, as chemist, F. S. Russell, a zooplankton specialist, and Marie Lebour[12] to join him in a year's study of the plankton cycle near Plymouth. Their work began in January 1934 at the old ICES station L4, about five miles southwest of the Plymouth breakwater, at a depth of 50 meters. There, throughout 1934, until the end of the spring bloom in 1935, they took vertical hauls with Harvey's new net for phytoplankton and zooplankton along with water samples to determine temperature, salinity, plant pigments, and nutrient salts, especially phosphate and nitrate. Phosphate was the standard unit of plankton abundance, but Russell and Lebour's painstaking work enabled a picture to be constructed of the biological changes of species and their volumes that occurred during the year and their relation to the change of plankton stock expressed as phosphate or as pigments.[13]

Once the zooplankton had been analyzed, a lengthy and tedious task, Harvey summarized the results in a notable paper under all their names, titled "Plankton Production and Its Control," published in 1935. Its

10. Harvey 1934a, p. 790.

11. L. H. N. Cooper, interview at Plymouth, May 25, 1982.

12. On F. S. Russell (1897–1984) see Blaxter et al. 1984 and Denton and Southward 1986. Russell joined the Plymouth Laboratory in 1924 after a career in the Royal Air Force and as a fishery biologist in Egypt. He was noted for his documentation of zooplankton vertical migration and for the discovery that the production of the English Channel was linked to changes of nutrients and hydrography. In 1928–1929 he joined Sheina Marshall and A. P. Orr on the Great Barrier Reef Expedition under C. M. Yonge. Russell became director of the Plymouth Laboratory in 1945, succeeding Stanley Kemp. He was knighted upon retirement in 1965. Marie Lebour (1876–1971), after joining the laboratory staff in 1915, quickly built a reputation for taxonomic studies of diatoms and dinoflagellates, but particularly for her exceptional ability to rear marine larvae, allowing the life histories of many species to be described (see F. S. Russell 1972).

13. Marine Biological Association 1935, pp. 468–470; Harvey et al. 1935, pp. 407–408.

ostensible aim was "to test the broad conclusion arrived at that the plant population is regulated by the intensity of grazing by animals."[14] In fact, it summarized implicitly much of the evidence that the Plymouth scientists had gathered since 1921 for the control of plankton abundance by light and nutrients, paying special attention to Harvey's insight that zooplankton grazing held the phytoplankton below the potential level allowed by light and nutrient abundance during the spring bloom.

Harvey, Cooper, Lebour, and Russell's quasideductive analysis of plankton abundance was based on the axiom that the main controls of the bloom were light and nutrients, a view that had originated at Kiel, but had been documented copiously from 1921 onward by the work of Atkins, Harvey, and Cooper. Light in particular could now be quantitatively related to photosynthesis in the sea as Penelope Jenkin had shown. Nutrients too bore a quantitative relation to plankton abundance once it was firmly established by Cooper how much phosphorus was present in representative plankton organisms.[15]

With this background, Harvey's argument had five main elements.

1. Observation in the Channel showed that there were complex fluctuations of phytoplankton and zooplankton during the year, but in general there was an inverse relation between phytoplankton and zooplankton (Figure 9.2A).[16]

2. Lebour's counts of phytoplankton made it possible to relate plant pigment units (HPPUs), a measure of the plankton crop, to the actual numerical abundance of phytoplankton (Figure 9.3A). Furthermore, both cell number and HPPUs could be converted to phosphate amounts, the common currency used to assess standing stock. Thus cell number, pigment content, and phosphate content became readily interconvertible. Zooplankton too could be fitted into the scheme, for their phosphorus content was reasonably well correlated with their abundance (volume) per cubic meter.[17]

3. There was a close relationship between phytoplankton abundance and the food consumed by zooplankton. Abundance of fecal pellets of the zooplankton, which were collected along with the organisms in Harvey's fine net, was closely correlated with the abundance of phytoplankton. As

14. Harvey et al. 1935, p. 407.
15. Cooper 1934a.
16. Harvey et al. 1935, p. 417. Because of the wide range of sizes of phytoplankton cells, Harvey established a common unit, the "diatom of average cell contents," allowing several small cells to be considered equivalent to one larger one. Figure 9.3A shows the results of this conversion.
17. Harvey et al. 1935, pp. 420, 423.

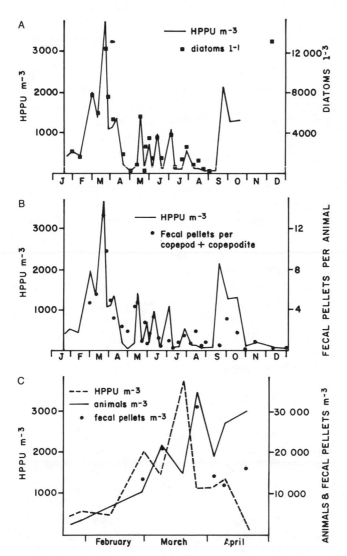

FIGURE 9.3 The evidence leading Harvey, Cooper, Lebour, and Russell (1935) to the conclusion that zooplankton grazing controlled the abundance of diatoms in the sea, especially during the spring bloom. *A*, Comparison of plant pigments (HPPU) with the number of diatom cells counted, showing that pigment content was a quick, accurate index of the phytoplankton population. *B*, The relationship between phytoplankton abundance (HPPU) and the number of fecal pellets produced by zooplankton grazers, showing a close linkage throughout most of the year between the abundance of plants and the rate at which animals fed on them. *C*, A close examination of the spring bloom, showing that although animals continued to increase in the spring, the phytoplankton population (HPPU) and number of fecal pellets produced by the grazers declined sharply late in March, indicating that the grazers had reduced the number of phytoplankton. Redrawn from Harvey, et al. 1935, pp. 417, 424, 426, with permission of Cambridge University Press.

the bloom decreased, so too did the fecal pellets of the zooplankton (Figure 9.3B).[18]

4. As he had done in 1933, Harvey calculated that the amount of phytoplankton present during the bloom was much less than the amount that had been produced during the preceding days (based on the decrease of phosphate). As he said, "Again in 1934 the estimated production of diatoms during the period of the spring outburst exceeded by many times the population present on any day when samples were taken."[19] A large amount of the spring bloom phytoplankton had disappeared during the season, both off Plymouth and in Loch Striven (see Figure 9.2B), which he used for comparison.

5. The missing phytoplankton cells could not be found near or on the bottom, nor was there evidence that they were dying above the bottom. Their abundance had not been limited by nutrients because appreciable phosphate was still present. Moreover, grazing animals continued to increase as the phytoplankton decreased, although their fecal pellets decreased in abundance and became brown, not green (Figure 9.3C). The conclusion was clear—grazers were removing much of the phytoplankton crop, keeping it well below the levels that might have been established on the basis of light and nutrients. Zooplankton grazing controlled phytoplankton abundance.[20]

In this way another major factor controlling the extent and timing of the annual plankton cycle fell into place. Showing by calculation and by empirical evidence that the rate of diatom division matched their production and that copepods could graze fast enough to affect the bloom,[21] Harvey concluded that "the survey has in general indicated that a change in diatom population is brought about by a change in one or both of two opposing factors—the rate of growth of the diatoms (depending upon illumination and probably concentration of nutrient salts) and the rate at which the diatoms are eaten (depending upon the number and kind of herbivorous animals)."[22]

Major problems still remained. For example, not all the phosphorus that disappeared from solution could be accounted for by phytoplankton crop and grazing; in particular it was not clear where the phosphate that regenerated during the summer came from without major mortalities of plants and animals, which had never been observed. And what happened

18. Harvey et al. 1935, pp. 423–424.
19. Harvey et al. 1935, p. 425. The original observations are discussed in Harvey 1934a.
20. Harvey et al. 1935, pp. 425–428.
21. Harvey et al. 1935, pp. 431–432, 433–434.
22. Harvey et al. 1935, p. 439.

to the apparently undigested food in the copious fecal pellets produced by zooplankton during the bloom? Harvey suggested that it was returned to the food web by microzooplankton, which were probably eaten in turn by larger grazers.[23]

The collaboration of Harvey, Cooper, Lebour, and Russell was an extension of their work as individuals at Plymouth, which also drew on the achievements and intellectual stimulus provided by Atkins. Their work, which established the importance of grazing in the plankton cycle, summarized, in effect, all the work begun by Atkins and Harvey in 1921. Harvey clearly saw that one summit of achievement had been reached, for "several factors which influence, and from time to time control, the production of phytoplankton can now be enumerated with some degree of certainty—the concentration of phosphate, and of nitrate, the illumination and temperature, the rate at which the organisms are being eaten by zooplankton, and the extent to which vertical currents carry them down beyond the level of sufficient light."[24]

"Plankton Production and Its Control" soon came to be regarded as the epitome of work on the plankton cycle at Plymouth, the outcome of foundations laid by Brandt, Raben, Lohmann, and Nathansohn. It played an important role in stimulating new research, both in Plymouth and other laboratories, but it did not, as might have been expected, lead to new syntheses of ideas about plankton dynamics by the group at Plymouth. Instead Harvey alone continued to work toward new, broadly based ideas about the marine ecosystem of the English Channel applicable to the seas in general. Russell, Cooper, and Lebour returned happily to their own more narrowly oriented research early in 1935 after barely more than a year working together.

In Plymouth after 1935 two factors had a dramatic effect on the direction taken by the well-established individual research programs of Harvey, Cooper, and Russell. The first was the difficulty that Harvey and Cooper found in relating nutrient supply to the annual plankton cycle. The problem of nitrogen supply (especially the process of nitrification) remained an enigma; the phosphorus cycle could not be closed, perhaps because they lacked information on the conversions of organic phosphorus; and other elements, such as iron and manganese, appeared to play some role in controlling phytoplankton abundance. Harvey was never convinced that light and nutrients alone controlled the onset of the bloom; other factors, such as trace elements, might be involved in creat-

23. Marine Biological Association 1935, pp. 469–470; Harvey et al. 1935, pp. 436–437.
24. Harvey 1937b, p. 205.

ing anomalies in time and space like those that had been observed earlier by Gran and Braarud and Klem off Norway (see Chapter 5).[25]

The second factor, which reinforced the first very quickly, was Russell's observation, based on his regular collections of zooplankton since 1924, that a major change had occurred in the English Channel in 1930. Then, or shortly afterward, zooplankton abundance fell, winter phosphate values dropped significantly, the abundance of many fish and then larvae changed, and a new group of plankton organisms appeared, characterized by the presence of the chaetognath *Sagitta setosa* replacing *Sagitta elegans*, which had been abundant before 1930.[26] A major change of the whole marine system (later named the *Russell Cycle*)[27] had occurred; its main practical outcome was the failure of the winter herring fishery in the Channel and the replacement of herring by pilchards. Russell showed, by drawing on the long series of phosphate analyses made by Atkins and Cooper, that there was a correlation between winter phosphate levels and the abundance of larval fish. Somehow fish recruitment was positively correlated with nutrients. Fortuitously, the long-term biological and chemical work that had occupied Atkins, Harvey, and later Cooper took on a sudden practical significance, which dominated work at the laboratory not only after Russell's appointment as director in 1945 but as early as 1935 when this remarkable natural change was fully recognized.

Cooper, an analytical chemist, returned quickly to nutrient analyses and to broader problems of nutrient supply in the Channel during and after 1935. Iron presented an important problem, for it appeared to have a growth-promoting effect on phytoplankton despite the fact that it was highly insoluble in seawater unless it was in complex with organic matter. For several years, Cooper wrestled with the difficult problem of quantitatively estimating the iron available to phytoplankton in seawater. Collaborating at times with Harvey, he had reached an analytical impasse in 1939 when war broke out; although he summarized his work later, he never returned to the problem in a major way.[28]

The nitrogen cycle, crucial to phytoplankton production, also got attention from Cooper. In an important review, he outlined the whole cycle, describing the physical chemistry and thermodynamics of its constituent processes and showing the probability of various reactions based on chemical principles. The significance of nitrogen fixation and the details of nitrification in the sea remained unknown in 1937, but Cooper

25. Harvey 1942, pp. 242–243; Harvey 1945, p. 129.
26. F. S. Russell 1935a, b.
27. The background and recent changes are summarized by F. S. Russell et al. 1971 and Southward 1980. The Channel returned to something like its pre-1930 state in 1965.
28. Cooper 1935a, 1937c, 1948a, c; and Cooper, interview at Plymouth, May 25, 1982.

confirmed that ammonia and urea could be used as nutrients by phytoplankton. In addition, he demonstrated that denitrification was not as poor a way of oxidizing organic matter as most authors (from the time of Gran's work in 1901; see Chapter 2 at n. 75) had assumed.[29]

Phosphorus too received Cooper's attention. At Harvey's suggestion he followed the release of phosphate from decomposing plankton. Phosphate was released in significant amounts within a few hours after death from decomposing animals; even after several days inorganic phosphate continued to be released. Phytoplankton cells also released phosphate as they decomposed, but more slowly and to a lesser extent.[30] Cooper's results supported, at least in part, Atkins's suggestion, made several years earlier, that the rapid renewal of phosphate during the summer was due mainly to cell lysis, which did not require extensive, slow chemical transformations of the nutrient. Thus phosphate could be returned as an inorganic nutrient far faster than nitrite and nitrate, which resulted from more complicated transformations.

Throughout the 1930s, Cooper maintained the regular phosphate analyses of Channel water taken several times a year. These, combined with his other nutrient analyses, enabled him to examine Harvey's ideas, up to 1926, that phytoplankton removed nitrogen and phosphorus from the sea in constant proportions, an idea extended by A. C. Redfield to include the constant elemental composition of the organisms.[31] Cooper's finding that the N:P ratio of seawater lay between 20:1 and 16:1 led him to suggest that the ratio could be used to track water masses.[32] A little later, correcting his (and many earlier) phosphate analyses for the presence of salts, he changed the value in seawater to 15:1 and that in plankton to 16:1, ratios still widely used in biological modeling of plankton-nutrient systems.[33] The phosphate analyses themselves allowed him to recalculate his estimates of production in 1931 to new values based on a far longer series of chemical data. The results showed that the English Channel produced from 1000 to more than 2000 metric tons of organic matter per square kilometer each season[34] and that production could vary in previously unexpected ways if nutrient levels changed. In brief analyses, he showed how varied amounts of phosphate could be expected to have different effects on fish with and without pelagic larval stages.[35]

29. Cooper 1937a, p. 198.
30. Cooper 1935b.
31. Harvey 1926, p. 87; Redfield 1934.
32. Cooper 1937b.
33. Cooper 1938b, c. On the use of elemental ratios in "biochemical modeling" see Redfield et al. 1963.
34. Cooper 1933b, 1938a.
35. Cooper 1948b.

After the war, which seriously interrupted his work, Cooper, in his own words, "went physical" in an attempt to determine the factors controlling major changes in marine production such as the Russell Cycle. Working under considerable difficulties (e.g., he had only one water bottle in 1948 and could not get the five more reversing bottles he needed from the MBA until the Royal Society provided him with a grant), he investigated the processes by which deep water formed or was restored to the surface in the Celtic Sea, the Bay of Biscay, and the North Atlantic as a whole.[36] Despite his cherished independence, it is clear that Cooper kept his work, by choice, within the conceptual bounds suggested by Harvey's ideas and the Russell Cycle. All his major research, both on nutrients during the 1930s and on physical oceanography later, fell within the framework established in his collaboration with Harvey, Lebour, and Russell in 1934, adapted later to the broad-scale problem of how oceanic production is controlled.

On Harvey's part, the problems opened up by changes in the Channel's marine ecosystem during the 1930s reinforced his belief that light, nutrients, and grazing, though clearly important, were not the only controls on such complexly balanced biological systems in the sea. Despite Atkins's belief that light, variations in the initial stock of phytoplankton, and differences in the stability of the water column were solely responsible for the time when the bloom began, Harvey was not convinced that the problem was so simple.[37] Before this disagreement with Atkins crystallized, Harvey referred to "the mosaic of conditions controlling the population," emphasizing that "we are dealing with *living organisms* which are subject to the resultant of so many physical laws that their behaviour is extremely complex."[38] He believed intuitively that the biological problem of the bloom had to be sought in biological processes operating within a physical framework.

Harvey's biological approach resulted, at least in part, from his failure to understand, or his choice not to emphasize, the control of the bloom by the interaction of vertical stability and mixing. Atkins clearly understood the need for both stability and nutrient renewal in 1929. Gran, Braarud, and Klem had shown the importance of these processes in the production cycle as early as the 1930s, but Harvey's publications as late as the mid-1950s give no indication that he understood the interplay of mixing and stratification. Even when H. U. Sverdrup's quantitative analyses of

36. L. H. N. Cooper, interview at Plymouth, May 25, 1982. The published results included Cooper and Vaux 1949; Cooper 1952, 1955a, b.
37. Atkins 1945. The difference appears to have been more than a scientific one. At some time, probably in the mid-1940s, Atkins and Harvey stopped speaking to one another.
38. Harvey 1928a, p. 164.

stratification and mixing appeared in 1953 Harvey did not give these physical processes star billing.[39]

Whether Harvey chose not to emphasize the primary roles of mixing and stability or did not understand them, his belief that "there is a reasonable *suspicion* that something else besides light, turbulence and grazing plays a part" in controlling the bloom directed his work away from physics toward growth, trace chemical constituents of seawater, and the nutrition of phytoplankton cultures.[40] Before the group's collaborative work ended in 1935 the annual report of the MBA's council stated that "Mr. Harvey has started an investigation on the effect of varying light and temperature on the growth rate of diatoms. Some quantitative knowledge of this is required to help in interpreting the causes of changes of diatom population in the sea, and is needed before further investigating the effect of varying concentrations of nutrient salts, iron and 'auxin' on their growth rate."[41] These subjects, especially the role of growth substances, including micronutrients, occupied him for more than twenty years.

Iron was the first of the "nonclassical" nutrients that Cooper and Harvey investigated, spurred on by Gran's belief that it could control the timing of the bloom near shore.[42] The problem was that plankton cells contained far more iron than was held in solution in seawater. While Cooper struggled with the analytical methods, Harvey provided evidence that diatoms could take up iron as colloidal particles, perhaps via the cytoplasm on their test surfaces.[43] But the problem of determining available iron proved too much for them. Their work was postponed in 1937 while Cooper worked on methods; new work and the war brought a permanent end to that line of research.[44]

In 1938 Harvey began working on the growth of the diatom *Ditylum*, which proved to grow well in phosphate-, nitrate-, and iron-enriched winter seawater but not in similarly enriched summer seawater. Adding manganese to summer water resulted in growth of the diatom. But the cells grew in artificial seawater that contained manganese only if soil extract was added. Harvey demonstrated that the active ingredients appeared to be organic acids. Their activities were mimicked by sub-

39. Sverdrup 1953; Harvey 1955, pp. 101–102.

40. The quotation was given by Atkins (1945), suggesting that they were still on speaking terms then, although by 1948 their relations were distant (F. A. J. Armstrong, personal communication, 1985).

41. Marine Biological Association 1936, p. 452.

42. Gran 1931, p. 41.

43. Harvey 1937a, b.

44. Marine Biological Association 1938, p. 258; L. H. N. Cooper, interview at Plymouth, May 25, 1982.

stances such as sulfur-containing amino acids and biotin, whose role might be to keep iron and manganese in solution and thus available to the cells.[45] The "mosaic of conditions" was, as he had predicted, important, but it was difficult to characterize because it changed both from time to time and from place to place. Moreover, the substances involved could be inorganic salts, unknown organic compounds, or even vitamins.[46]

Harvey never underestimated the importance of the "classical" nutrients, salts of nitrogen and phosphorus, which had been studied at Plymouth for decades. During the early war years, before his experimental work was forced to a halt, he verified the importance of ammonia as a nutrient, showing that it was absorbed in preference to nitrate. Urea and uric acid too could be used as a source of nitrogen, but, as he showed later, although organic phosphorus could be absorbed by phytoplankton cells, organic nitrogen (except for urea and uric acid) could not.[47]

Phosphorus was especially interesting because of its involvement in the Russell Cycle. In 1940 Harvey began to examine the idea that the total phosphorus in water masses, convertible to inorganic salts from other forms, might be used as an index of the potential fertility of the water, as the work of A. C. Redfield, H. P. Smith, and B. H. Ketchum in the Gulf of Maine suggested.[48] When the laboratory was bombed in 1941 this work lapsed, but Harvey took it up again in 1943–1944, aided, beginning in 1948, by F. A. J. Armstrong, who had joined the staff as an analytical chemist. Their results, which appeared in 1950, showed that total phosphorus was relatively constant throughout the year in the English Channel. Moreover, the mean phosphorus present in the water, which varied from water mass to water mass, appeared to be directly related to the amounts of phytoplankton, zooplankton, and larval fish that were present. Total phosphorus, though not the only determinant of the level of production, could be used as an index of the production possible in the sea.[49]

Harvey's wide-ranging, subtle studies of the factors influencing phytoplankton dynamics, which ended when he retired in 1958, always took

45. Marine Biological Association 1940, pp. 441–442; Harvey 1939. Harvey returned to the study of manganese after the war (Marine Biological Association 1947, pp. 668–669; Harvey 1947, 1949).

46. Harvey demonstrated the effect of vitamin B_{12} and sulfur deficiencies on diatoms in 1954 (Marine Biological Association 1955, pp. 663–664).

47. Harvey 1940, 1953, 1957; Marine Biological Association 1956, p. 654.

48. Marine Biological Association 1941, p. 428; Redfield, Smith, and Ketchum (1937) developed an elaborate box model of the phosphorus cycle in the Gulf of Maine based on interconversions of organic-inorganic phosphorus compounds and renewal of phosphate by eddy diffusion to the surface waters.

49. Armstrong and Harvey 1950, pp. 145, 159–161.

place within a larger framework. The "mosaic of conditions" controlling plankton which he referred to in 1928,[50] still occupied his thoughts while he was writing the review *Recent Advances in the Chemistry and Biology of Sea Water*, which was published in 1945.[51] But by then he was beginning to think of marine systems less in terms of complex, multivariate systems of control than as the result of intricate balances among nutrients, plants, and animals.[52] This concept, as he expressed it in 1955, three years before he retired from active work at the Plymouth Laboratory, embodied his belief that marine ecosystems were homeostatic, constantly shifting around some maximum level of production that could be achieved only when the transfer of nutrients to phytoplankton, or of phytoplankton cells to grazers, was closely coupled.

> In nature an equilibrium between the standing crop of plants, herbivores and carnivores is continually passing in and out of balance. Light supply changes from day to day, and generation follows generation of zooplankton. An excess of carnivores, which can only be temporary, leads to a deficiency of herbivores and an abundance of plant food. While an abundance of plants lasts, individual herbivores eat more and void more partially digested food, from which much organic matter passes into solution. Likewise an excess of herbivores, which again can only be temporary, will lead to an accumulation of nutrients in the photosynthetic zone and an abundance of plant food at a subsequent date. The extent and frequency with which the *balance of life* in the sea passes in and out of equilibrium increases the proportion of organic matter destroyed by non-biological and bacterial respiration.[53]

Although the Plymouth scientists agreed that the renewal of nutrients, a process championed first by Nathansohn and Brandt, governed the level of production of planktonic systems within limits set by temperature and light, their research added greatly to the sophistication of plankton dynamics. Their first major contribution, indisputable evidence that nutrients behaved as controlling substances in the sea, was followed by the first complicating factor from the biological realm, evidence that grazing, in addition to light and nutrients, governed both the level of primary production and the end of the spring bloom. They showed too that a large number of chemical factors had to be considered as determinants of production in addition to light, phosphorus and nitrogen, and grazing.

50. Harvey 1928a, p. 164.
51. Harvey 1945. This book, an elaboration of Harvey 1942 (which he revised as chapter X), was written after his experimental work was ended by the bombing of the Plymouth Laboratory in March 1941.
52. Harvey 1945, pp. 124–125.
53. Harvey 1955, pp. 114, 116.

Russell's startling observations of changes in the English Channel around 1930, after a time, reinforced Harvey's belief that whole systems must be studied in addition to their parts. In 1948, while he was compiling information on the phosphorus cycle in the English Channel, Harvey recognized the need for information on the standing crops and production of all the organisms found there, including fish and benthos, if his ideas on the production-governing role of total phosphorus were to be tested adequately. Fortunately Norman Holme had begun a survey of the benthos in 1949, and estimates of fish abundance could be obtained from the fishery. Even bacteria were included in Harvey's model of the whole production system of the English Channel, published in 1950.[54] The result, "a picture, crude and in part indefinite, of the quantitative relations between the different ecological groups and of how change of circumstance may affect the populations," was a direct, though not a logical, outcome of the tradition begun by Allen that the life of the sea must be examined as a whole. Partly by coincidence, perhaps partly through Allen's judgment of him, it was Harvey, building on the work of Atkins, Lebour, Russell, and Cooper, who came closest to developing quantitative models of marine ecosystems based on the mainly qualitative ones that they had taken over from Brandt and the Kiel school thirty years before.

When Harvey retired in 1958, research on plankton dynamics was being carried on in many places other than Plymouth. A large volume of new work, much of it in the United States, extended and amplified both the grazing model that Harvey, Cooper, Lebour, and Russell had presented and the work on phytoplankton growth that Harvey had carried on after 1935. The Plymouth Laboratory itself expanded greatly after 1946, taking on several new scientists and a new range of activities. But plankton dynamics, that is analysis and modeling of the factors controlling plankton production in the way they had been done in 1935, was not on the agenda, either for Harvey or his new colleagues.

It has been suggested—and is still heard in conversation—that work on plankton dynamics at Plymouth "declined" after 1935. There is a grain of truth in this, for Harvey's experimental approaches between 1935 and the 1950s show that he was in search of fresh insights into problems that had not been solved in 1935. More data and new experimental approaches were needed.[55] But a closer look indicates that "de-

---

54. Holme 1953; Harvey 1950, pp. 100–128.
55. J. E. Smith, a former director of the Plymouth Laboratory (1965–1974), has suggested that the need for new corroborative data stalled work in plankton dynamics after 1935 (Smith, interview at Plymouth, May 24, 1982).

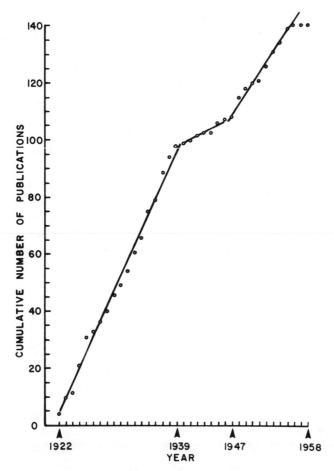

FIGURE 9.4 A cumulative plot of publications on plankton dynamics by the Plymouth group (Atkins, Harvey, Cooper, Russell, and their colleagues), 1922–1958. Note that the only anomaly in the rapid increase of cumulative number of publications ocurred from 1939 to 1946, a result of World War II.

cline" is not an appropriate word to describe the changes that began in Plymouth in the mid-1930s.

Even a gross indicator, such as the pattern of publication by the main actors in the plankton dynamics story at Plymouth—Atkins, Harvey, Cooper, Russell, and a few others—indicates that research continued at a high level from the early 1920s until Harvey's retirement in 1958 (Figure 9.4). Until the war years the Plymouth group's research production on topics related to plankton dynamics was high; after the war it resumed its rapid pace, outstripping the output of the Kiel school (see Figure 6.1). What occurred after 1935 was not a decline but a significant redirection of the work that took place in Plymouth. Each scientist continued work, but changed direction to study, for example, growth factors, improved methods of nutrient analysis, physical oceanography related to levels of production, and the consequences of major changes in marine ecosystems. Only Atkins held to his long-established research.

The effect of World War II was profound, yet in the long run it was surprisingly insignificant in the total pattern of work at Plymouth. After 1939, when their ship *Salpa* was requisitioned by the admiralty (it was never returned), until 1948, research work at the laboratory was progressively crippled. By 1940 Cooper was working for the Ministry of Supply, and Russell had joined the Intelligence Branch of the Royal Air Force. Atkins, first a member of the Home Guard, then an officer in the Royal Army Medical Corps, eventually became a meteorological officer in Plymouth. There he and Harvey worked on marine fouling and corrosion and other kinds of war-related activities. Only Harvey had the time or the facilities for pure marine research, though this soon ended when the laboratory was bombed in March 1941 and the library was removed to storage in Tavistock.[56] Forced out of laboratory work, he spent 1941–1942 reviewing the group's work on plankton dynamics in an essay that later grew into his well-known book *Recent Advances*.[57]

Most of the damage to the laboratory was soon repaired, and Harvey renewed experimental work in a small way late in 1943. But it took five years before there was enough money, equipment, a new ship, and a reconstituted staff to carry on full-scale marine research. When the work began again—at a steady pace characteristic of Plymouth—Harvey's work on micronutrients, photosynthesis, and growth was overshadowed, at least in volume, by the research of a younger, expanding staff. Russell, who succeeded Stanley Kemp as director in 1945, maintained the Allen tradition of individual work but kept the orientation of the laboratory turned toward processes underlying the Russell Cycle.

Quite clearly, if we judge by the pattern of publication and the quality of the work on plankton studies, the Plymouth group's work did not "decline" after 1935. What then has given us the misapprehension that a decline occurred? The answer: the pattern of work. If one focuses only on the results of the team made up of Harvey, Cooper, Lebour, and Russell after 1935, a decline did occur. Plymouth never again achieved the closely coordinated team effort that resulted in the near-paradigmatic paper of 1935, "Plankton Production and Its Control." But it was the team effort

56. Marine Biological Association 1941, pp. 423–425; Hardy 1946, pp. 232–234.
57. Harvey 1942, 1945.

itself that was anomalous and requires explanation, not the return of its members to their own scientific work.

The team's collaboration in 1934–1935 was the result of Harvey's belief that only by collaborating could a complete examination of the annual plankton cycle, including the role of grazing, be made. The "Allen tradition"—a norm of the laboratory—held that individuals were appointed on their merits, as near equals, to contribute freely in their own ways to a collective understanding of marine processes. This view sprang in part from a typically British individualism, but also from a widespread belief, still held in some quarters in science, that really difficult problems are best solved by individuals using their own intellectual talents rather than by teamwork, which is best suited to big, labor-intensive efforts.[58] F. S. Russell, as director of the laboratory, expressing his view of how complex research could be conducted with limited resources stated that "the most that can be done is to distribute this staff in a balanced manner so that there is one engaged in each of the possible major lines of inquiry. Some might argue that it would be better to concentrate all the energy onto one specific problem. This could only be done by the formation of a school and the interests of the leaders of this school might determine a one-tracked course for many years."[59] At Plymouth, a "one-tracked course" was not considered the way to study the mosaic of processes in the English Channel, or in any other scientific setting. A more flexible, adaptable way of doing research was needed.

In fact, teams were formed at Plymouth for specific purposes. One of the best examples of a long-lived team was the group established at Lowestoft in 1902 under Walter Garstang's direction, which was responsible for England's ICES research (much of it routine surveys) in the North Sea. Such a group, with one or two senior scientists directing the work of junior staff and technicians, was, and is, typical of many in marine science carrying out mission-oriented scientific work. And looked at broadly, the Plymouth staff itself, circa 1935, as in 1902, supposedly a group of co-equal individuals, each with his or her own work, was a de facto team, loosely structured, but able to profit scientifically from each others' research. Rank distinctions do not appear to have been very important at Plymouth, nor was there a pressing need to compete for improved status under the Allen system. The Plymouth scientists could meet the needs of their research and achieve recognition without exten-

58. Hagstrom 1965, esp. chapter III.
59. F. S. Russell 1948a, p. 768. See F. S. Russell 1948b on his views of pure versus applied research.

sive collaboration, but they were united loosely as a team by their work on marine processes and by the shared ideal of the Allen tradition.

Despite its lack of a formally established social structure, the Plymouth Laboratory was acknowledged by most who worked there to be not only a success but an inspiration. The Austrian zoologist Wolfgang Wieser summarized this widely held view in 1952 when he thanked F. S. Russell and the laboratory staff "not only for their readiness to help and for all kinds of advice, but also for the spirit prevailing at Plymouth which makes work so easy and the Laboratory itself the most excellent of its kind in Europe."[60]

The "spirit prevailing at Plymouth," which so impressed Wieser and generations of earlier visitors, was based on the relatively unfettered intellectual talents of hand-picked scientists with roughly equivalent status in their professional setting. But this kind of research had its negative side—for example, the inherent instability of groups such as the one formed by Harvey, Cooper, Lebour, and Russell in 1934–1935. As Harvey arranged things, his ideas were given substance by the technical work of Cooper on nutrients, Lebour on phytoplankton, and Russell on zooplankton. Scientists of palpably equal status were unlikely to want to collaborate for long as, in effect, technicians under Harvey's direction. So, although they dispersed on the best of terms, it is no accident that this productive team was so short-lived.

The situation at Plymouth, in which scientists of equal status temporarily subordinated their work to that of another, has few similarities with the situation at Kiel three decades earlier. There the institutional structure of science compelled collaboration (see Chapters 3 and 6). Men in the junior ranks, such as Schütt, Apstein, and Lohmann, were dependent on the senior ranks (especially on Brandt) for resources, positions, and patronage. In turn, senior academics such as Brandt depended on their junior colleagues to carry on day-to-day research and administration. When the junior scientists could no longer stand the frustration of finding no senior positions (generally professorships) opening up, they left, draining the intellectual resources from what had been a noteworthy school of marine biological research.

Clearly the situation at Plymouth was different, more evolutionary than reactionary. Loose-knit or temporary team efforts came and went, but the most significant changes were brought about as scientific fashion changed, or as intractable problems were exchanged for apparently more profitable ones. Thus the study of environmental factors controlling the bloom, such as light, nutrients, and grazing, studied first at sea, gave way

60. Wieser 1952, p. 148.

to laboratory studies; these reached their climax in parallel with, rather than as precursors to, Harvey's models of the marine ecosystem in the English Channel. All were the result of an approach to doing science quite different from that at Kiel.

Despite the differences, the group at Plymouth inherited a theory of nutrient control and their approach to the estimation of seasonal production in the sea from the Kiel school. How may the Plymouth group's work on plankton dynamics after 1921 be assessed in comparison with the foundations laid at Kiel?

Always held back by their modest funding and sparse equipment, the Plymouth scientists had one important advantage over their predecessors at Kiel, the availability of ships. Their long, regular series of cruises, intended at first to fulfill the objectives of ICES, provided the details from which the group at Plymouth developed their knowledge of the hydrography and nutrient dynamics of the English Channel. Nothing comparable was available to Brandt, who struggled to get ship time for a few days each year at most.

Even with ship time, however, the work of the Plymouth group could not have proceeded as it did without improved techniques and instruments. Matthews's colorimetric technique of measuring phosphate was the first step in a technological revolution at Plymouth. Later, Atkins's and Harvey's new, highly sensitive colorimetric methods for phosphate, silicate, and nitrate, used on samples from scores of cruises, revealed the complex interplay of these nutrient salts, providing the first definitive evidence that nutrient depletion actually occurred in the sea.[61] Once highly sensitive techniques came into use for nitrite and ammonia as well as nitrate analysis it became clear, too, that changes involving the whole nitrogen cycle were going on in the water column. This evidence, imperfect but suggestive, could not have been obtained using the techniques that Brandt had available, even in 1927.

Atkins and Poole's skilled work with photocells, rapidly developed from the early work of Shelford and Gail, allowed the Plymouth group to link light and photosynthesis quantitatively, not only from place to place and season to season, but, most important, with depth, so that the total vertical extent of the productive layer could be determined. When F. S. Russell brought back from the Great Barrier Reef Expedition of 1928–1929 his knowledge of Marshall and Orr's light and dark bottle techniques, the stage was set for an important step forward: the measurement

---

61. Brandt had found that both phosphate and nitrate seldom, if ever, completely disappeared from the water. This result (which was the result of imperfect analytical techniques, although the best available) led him to suggest that a minimum level of nutrients existed below which they could not be absorbed by the phytoplankton.

of photosynthesis in relation to light intensity throughout the water column.[62]

The sensitive new techniques and instruments used after 1921 in Plymouth were indicators of profound changes in the way the Plymouth scientists were trained compared with their predecessors. Atkins and Harvey were both skilled in physics and chemistry, although their work was intended to be biological directly or indirectly. Their analytical, deductive, and mathematical abilities were important in the development of new work at Plymouth in the 1920s, as were Cooper's in the early 1930s. The first result was that Atkins and Harvey quickly applied physics to biological problems, deriving their physical oceanography mainly through ICES from Scandinavian physical oceanographers. But they used physics in quite original ways too, such as Atkins's work on temperature stratification and Harvey's on the importance of evaporation from the sea surface. The evidence is strong that Allen deliberately chose scientists with physical, chemical, and mathematical training to solve biological problems. As a result, a distinctive union of physics, chemistry, and biology became characteristic of work at Plymouth.

The laboratory culture of phytoplankton, never a priority with Brandt, evolved at Plymouth from the problem of studying plankton at sea. It became a distinctive trademark of the Plymouth style in plankton studies after 1935. Allen and Nelson set the pattern early when they developed laboratory techniques for growing phytoplankton in culture. After her arrival in 1915 Marie Lebour too acquired the reputation of having a green thumb for culturing marine organisms.[63] Ernst Schreiber at Helgoland added an array of new culture techniques during the late 1920s. Lebour and Schreiber set the stage for Harvey, whose work with cultures, beginning in 1935, was based on his belief that the complexities of blooms and of the annual cycle must be biochemical as well as chemical and physical. His experiments, showing the different growth-promoting effects of nutrient-enriched water taken in summer and winter, were the forerunners of many subtly designed laboratory experiments that showed the complex ways in which micronutrients controlled phytoplankton growth.[64] Each depended on analytical techniques and craft lore developed or accumulated after the 1920s.

Viewed broadly, the most important accomplishments of the Plymouth group were their model of plankton production, published in 1935, and

62. Russell believed that his contact with Marshall and Orr on the expedition provided an important link between the two groups (Frederick Russell, interview at Goring-on-Thames, England, March 31, 1983).

63. F. S. Russell 1972, pp. 780–783.

64. Summarized with other work by Harvey 1955, pp. 103–113.

Harvey's quantitative summaries of the marine ecosystems in the English Channel published in 1950. The first, which centered on the role of grazing in shaping the size and duration of the bloom, contained the intellectual seeds of Harvey's influential belief that marine ecosystems as a whole achieved a shifting balance of standing crops and production based on interactions of physical, chemical, and biological processes. His belief, by the 1950s, that marine communities were delicately balanced homeostatic systems fluctuating around mean values with high efficiency was one that Victor Hensen would have endorsed. But Harvey's inspiration was not Hensen; it was the purely scientific concern of trying to account for the passage of varying amounts of phosphorus in and out of the marine organisms in the English Channel. By chance, Hensen's ideas and the accomplishments of the Plymouth group converged between 1920 and 1950.

Basing its early work on the nearly forgotten achievements of the Kiel school, the Plymouth group accumulated major and minor successes using a loosely structured way of doing scientific work that was widely emulated or admired but which was greatly different from the modus operandi of the Kiel school. With modest physical resources but unusual human ones it added quantitative detail and new components to Brandt's general model of plankton dynamics. The Plymouth group, with Harvey at its center, provided the inspiration for a new way of modeling production in the sea based on the union of experimental and mathematical marine biology.

10 Appreciating Mathematics: The Origin of Plankton Modeling in the United States, 1934–1946

> The unfortunate biologist, even if mathematics are to him a
> closed book, as is the case with too many of us, must per-
> force take as keen an interest as do his physical confrères, in
> the modern applications of mathematics to oceanic dynamics,
> and hold as high an appreciation of them.
>
> H. B. BIGELOW 1931

The Plymouth group's view of how the plankton cycle was controlled in the sea crystallized during the late 1930s and 1940s into a virtually paradigmatic qualitative model involving biological, chemical, and physical processes that took place mainly at temperate latitudes in areas such as the Norwegian Coastal Current, the English Channel, the Gulf of Maine, and Long Island Sound. Admittedly a few skeptics claimed that grazing was not the only explanation for the frequently observed inverse relationship between phytoplankton and zooplankton.[1] Despite this, the Plymouth group's views became increasingly influential, partly through Harvey's books, but equally through new work on the mathematical relationship among factors such as light, turbulence, nutrient levels, and grazing, carried out in the United States beginning late in the 1930s.

Although Harvey never visited the United States and few American biological oceanographers visited Plymouth during the 1930s and 1940s, the Plymouth group's work inspired American scientists, most of them at

---

1. Notably A. C. Hardy, who put forward his animal exclusion hypothesis (with E. R. Gunther) in 1935. Hardy claimed, on the basis of observations in the Southern Ocean, that zooplankton avoided dense concentrations of phytoplankton, giving rise to the same inverse relationship attributed to grazing by Harvey. Hardy's student C. E. Lucas (1938) attempted to broaden the concept using W. C. Allee–type holistic concepts. A number of marine biologists, notably E. Steemann Nielsen (1937b) and G. A. Riley (1939a, pp. 68–69), opposed Hardy's idea before it quietly died of neglect in the late 1940s.

Yale University, later at Woods Hole Oceanographic Institution. They drew their energy from the work at Plymouth, from tensions within the American ecological community, and from the rapid expansion of institutional oceanography. Out of controversies over how data were to be used in ecological studies and out of the growth of mathematical physical oceanography in the United States came a powerful, influential new mathematical approach to the question of how phytoplankton and zooplankton were controlled in the sea.

Gordon A. Riley (1911–1985) arrived at Yale early in the autumn of 1934, while Harvey and his colleagues were at work in the English Channel, to begin graduate work on experimental embryology with Ross G. Harrison. After his early education in Springfield, Missouri, Riley had completed an M.S. degree at Washington University, St. Louis, working with Caswell Grave on the embryology of an ascidian. Grave had suggested that Riley go on in embryology, and helped him to enter Yale.[2] But Riley found embryology unexciting; far more fascinating was the new world of limnology he encountered in the lectures of G. Evelyn Hutchinson during the spring term of 1935. His conversion to limnology was rapid. Riley became Hutchinson's first graduate student, working on the copper cycle in Connecticut lakes as his Ph.D. research. At the same time he first encountered statistics, taught at Yale by Oscar W. Richards (b. 1902), whose Ph.D. thesis at Yale in 1931 dealt with the mathematical description of yeast growth, a topic closely linked to the emergence of mathematical population ecology in the United States during the late 1920s and early 1930s.[3] Influenced by Richards, Riley applied statistics to his limnological work, which began in September 1935 and continued through November 1936, when he began to write his Ph.D. thesis.

Richards provided the first mathematical framework for Riley's work on limnology, but it was Evelyn Hutchinson (b. 1903) who first aroused Riley's interest in ecological problems and who continued to inspire him as his work shifted from limnology to oceanography in 1937. When Riley arrived at Yale in 1934, Hutchinson had been there for six years, and had returned only two years before from the Yale North India Expedition of 1932, which had studied the prehistory, geology, limnology, and zoology of the Indian-Tibetan border region.[4]

Hutchinson was a born ecologist, brought up among the intellectual

2. Riley's unpublished memoirs, *Reminiscences of an Oceanographer* (Halifax, Nova Scotia, 1984) provide the most detailed record of his career. See also essays by Hutchinson and Deevey in Wroblewski 1982.

3. Kingsland 1985, pp. 149, 151, 178; Riley 1984, p. 8.

4. On Hutchinson's early career and his arrival at Yale see G. E. Hutchinson 1979, chapters 7 and 8; Hutchinson 1936 also gives insights into the personality of this remarkable ecologist.

and environmental riches of Cambridge, England, during the early years of the century. Very early he became interested in aquatic insects, especially waterbugs, a study he carried on while he was quite unhappily employed as a lecturer in South Africa at the University of the Witwatersrand between 1926 and 1928. During his stay in South Africa, Lancelot Hogben, newly appointed professor of zoology at Cape Town, suggested that Hutchinson study the chemistry of a number of small lakes near Cape Town. The effect was instantaneous: "Having got started, I realized that at last I had found what I wanted to do."[5] Coincidentally, when Harvey's *The Biological Chemistry and Physics of Sea Water* appeared in 1928, Hutchinson read it at once. The time was ripe for his new career, for as he recalled,

> When I returned to Cambridge on my way to America [in 1928], Penelope Jenkin,[6] who I had known as a student . . . showed me August Thienemann's newly published *Die Binnengewässer Mitteleuropas*. Reading this, I learned how the Lake Chrissie pans [shallow ponds] fitted, as new categories, into the scheme of lake classification that was being developed in Europe. I also came to see, reading this book after Elton's *Animal Ecology*, how all the ways of looking at nature that I had acquired in a random, disorganized way could be focused together on lakes as microosms. I had, in fact, become a limnologist.[7]

By 1934, when Riley arrived at Yale, Hutchinson had published nearly 40 papers, most of them on aquatic insects, and his papers on African and North Indian limnology were beginning to appear. His reputation as an iconclast began to blossom. As he recalled,

> During my first year I had given a departmental seminar—always called a journal club in those days—on my research. I talked about the pans of the Transvaal. The whole approach, based primarily on Thienemann's point of view, was, I suspect, utterly unlike anything that had hitherto been presented to the department. . . . To the limnological community of North America, had any of its members been present, it would have given the same impression as a Dominican or Jesuit sermon would have done to a seventeenth-century Puritan congregation.[8]

It was more than Thienemann's unfamiliar approach to the classification of lakes that made Hutchinson so unusual in the American academic

5. Hutchinson 1979, p. 206; see Hutchinson 1979, pp. 202–208, on the origins of his interest in limnology.

6. Penelope Jenkin (see Chapter 8, n. 53) began her career as a limnologist. In 1928 she was about to leave on the Percy Sladen Expedition to the Rift Lakes in Kenya.

7. Hutchinson 1979, p. 208.

8. Hutchinson 1979, p. 240.

world in the 1930s. His unorthodox approach to the use of data and the testing of hypotheses, rapidly taken up by Riley with encouragement from Richards, ran counter to the strongly empirical tradition of academic ecology in the United States.

Hutchinson's approach to scientific questions was rigorously mathematical and deductive from the first; fact gathering had its place mainly as support for preexisting concepts on theoretical structures. This characteristic of Hutchinson's thought, latent until 1931, was fully expressed after the publication of Harold Jeffreys's book *Scientific Inference* in 1931, as Sharon Kingsland has shown.[9] The question centered around the use of deductive logic. As Jeffreys expressed the problem, syllogistic reasoning per se was uninteresting if the major premise was known with certainty—but if it was only a probability, new knowledge became possible. Thus, as Jeffreys said, "the same syllogism may or may not provide new knowledge, according to the means of knowing its premises."[10] The consequences of any scientific hypothesis could be deduced, but both the major premise and the conclusion were probabilistic, not certain. By exploring a range of probable outcomes of various scientific hypotheses, one could determine the most probable causes of natural phenomena within the limits of statistical probability. Kingsland has explored how Hutchinson used this approach in population ecology; here I explore how Gordon Riley assimilated these ideas, using them to transform biological oceanography from a subject that was mainly qualitative to one that was statistically based and quantitative. This transition, in the cases of both Hutchinson and Riley, occurred in opposition to the ideas of a powerfully conservative group of American scientists, terrestrial and aquatic ecologists, many of them brought up academically in the traditions of the great schools of ecology at the Universities of Nebraska and Chicago.

In the mid-1930s, Evelyn Hutchinson, a mathematically able, hypothesis-testing naturalist, began to play a powerful, though at first lonely, role as an exponent of a new approach to ecology. Riley recalled his early days with Hutchinson at Yale:

> There were really very few people in those days who were interested in that aspect [mathematical ecology] of the subject or knew how to handle it, and he was one of the leaders at that time. . . .
> I think Evelyn, certainly I, was a little arrogant about this. I felt that this was the way we were going and damn it we were going to do it even if it wasn't appreciated, and we had a rather snooty attitude toward some of

9. Kingsland 1985, pp. 178–180.
10. Jeffreys 1931, p. 4.

these old boys that were still content to just publish species lists and call it ecology.[11]

Their "snooty attitude" was a reaction to the overwhelming influence of the description and classification of communities that permeated ecology during the early decades of the century. Population ecology had begun to emerge in the hands of demographers and animal ecologists during the 1920s, bringing mathematical population modeling, especially the logistic equation, into scientific consciousness. But, as Riley suggested, most ecologists were preoccupied with the biotic composition of communities, an adequate terminology for assemblages of organisms, and frequently, loosely drawn analogies between biotic communities and organisms.[12]

Symptomatic of American ecology in the 1930s was a symposium titled "The Conference on Plant and Animal Communities," held at the Biological Laboratory in Cold Spring Harbor, New York, in late August–early September of 1938. Its proceedings, with a heavily edited transcript of the discussion, were published in the *American Midland Naturalist* the following year. They were devoted, in the words of the editor, Theodor Just, to resolving "the status and delimitation of the fundamental entity of community study."[13]

The Cold Spring Harbor symposium, which summarized explicitly or implicitly the views of many influential American ecologists, evoked reactions that ranged from mildly approving to derogatory. Even one of the participants, the critically minded Thomas Park, claimed that ecology was "in need of a thorough housecleaning followed by a well-planned program of refurbishing," as a result of its "careless observation and experiment, misdirected overenthusiasm and lack of intellectual focus." He concluded that "ultimately ecology must cease to be a descriptive, qualitative science and become quantitative in character."[14] But it was not clear to Park how that transformation was to occur except by the accumulation of more quantitative data. His prescription is quite unlike the hypothesis-testing approach that was being applied by Riley and Hutchinson at exactly the same time outside the traditional boundaries of American ecology.

The English ecologist Charles Elton (b. 1900) was less complimentary, referring to the obsession with terminology at Cold Spring Harbor (and

11. G. A. Riley, recorded interview, October 27, 1980.
12. These aspects of American ecology in the 1930s can be reconstructed using Cooke 1977; Kingsland 1985, 1986; McIntosh 1976, 1980, 1985 (esp. chapters 3–5); Nicolson 1983, chapters 2–4; Tobey 1981; and Worster 1985, chapters 10, 11, 14, 15.
13. Just 1939, unpaged editor's preface.
14. Park 1939b, pp. 332, 336.

elsewhere) as "scholasticism in ecology."[15] Elton himself, in his book
Animal Ecology published in 1927, had tried to give a natural historian's
order to ecology, using the organizing concepts, originating in sociology
and economics, of food chains, food size, pyramids of numbers, and
ecological niche. Elton, never a theoretician, attempted "to turn natural
history into science" using observation of nature rather than sophisticated concept formation or hypothesis testing.[16] The result, almost paradoxically, was a simple but very useful way of viewing organisms in terms
of their size, food supplies, and habits, that appealed to Hutchinson, in
creative exile in South Africa, when it appeared in 1927. Elton claimed
quite justifiably that the "real shape of nature had to be discerned by field
studies," not through what he regarded as "the growing rigidity of
ecological concepts"; he was more interested in the natural world than in
constructing new edifices of theory.

To Thomas Park, the most perspicacious of the American ecologists at
the Cold Spring Harbor symposium in 1938, the main failing of ecology
was that "the facts have not been adequately assembled into principles
and concepts."[17] He envisaged mathematical population ecology being
used to amplify and clarify community ecology. Implying that ecological
communities were the sums—or perhaps the products—of individual
competitive interactions of a Lotka-Volterra kind, he attempted to sketch
out a way of making community ecology operational rather than defini-
tional or synthetic by using the clarity contributed by experimental,
quantitative population dynamics. Ecology could be reformed, according
to Park, if the mathematical ecologist and the field ecologist cooperated
by applying observations from the field to the simple models devised for
mathematical population ecology in the laboratory. The refined models
then, if they were realistic, could be used as analogues of real processes,
despite their oversimplification of nature.

Park believed that ecology did not lack "principles," that is theoretical
or observational generalizations, of various logical kinds and qualities.
He and another Chicago ecologist, W. C. Allee, assembled a number of
these in 1939; most were generalizations based on field observations,
describing the reactions of animals to physical environmental factors or
to each other. According to Allee and Park, groups of organisms were
given unity by the Eltonian factors of food-chain relationships, pyramids
of numbers, and niche diversification, and their populations were governed by competition with each other for resources, according to the

15. Elton 1940b, p. 151. Elton's role in ecology is discussed in Cox 1979.
16. Hardy 1968, p. 3; Worster 1985, pp. 293–298.
17. Park 1939a, p. 250.

principles outlined by Lotka, Volterra, and Gause.[18] When Allee and Park concluded that "there is no dearth of major and minor ecological principles about which to orient" the facts and that "focussing attention on a theoretical framework will lead to more important work in ecology," they expressed a sentiment of growing popularity among American ecologists. But their approach was still largely descriptive, despite Park's increasing reputation as an experimental ecologist. It was far from clear to most ecologists in 1939 how the theoretical respectability of ecology could be improved if not by fact gathering. Many claimed that they expected theoretical principles to emerge virtually unbidden from data collected in an unbiased way. The ideal of testing hypotheses had not fully entered ecology, in spite of Thomas Park's tentative approaches to the quantitative description of community-level processes through mathematical population dynamics.

The kind of hypothesis-testing approach that Hutchinson and Riley were using in the late 1930s was a regular feature of scientific investigation only in agriculture and in fisheries research. R. A. Fisher expressed very clearly the philosophy—and power—of testing hypotheses in his book *Statistical Methods for Research Workers*, first published in 1925. Fisher, chief statistician at the Rothamsted Experiment Station, was involved daily in interpreting the results of complex experiments on crops and livestock. Decisions about the effect of environmental changes or experimental treatments took on direct economic importance in such a field. While ecologists were using statistics almost entirely to summarize the abundances of plants and animals, or to compare plots or populations, Fisher was expounding techniques for assessing the validity of hypotheses. With beautiful clarity he expressed the approach familiar to Hutchinson and Riley through the expositions of Harold Jeffreys and Oscar Richards. "A hypothesis is conceived and defined with all necessary exactitude; its consequences are deduced by a deductive argument; these consequences are compared with the available observations; if these are completely in accord with the deductions the hypothesis may stand at any rate until fresh observations are available."[19]

An important focal point of the tensions that existed in American ecology was Frederic Clements and Victor Shelford's book *Bio-ecology*, which appeared in 1939. Clements and Shelford attempted to unify plant and animal ecology within the structure of large, supposedly natural organic units, biomes, with organismic properties. These ideas had played a

18. Allee and Park 1939; see also McIntosh 1976, pp. 358–359.
19. Fisher 1925, p. 9. H. J. Buchanan-Wollaston of the Fisheries Laboratory at Lowestoft introduced similar ideas to fisheries scientists in three papers between 1933 and 1936; see esp. Buchanan-Wollaston 1935, pp. 251–252, 253–254.

large part in Clements's thinking since early in the century; now he and Shelford applied the biome concept to a large part of the living world, including the seas.[20] Both Elton and Hutchinson, who reviewed *Bio-ecology* in 1940, recognized it as an exercise in ecological taxonomy or semantics. Hutchinson commented that, according to Clements and Shelford's approach, "the general principles of ecology . . . appear as a set of rules for the construction of a language."[21] But they ignored mathematics, a more powerful language, which had the virtue of allowing ecological hypotheses to be tested. Clements and Shelford's distrust of mathematics excluded a method that would allow their complex classification of the natural world to be examined critically.

There were other grave defects to the approach taken in *Bio-ecology* according to Hutchinson. If biotic communities were truly organisms, "it should be possible to study the metabolism of that organism"[22] using the techniques of mathematics and chemistry. Hutchinson did not debunk the idea of supraorganismic unity; on the contrary, he found it a useful operational metaphor for measurable processes occurring within ecosystems such as lakes. In 1941 he described the "intermediary metabolism" of stratified lakes, analyzing the phosphorus cycle using mathematical techniques to express the eddy diffusion of phosphorus and its physico-chemical transfers between mud, water, and organisms. But too great a stress on the organism-like nature of ecological assemblages had its problems; Clements and Shelford, by separating the community too much from its environment, neglected biogeochemical relations linking organisms to their physical and chemical surroundings.[23]

Hutchinson's brief review of *Bio-ecology* indicates how far his thinking had moved beyond the approaches taken by classical ecologists. Although Elton and Hutchinson agreed that Clements and Shelford had provided a largely empty classification of community types, only Hutchinson suggested that mathematics could provide a powerful approach to the description and testing of hypotheses about nature. Elton, in his review, viewed statistics only as a way of validating preexisting ideas about community composition; Hutchinson's view was that mathematics could be used to develop and test new ideas about the interactions

---

20. On the development of Clementsian ecology, see esp. Nicolson 1983, chapter 2; Tobey 1981, chapters 4, 6; and Worster 1985, chapter 11.

21. Hutchinson 1940, p. 267.

22. Hutchinson 1940, p. 268.

23. Hutchinson's interest in biogeochemical relationships began early; he knew of the work of Viktor Goldschmidt from his father, a mineralogist at Cambridge, and of V. I. Vernadsky from his colleague at Yale, Alexander Petrunkevitch (see Kingsland 1985, pp. 178, 252; Hutchinson 1979, pp. 232–233).

between organisms and the physical environment. His approach was mechanistic and reductionist, but his sympathies were broader, based on a belief in the unity of natural and intellectual processes. His powerful, sophisticated approach to nature was quite unlike any other in evidence in the United States during the 1930s.

The small lakes around New Haven, Connecticut, were ideal subjects for a small group interested in extending European limnological concepts to North America and in developing new approaches to limnology and ecology. Riley began a general survey of three lakes during the autumn of 1935. He attempted to characterize the physical, chemical, and biological features of each, and to add something new, a study of the copper cycle to determine if copper was limiting, how it changed from season to season, and if copper governed the kinds of organisms found in each lake.[24] Later more students, including E. S. Deevey and W. T. Edmondson joined the group; Linsley Pond became one of the best known small lakes in the world as the result of work by Hutchinson and his colleagues during the late 1930s and 1940s.[25]

Riley's thesis, which was begun late in 1936, described the limnology of the three ponds, and discussed the role of copper in shallow, productive ponds. Statistics were an important part of Riley's approach. In a paper based on his thesis he showed that regression analyses could be used to relate the amount of chlorophyll in the water to the amount of organic matter, from which the amount of organic matter contributed by the phytoplankton could be calculated. In a more complex analysis, he showed, using multiple correlation techniques, how the copper in lake water was a function of its alkalinity, precipitation, the amount of organic matter, and the attached vegetation. Some of the relationships were nonlinear, calling for the use of special techniques to calculate the correlation coefficients.[26] It is clear, although Riley did not say so explicitly, that he was using multiple correlation techniques to test the adequacy of hypotheses about factors controlling the copper cycle. This approach became the most significant feature of Riley's work for nearly the next decade.

Doing the research and writing the thesis were relatively straightforward, but getting the thesis published was more difficult. Riley sent a manuscript based on his thesis to Chancey Juday of the University of Wisconsin, a member of the editorial board of *Ecological Monographs*. It was returned with the request that the theoretical statistical part be

24. Riley 1937a; Riley 1939b, pp. 66–69.
25. Hutchinson 1982 and Riley 1984, pp. 6–9, describe the early work.
26. Riley 1939a, pp. 63–67.

removed; only the descriptive sections were deemed satisfactory for publication.[27] After some delay, the descriptive part was published, the first of a series called "Limnological Studies in Connecticut," in 1939.

Hutchinson and Albert E. Parr, the director of the Bingham Oceanographic Laboratory at Yale, encouraged Riley to publish his theoretical work; Parr suggested that a short paper could easily be published in Yale's *Journal of Marine Research*. Incorporating some new results from Linsley Pond gathered in 1938, Riley published a paper in June 1939[28] with the title "Correlations in Aquatic Ecology." This paper proved to contain a statement of the scientific principles that governed all Riley's later work on aquatic ecology.

The essence of Riley's approach was to test the validity of hypotheses about causal relationships in nature using the newly developed technique of multiple regression analysis. Nature was both highly variable and highly complex; simple inspection of data could not lead to firm conclusions about the causal relationships at work. As a result

> in order to arrive at an accurate statement of the biological significance of a factor, it is necessary to isolate simple statements of cause and effect from the maze of inter-relationships. Such isolation is effected by developing a theory (on a basis of extrinsic biological probability) to explain the interaction of various phenomena, and then testing the theory by the multiple correlation method. The multiple correlation coefficient obtained from the equation does not prove the correctness of the theory, for it implies no causal significance; however, it shows whether or not the facts are consistent with the theory, and it permits the calculation of errors that might be due to undetermined causes.[29]

The result of a multiple regression analysis of ecological factors was a provisional model of the causal factors at work in the environment. It was the ecologist's responsibility to devise hypotheses that were reasonable, based on his observations of nature, then to determine the significant factors that caused variations in the factors being observed. This approach was the reverse of what Thomas Park advocated, the assembly of facts into principles and concepts that would express ecological reality. Instead, well-founded hypotheses came first. Their analysis yielded fac-

27. Riley 1984, p. 24. The thesis itself (Riley 1937a) contained no statistical section and little theoretical discussion. The nondescriptive parts that Juday objected to were written after Riley's thesis was submitted early in 1937.

28. The incident was similar to the fate of Raymond Lindeman's classic paper on trophic dynamics in 1941; see Cook 1977; Hutchinson 1979, pp. 246–248; and Reif 1986. Lindeman, another of Hutchinson's students, also had his work rejected by *Ecological Monographs*.

29. Riley 1939a, p. 59.

tors that provisionally (until hypotheses or data were improved) could be regarded as causal, often in complex association with each other.

Riley believed that his statistical approach to ecology, using multiple regression techniques, had the kind of power that was needed to resolve the intellectual tangles faced by ecologists. Biological meaning and biological significance could be teased from nature using the techniques, very similar to those advocated by Fisher, that had been introduced to him by Oscar Richards in the intellectually eclectic environment surrounding Evelyn Hutchinson at Yale. By chance, early in 1937, he got the first opportunity to test his approach on a large scale in the oceans. They soon proved to be just as fertile areas as Linsley Pond for a questioning mind armed with quantitative techniques for testing ecological hypotheses.

Since 1933 Albert Parr had been using Woods Hole Oceanographic Institution's new ketch *Atlantis* for a series of cruises involving Bingham Oceanographic Laboratory personnel in the Gulf of Mexico and Caribbean Sea.[30] Aimed first at collecting deep-sea fish, they soon became more broadly oceanographic. Riley, who was at loose ends after writing the first draft of his thesis, was invited to join the spring cruise of 1937, departing Mobile, Alabama, to study the effect of the Mississippi River on the production of the northern Gulf of Mexico. During the cruise, in mid-March, Riley showed that nutrient levels increased near the mouth of the river and that phytoplankton chlorophyll increased as a result, just as would be expected.[31] He also discovered that he liked working at sea.[32] Never prone to seasickness, he could use the ocean to test the ideas he was beginning to formulate about how environmental factors governed geographical and seasonal variations of plankton production.

In the depths of the Great Depression it was not easy for a newly qualified ecologist to find a job. But a Sterling Fellowship, arranged by Hutchinson, was awarded to Riley beginning in September 1937, to continue work on Linsley Pond. Riley's thesis had shown the general features of the annual cycle in the pond; now he and Hutchinson intended to concentrate on estimating the production of the pond and its relationship to variations in light, temperature, and nutrients. For the first time they used light and dark bottles, incubating them for up to a week to estimate production throughout the year.[33] Their aims were thwarted by the residents around the lake, who poisoned it with copper sulfate in June 1938 to remove nuisance algae, but the results showed

30. Schlee 1978, pp. 63–67, 73–77; Schlee 1980, pp. 53–54.
31. Riley 1937b, pp. 67–72.
32. Riley 1984, pp. 9–16.
33. The long incubation times used in Riley's early studies were severely criticized later. Many more recent studies have shown that long incubations give unreliable results.

that the light and dark bottle technique could be used in lakes, and that there were statistically significant links between gross production, light, and temperature, just as the Plymouth work had suggested there should be.[34] Despite the failure of the project as a whole, Riley believed by 1938 that his experimental and statistical techniques could add new levels of understanding to plankton dynamics.

The opportunity to carry his ideas forward in the sea came when Parr hired Riley as a marine biologist for the Bingham Oceanographic Laboratory in June 1938. Immediately he began to study the control of production in Long Island Sound, basing his work at Milford, Connecticut. For three weeks in July–August 1938, with a grant from the Carnegie Institution, he worked at its Tortugas Laboratory on the control of production in the tropics.[35] Both had profound consequences for Riley's work in the open sea later.

Riley expected his plankton work at the Tortugas Laboratory, located west of the Florida Keys, to be significant because the very low abundance of plankton in the subtropics should be easy to relate to environmental factors. Plant biomass was indeed low—about 1% of the level in the English Channel—but production (estimated using 5- to 7-day incubations of light and dark bottles) was surprisingly high, up to half that in a temperate location, probably because the euphotic zone was much deeper in the clear waters of the subtropics than at temperate latitudes.[36] For the first time he used a multiple regression equation to express the factors governing production in an aquatic system, a technique he refined and extended into a predictive system during the following years. At the Dry Tortugas, according to his analysis,

$$O_2 \text{ production } (g \cdot week^{-1}) = 2.18 \text{ chlorophyll } (mg \cdot m^{-3}) + 0.75$$
$$\text{phosphate-phosphorus } (mg \cdot m^{-3}) - 0.00035 \text{ zooplankton}$$
$$(animals \cdot m^{-3}) - 0.09.$$

Although production could be predicted without considering the abundance of zooplankton grazers, including their abundance in the equation above along with phosphate resulted in a very good fit between the variations observed from week to week in nature and those predicted by the regression equation.[37] For the first time the Plymouth model was expressed quantitatively in a predictive equation.

Long Island Sound, close to home as was Linsley Pond, was a perfect

34. Riley 1940, pp. 299–305.
35. Riley 1984, pp. 19–21.
36. Riley 1938b, p. 338; Riley 1938c, p. 98.
37. Riley 1938b, pp. 344–346.

setting for a lengthy study of the factors controlling plankton production. In particular, it gave Riley a chance to evaluate the light and dark bottle technique, which had been in regular use in the sea since Marshall and Orr's work in Loch Striven in 1927.[38] Were the light and dark bottles true microcosms, reflecting changes in nature, or did they produce fatally flawed artifactual results? Riley's work from June 1938 through August 1939, interrupted once or twice for work farther from shore, established the validity of the technique for studying highly productive near-shore ecosystems. Furthermore it gave a rich harvest of new results based on experiment, amplifying the Plymouth model and making it applicable to a new range of environments by concentrating on plankton production as a balance between accumulation and degradation of biomass and consumption and regeneration of nutrients.[39]

Riley's approach was to use the light and dark bottle experiments "as a stepping stone in the gap between purely descriptive studies of natural waters and the rigidly controlled laboratory experiments" on plant growth that he carried out in the Milford laboratory. But he did not expect unequivocal results; there would be, and were, variations from his expectations. Multiple regression analysis could be used here to make sense of at least some of the factors complicating the results of the experiments. Comparison of results from the unconfined Sound and from the bottles was "never a matter of seeking identities but rather of determining whether the similarities are statistically significant and whether the differences are of the kind that would logically be expected."[40]

The essence of Riley's technique was to use light and dark bottle experiments to separate the effects of such factors as nutrient supply, grazing, light, and temperature on primary production and the increase of chlorophyll. In this, the results of the light bottle incubation represented unconfined nature, that is, the balance of photosynthesis, respiration, and grazing by herbivores. In the dark bottle, photosynthesis stopped, but nutrient regeneration, plankton respiration, herbivore grazing, and bacterial metabolism continued. The difference between results from the two bottles could be used to estimate oxygen and chlorophyll production as well as nutrient uptake by phytoplankton. Although perfectly respectable multiple regression equations could be obtained relating primary production to nutrients, phytoplankton standing crop, tem-

38. See Chapter 8 at n. 52. The Danish scientist Einar Steemann Nielsen (1937a) first used the technique to measure annual primary production in the sea, working in the Sound off Helsingør between October 1932 and September 1933.

39. Riley 1941a, p. 2.

40. Riley 1941a, pp. 20, 27, 64–65.

perature, light, and grazers,[41] the bottles gave him much sharpened insight into the nature of the processes occurring in the Sound.

One startling result of the new analyses was that nearly half the nutrients taken up by the phytoplankton came from nitrogen and phosphorus regenerated near the surface. Deep water had its role as a source of nutrients, as had been suspected since Brandt and Nathansohn's day, but the role of zooplankton and bacteria near the surface was far greater than even Harvey and Cooper had suspected.[42] In Long Island Sound the production that resulted from regeneration of nitrogen and phosphorus was very high—the range of net production based on oxygen production was 400 to 700 grams carbon per square meter per year. The average was about 375, roughly three times the production of the English Channel.[43] By estimating production using changes of nitrate, phosphate, and chlorophyll (each of which gave different results), Riley was able to compare his results with work, such as that done in the English Channel, that had not used the light and dark bottle production technique.

Long Island Sound differed from the Channel in another way. Although Riley agreed that Harvey had demonstrated the importance of grazing in controlling phytoplankton blooms, in the Sound, he found that more phytoplankton died naturally than was eaten. The rather sparse crop of zooplankton could not control the high phytoplankton production in the Sound. Local factors—such as the high nutrient level and shallow, turbid euphotic zone of the Sound, the limited nutrient supply and deeper euphotic zone of the English Channel, or the very deep, nutrient-starved euphotic zones of the tropics—outranked latitude in controlling production. Although it is unlikely that Riley knew of Brandt's work at first hand in 1939, his studies of Long Island Sound and of the northwestern Atlantic (discussed later) resulted in a significant challenge to the Kiel school's model, which postulated low production in the tropics due to the slow renewal of nutrients from deep water, as well as to Harvey's conclusions about grazing.

But what of the light and dark bottle method on which these revisions depended? Was it an adequate representation of nature? The answer was clearly yes, though with qualifications. First, Riley showed that biological changes in the unconfined Sound waters were correlated with environmental variables such as light, temperature, and nutrients. Just such changes occurred in the bottles. But were they concurrent with events in the Sound? Again the answer was yes. Chlorophyll, oxygen, and phos-

41. Riley 1941a, pp. 27–64.
42. Riley 1941a, p. 23.
43. Riley 1941a, pp. 73–85.

phate varied in much the same way and at nearly the same time in the experimental bottles as in the Sound. But enclosing Sound water in the bottles did smooth out some of the natural variation by reducing the mixing caused by wind, currents, and convection.[44]

Any experimental technique is artificial and produces effects not usually seen in nature. Riley recognized this and attempted to analyze and reduce the sources of error in his studies of Long Island Sound. Even with the uncertainties that inevitably remained, Riley's analysis was of the highest importance for his own work and for the development of biological oceanography. He readily accepted the verisimilitude of the light and dark bottle method and its use to determine zooplankton grazing, nutrient regeneration, and bacterial metabolism in the sea. After the publication of his paper on Long Island Sound in 1941 there were few truly significant criticisms of the potential usefulness of his technique as an experimental means of simplifying studies of primary production, although the way it was applied was frequently criticized. Riley continued to use the light and dark bottle technique, combining it with multiple regression analysis, in increasingly ambitious studies of production in the open sea.

It was quite natural that Gordon Riley would go to sea again and that limnology would lose him. His interest in the quantitative approach, initiated and fostered by Evelyn Hutchinson, could be directed just as effectively at sea as in lakes. Parr's cruise in 1937 had captured Riley's imagination; he was challenged by the difficulties of working on a ship and entranced by the oceanic world it revealed to him. His position at the Bingham laboratory and Parr's connections made it natural for him to expand his scientific horizons beyond Long Island Sound. In the spring of 1939, at Albert Parr's suggestion, he asked Henry Bryant Bigelow, director of Woods Hole Oceanographic Institution, for research time on *Atlantis*, which would be returning from Cuba to Woods Hole in May. A series of stations from the tropics to temperate waters could add a great deal to his increasing knowledge of the factors governing regional variations in plankton. The arrangements were soon made, but Bigelow, always blunt, was not reluctant to state his reservations about Riley's approach. As Riley recalled much later,

> Henry and I [became] good friends, but we did not see eye to eye. I was very much under the influence of the oceanographers at the Plymouth Laboratory—Atkins and Harvey and the others—who were trying to develop quantitative measurements of plankton and ecological variables and to

44. Riley 1941a, pp. 66–73, esp. figures 12 and 13.

analyze seasonal cycles and regional variations in a quantitative way. To Bigelow this was a profound and distasteful oversimplification. I recall going to see him in Cambridge . . . to request ship time on the *Atlantis* on a cruise to the subtropical Atlantic and to discuss plans for a survey of Georges Bank that was to start in the fall of 1939. With the brashness of youth I had some grandiose plans for doing the same sort of thing in our part of the ocean that the Plymouth people were doing. Bigelow listened patiently and then, with a combination of bluntness and tolerance that was characteristic of the man, he said, "Anybody who thinks he can predict more than 10% of plankton variability is a damn fool, but good luck."[45]

Bigelow's view of Riley's work was hardly atypical. But despite his skepticism, a manifestation of his highly traditional approach to biology, Bigelow's "Oceanographic," as the Woods Hole Oceanographic Institution, established in 1931, was called, gave Riley the opportunities he needed to expand his research.[46] Bigelow himself, despite his classical upbringing, which went back to Alexander Agassiz, was an exponent of a unified science of oceanography, which called for an attack on the complexities of nature and which appears to have owed a good deal to the philosophy that pervaded much of American ecology in the 1930s.[47] In an influential report to the National Academy of Sciences in 1929, he wrote of "our general ignorance of the inter-relationships in the very complex chain of events in the sea that govern the comparative success or failure of its inhabitants in the struggle for life. Nothing in the sea falls haphazard; if we cannot predict, it is because we do not know the cause, or how the cause works."[48]

Bigelow and Riley had the same aims; it was merely the means to achieve them that was the point of contention. Bigelow did not see how complex, interdisciplinary scientific problems could be solved by setting up teams of scientific specialists along disciplinary lines. Instead, he built the Oceanographic by selecting individual scientists for their merits, avoiding the formation of schools or groups, much as Allen had done at Plymouth. Into this individualistic environment Riley fitted very well,

45. Riley 1979, p. 74. See also Riley 1984, pp. 23–24.
46. On the founding of Woods Hole Oceanographic Institution (WHOI), see esp. Burstyn 1980 and Schlee 1980. Bigelow's importance should not be underestimated. His report (1931) on the state of United States oceanography ca. 1930 is a classic (chapter V reveals his biological conservatism), his efforts for the foundation of WHOI met with success (given a great deal of external help; see Burstyn 1980, also Brosco 1985), and his inspiration permeated the Oceanographic while he was director.
47. The approach was not unique to Bigelow; his counterpart at the Scripps Institution of Oceanography, T. Wayland Vaughan, shared the viewpoint that oceanography should synthesize the factors governing life in the sea (Vaughan 1930).
48. Bigelow 1931, p. 206.

first as a regular visitor to Woods Hole, then from 1942 to 1948 as a staff member.

The ship too selected its men. *Atlantis*, which was built specially for the new Oceanographic, was crowded, frequently uncomfortable, often un-hygienic, and not particularly seakind.[49] Susan Schlee, quite correctly, called *Atlantis* "the tail that wagged the entire dog" at the Oceanographic during its early days. Its capabilities governed the kind of research that was done and the kind of people who were prepared to put up with its discomforts.[50] Riley, by his own account, found the ship admirable. Between the spring of 1939 and the early summer of 1942 it was the sea-going medium by which he investigated and quantified the environmental controls of the spring bloom (Figure 10.1).

Atlantis cruise 83 left Havana, Cuba, on May 15, 1939, and arrived at Woods Hole on June 3. It was a relatively relaxed affair involving only one station a day,[51] at which Riley measured the temperature profile and took water samples to determine chlorophyll, total plant pigments,[52] nitrate, and phosphate. At each stop he began a three-day light and dark bottle experiment that was run in a tub on deck, maintained at surface temperatures by changing the water frequently. The results added significantly to his conclusions about the tropics, based previously on only a few weeks in the Dry Tortugas. Because the euphotic zone was deep in the south, tropical production and standing crop throughout the water column proved to be equal to or slightly higher than in the shallow temperate seas, (production estimates being based, admittedly roughly, on fitting surface production values to the presumed extinction of light at each station[53]). Just as his work at the Tortugas in 1938 had suggested, experimental methods cast some doubt on the prevailing picture, derived from Brandt's and Lohmann's work, that tropical standing crops and production were very low. Broad generalizations about the distribution of biomass and production had to yield place to experimental studies involving the causal factors at work in each location. In September 1939 Riley began to study Georges Bank, leaving aside for a couple of years the questions that still remained about geographical variation of plankton abundance and production.

49. Schlee (1978) gives an extensive account of *Atlantis* and its expeditions.
50. Schlee 1980, p. 49.
51. Riley 1984, pp. 26–30.
52. During his work on Linsley Pond, Riley discovered that Harvey's colorimetric method of estimating phytoplankton was difficult to use because carotenoids were abundant in freshwater algae. He devised a technique for separating and colorimetrically estimating chlorophyll as well as total pigments (Riley 1938a).
53. Riley 1939c, pp. 153–154.

FIGURE 10.1 Gordon Riley (far right) at sea on *Atlantis* in May 1941 (cruise #114). His companions (left to right) are L. V. Worthington, Dean F. Bumpus, and Fritz Fuglister. Photograph printed with the permission of the Woods Hole Oceanographic Institution Archive.

Georges Bank was the natural focus of work by scientists at Woods Hole Oceanographic Institution. Its rich fishery had been heavily exploited for more than a hundred years,[54] but in the 1930s relatively little was known of the biological or physical processes controlling its production. Henry Bigelow began to study the Gulf of Maine, including Georges Bank, beginning in 1912. His work continued, with interruptions due to World War I and the lack of a ship, until 1924.[55] Bigelow's three monographs on the Gulf, published in the 1920s, provided a background of descriptive physical oceanography and biology from which new, ecologically oriented research was a natural development in the late 1930s. Bigelow himself took visiting scientists to the area whenever he could once *Atlantis* was available. For example, in 1932 he took H. H. Gran and Selman Waksman to sea in the Gulf. They, Norris Rakestraw, and Bigelow's colleague at Harvard, Alfred Redfield, all began to contribute new information on the biology and chemistry of its waters using the opportunities provided by the Oceanographic and *Atlantis*.[56]

Bigelow's early surveys showed that Georges Bank was isolated from the rest of the Gulf of Maine. He noted in particular that its phytoplankton production season was longer than that of the open Gulf, perhaps because of intense vertical circulation bringing nutrients to the surface waters.[57] Whatever the mechanisms, the bank was rich in plankton and fish, particularly haddock, which spawned in the area and were heavily fished there. By 1939 Bigelow began to direct the attention of his staff at the Oceanographic to Georges Bank, using the question of what made the bank so suitable for haddock. This was the underlying reason for a series of cruises scheduled for the period from September 1939 to July 1940. George L. Clarke of Harvard took charge of the plankton work; Riley was asked to examine the plant pigments and nutrients.[58] Thus, just as the work of the Plymouth Laboratory ended because of the outbreak of war, some of Riley's most significant work began on Georges Bank.

54. Merriman 1982, pp. 23–27. On scientific knowledge see Wright 1987 and other essays in Backus 1987.
55. Brosco 1985, pp. 15–34; Schlee 1973, pp. 256–261. Also very helpful is Susan Schlee's unpublished manuscript "American Oceanography—1900–1950," esp. chapter III, "Henry Bigelow and the Gulf of Maine." I am grateful to her for the chance to read it. Bigelow's own accounts of the work were published for many years in *Science*, and the *Bulletin of the Museum of Comparative Zoology* beginning in 1913.
56. See Rakestraw 1933; Gran 1933b; Waksman, Reuszer et al. 1933; and Redfield et al. 1937.
57. Bigelow 1926a, pp. 485–486.
58. Iselin 1942, pp. 14–15. Columbus O'D. Iselin succeeded Bigelow as director of WHOI early in 1939. See also Schlee 1978, pp. 86–92; Clarke 1942, 1946; and Riley 1984, pp. 30–36.

Six *Atlantis* cruises between September 1939 and July 1940 and another four between March and June 1941 enabled Riley to compare this rich offshore marine production system with the more depauperate western North Atlantic and with the subtropical waters of the Gulf of Mexico. The spring cruises, in particular, gave him the opportunity to look closely at the onset of the bloom on Georges Bank, both to assess the Plymouth group's ideas and to try out the usefulness of his multiple regression techniques with data from a large number of stations in a hydrographically complex area. The four cruises in 1941, just before it became too dangerous to go to sea,[59] were devoted particularly to the beginning of the bloom using more detailed analyses than on the earlier cruises.[60]

One of the most striking features of Georges Bank was evident quickly. There was no indication that nutrients became depleted. Its shallow waters were well mixed year-round, so that phosphate and nitrate could always be detected. The result was a summer-long bloom. And, equally interesting, the spring bloom began near the center of the bank in March and spread radially to its edges in April. With statistical analyses Riley showed that the plant pigments in the water column could be predicted with assurance knowing only the temperature and the amount of phosphate and nitrate. Production too (as oxygen) could be predicted knowing only the values of a few variables, in this case light, temperature, plant pigments (equivalent to phytoplankton abundance), and the abundance of grazers.[61]

Using his observations and inferences from regression equations, Riley concluded that the low production in winter on Georges Bank was caused mainly by low light and turbulence, which mixed the cells out of the euphotic zone. Once light began to increase in the spring, deepening the euphotic zone, a bloom began in the shallow central area of the bank where the bottom prevented cells from being mixed below their compensation depths. On Georges Bank the bottom had the same effect as the stability imposed on less well-mixed waters, which led to the blooms characteristic of the English Channel or Norwegian coastal waters. Mixing had to be the main controlling factor on Georges Bank, because the bloom always began in the shallow central area despite the uniformity of light, temperature, and nutrients in the whole region.

The results of his 24-hour light and dark bottle experiments allowed

59. The Georges Bank cruises ended when German submarines made the work too dangerous to continue. *Atlantis* was withdrawn from use in September 1942 and drydocked until September 1946.
60. Riley 1984, pp. 40–44.
61. Riley 1941b, pp. 33–38, 52.

Riley to analyze the onset of the bloom with great care. Using growth rates of the phytoplankton from the bottles, he showed that when the cells were growing relatively slowly in January, storms occurred frequently enough to mix them out of the euphotic zone, preventing a bloom, even though other factors near the surface (nutrients and light) would have allowed one. In March, when the frequency of storms decreased and light was greater, mixing decreased and the spring bloom began. Then, as light continued to increase and the surface waters warmed, the bloom spread rapidly to the deeper water around the bank. Next, near the center, nutrients declined (though not to depletion), the turbidity increased (because of the growth of cells), reducing light, and grazers began to consume the phytoplankton, leading to a decline of the bloom in the shallowest areas. By summer, lower nutrient levels, high grazing, and high phytoplankton respiration kept net production well below the spring level.[62]

The first cruises yielded a clear-cut picture of the plankton cycle, corroborated by multiple regression analyses and a simulation of the growth of the phytoplankton under varied conditions of mixing. From the work of the first cruises, Riley concluded that "during the late autumn, winter, and early spring the chief limiting factors are solar radiation and vertical turbulence. The depth of the water is also important when it limits turbulence. This is believed to be the reason why the spring diatom burst began in the shallow central part of the bank."[63]

Multiple regression analyses allowed Riley to select the important ecological variables; then his numerical analysis of the growth of the cells, based on observations in nature and on experiments with light and dark bottles, enabled him to calculate the quantitative effect of environmental factors on the bloom. The results were highly suggestive, but they were still based mainly on statistical analyses rather than on deductive inference. His next step was to deductively buttress this inductive model of the plankton cycle on Georges Bank.

One important step in this procedure was to look at the cells themselves. He began laboratory cultures of the diatom *Nitzschia* during 1941, working in the basement of the old New Haven mansion occupied by the Bingham Oceanographic Laboratory.[64] Cells cultured in the laboratory showed many of the features that he had observed in the productivity bottles and in the sea, that is, slow initial growth, followed by rapid cell division (corresponding to the bloom), then eventually slower growth

62. Riley 1941b, pp. 38–54.
63. Riley 1941b, p. 56.
64. Riley 1984, pp. 42–43.

because of nutrient depletion or physiological changes of the cells.[65] Once again the bottles mimicked nature; results from light and dark bottles in the sea and culture bottles in the laboratory were similar enough to nature to contribute to knowledge of the bloom.

But if the general control of the complexities seen in the bottles lay in nature, it ought to be possible to find simple relationships between plant cell abundance and the physical features of the water column, particularly vertical mixing. In 1941, when his last four cruises took place, Riley was familiar with Gran and Braarud's idea that net production of phytoplankton could occur only if there were enough cells in the euphotic zone exposed to light to balance the respiration of all the cells in the water column above and below their compensation depths. According to Gran and Braarud's analyses of production in the Bay of Fundy, because phytoplankton respiration was about one-fifth of their photosynthetic rate, the cells could not be mixed below about five times their compensation depth without respiration exceeding photosynthetic production.[66] Under those conditions, of course, no bloom could occur and the population would decline. Could these relationships be demonstrated on Georges Bank? Frequent cruises should show some link between the abundance of plant cells and increasing stability in the spring.

The stability of the surface water could be estimated by the difference in density between the surface and 50 meters. Winter mixing should create a thoroughly mixed water column; as spring proceeded, the density of the warming surface water would decrease, reducing mixing of the water. Using this simple measure, Riley found that there was no correlation in March between the abundance of phytoplankton and stability, but in April a weak positive relation was detectable (Figure 10.2A). By May the relationship was negative. His monthly cruises had not been frequent enough to show the relationship just before the bloom began. But the results were good enough to indicate that "the relationship between surface phytoplankton and stability is zero in the early spring, becomes a direct relation during the early part of the spring flowering, and shifts to an inverse relation after the diatoms pass their peak," just as theory predicted.[67] This result led to a testable hypothesis, based directly on Gran and Braarud's concept, that, under conditions when nutrients were not limiting,

65. Riley 1943, pp. 46–50.

66. Gran and Braarud 1935, pp. 421–426. The depth at which the total respiration of the phytoplankton in the water column was equal to their photosynthetic production was called the *critical depth* by H. U. Sverdrup (1953).

67. Riley 1942, p. 71.

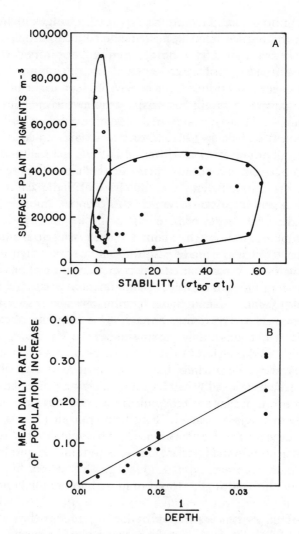

FIGURE 10.2 *A,* The relationship between the phytoplankton population (surface plant pigments) and stability (difference between the density at 50 meters and the surface) on Georges Bank in April 1941. Open circles, the constantly mixed region at the center of the bank; closed circles, stations where the depth was greater and temperature stratification developed in the spring. Because the cruise missed the beginning of the bloom in stratified waters, Riley believed that the positive correlation between phytoplankton and stability was diminishing due to nutrient depletion and grazing. Redrawn from Riley 1942, figure 26. *B,* The relationship between phytoplankton growth and the depth of the mixed layer on Georges Bank in March–April 1941. Redrawn from Riley 1942, figure 29. With permission of Mrs. G. A. Riley, Halifax, Nova Scotia.

(rate of increase of the phytoplankton population)

$$= (\text{photosynthetic rate}) \times \frac{(\text{depth of the euphotic zone})}{(\text{depth of the mixed layer})}$$

$$- (\text{respiratory rate of the cells}).$$

If photosynthetic rate, the depth of the euphotic zone, and respiratory rate were held constant for simplicity, the simple relationship

(rate of increase of the phytoplankton population)

$$\propto \frac{1}{(\text{depth of the mixed layer})}$$

was the result. By calculating the growth rate of the phytoplankton (based on its increase from one cruise to another, or since the water stratified[68]), he showed a very clear negative relationship between the growth of the phytoplankton and the depth of the mixed layer (Figure 10.2B).[69] Finally, after a long history of verbal formulations (described in Chapter 5), the effect of vertical circulation could be expressed quantitatively. Vertical circulation became a measurable variable along with light and nutrients in increasingly quantitative, predictive formulations of the plankton cycle.

By the middle of 1941, after only six years as an aquatic ecologist, Riley believed that he had developed the basis for a predictive approach to marine ecosystems as diverse as Long Island Sound, Georges Bank, the western and tropical Atlantic and the English Channel. Summarizing the results of his theoretical and statistical analyses of plankton production he stated that, at the conclusion of his work on Georges Bank, "the results obtained so far indicate that eventually the quantitative aspects of plankton growth can be predicted with considerable accuracy."[70] This sanguine hope was the result of his attempts to predict seasonal patterns of production throughout the western North Atlantic Ocean.

Qualitatively, according to Riley's summary of his work, the relationships among the physical, biological, and chemical factors that govern plant populations and production were weblike (Figure 10.3). In this

68. In the well-mixed waters over the center of the bank no thermocline formed. At deeper, less-well-mixed stations, where the water stratified, Riley had to calculate when the density discontinuity had formed so that phytoplankton growth rate could be calculated. See Riley 1942, pp. 73–75.

69. Riley 1942, pp. 75–79, esp. figure 29.

70. Riley 1941c, p. 171.

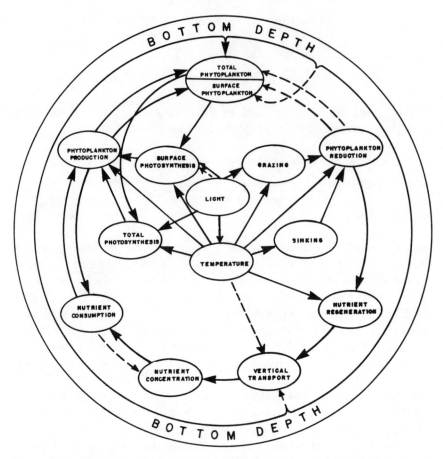

FIGURE 10.3 The mosaic of relationships among environmental variables, physical processes, chemical concentration, populations, and biological processes in marine ecosystems as varied as Long Island Sound and the subtropical Atlantic according to Riley's analyses in 1941. Solid lines indicate positive influences, dotted lines negative ones. Redrawn from Riley 1941c, figure 36, with permission of Mrs. G. A. Riley, Halifax, Nova Scotia.

Eltonian scheme, variables such as light, nutrients, temperature, vertical transport, and depth were intricately linked to plant production and populations—but in such a way that the patterns of positive and negative relations could be analyzed in step-by-step fashion, either qualitatively using tables of relations or quantitatively using statistics.[71] His qualitative technique, quite similar to the technique of loop analysis developed twenty years later,[72] was to explore the effect of positive or negative

71. Riley 1941c, pp. 164–167, explains the basis of the qualitative analysis.
72. Hutchinson 1979, pp. 217–219.

changes in such driving variables as temperature, light, and depth on the phytoplankton population and its photosynthesis. Then, adding detail, he explored the effect of changed nutrients, plant pigments, and production on the whole mosaic. Thereby, predictions could readily be made about the effects on the phytoplankton, of, for example, reducing the bottom depth, which would have the effect of increasing nutrient supply to the system.

The next step in Riley's synthesis was to use multiple regression analyses to evaluate the qualitative predictions that resulted from the model. Using broad seasonal averages of temperature, light, depth, phosphate, plant populations, and photosynthesis, he found partial correlations that were, in general, either positive or negative as suggested in the qualitative analysis, and of magnitudes consistent with what he knew about the control of production.[73] Statistics gave a new power and considerable verisimilitude to the simple mosaic of variables and processes involved in plankton production provided one focused on the patterns and general seasonal changes rather than on short-term variations.

The pay-off of this approach, using a simple diagram of ecological relationships superficially similar to an Eltonian food web, was that predictions were possible once a basic minimum of information was available about the planktonic ecosystem. Conventional ecological wisdom and unconventional statistical approaches could be married to allow predictions about how the phytoplankton populations of diverse marine systems were controlled. In fact, the statistical relationships were so good that Riley used the seasonal averages of relationships among nutrients, temperature, light, and depth to develop a set of equations that quite accurately predicted phytoplankton populations and their production from the Dry Tortugas to Georges Bank. Even the English Channel could be fitted into the system by making a few estimates of the relevant variables.[74]

Riley did not overestimate the power of his multiple regression equations. There was always a large residuum of unexplained variation, and short-term changes of population or production (on the scale of weeks rather than seasons) were difficult to predict. But by the time his active, sea-going research was forced to an end by World War II, he believed that he was hot on the trail of an ecologically based predictive theory of quantitative plankton dynamics.

73. Riley 1941c, pp. 168–169. In fact, 14 out of 20 of his qualitative predictions were borne out by multiple regression analysis.
74. Riley 1941c, pp. 169–171.

11 Disciplined Thinking in Biological Oceanography: Plankton Dynamics, Physical Oceanography, and Riley's "Synthetic Method"

> To be sure, I have never pretended that [models] are truly re-
> alistic. My only defense has been that they help us to
> think. . . . They frequently yield results that are not intuitively
> obvious, and they teach us caution about drawing conclu-
> sions that seem to violate mathematical logic. That is the way
> physics and astronomy have grown. Biological oceanography,
> messy though it may be, needs the same kind of disciplined
> thinking.
>
> G. A. RILEY 1984

The bombing of the Plymouth Laboratory in March 1941 had ended H. W. Harvey's hopes of continuing his experimental work throughout the war. Similarly, the Japanese raid on Pearl Harbor in December 1941 changed the course of Gordon Riley's work for several years. All work at sea came to an end. He continued to think about his models of how production varied temporally and spatially, especially about the unsolved food-chain relationships, such as the effect of herbivore grazing on phytoplankton populations, but he had no time to do research on plankton between the time of his last *Atlantis* cruise in September 1941 and the middle of 1946. When Riley returned to full-time research in 1946, oceanography had been transformed.

For Gordon Riley the transformation of oceanography began in 1942 when he moved to Woods Hole at Columbus Iselin's (then director of WHOI) request to work with B. H. Ketchum on the problem of marine fouling for the U.S. Navy. A year later, when that project ended, he joined

Maurice Ewing's underwater sound group, locating and charting sunken ships using echolocation. Shortly thereafter, given the resounding title Oceanographer of the Gulf Sea Frontier, he went to Miami and Key West to train naval officers to use bathythermographs and to teach submariners and sub-chasers principles of underwater sound. After a few more months with Ewing working on long-distance transmission of underwater sound, he returned to Woods Hole late in 1944; there he began to work on his Georges Bank data again when a variety of other jobs would allow. Finally, after five months in the western Pacific working on pre- and post-bomb surveys at Bikini Atoll in 1946, he was able to give his whole-hearted attention to research and to resume cruises into the Sargasso Sea on *Atlantis*.[1]

During the war the Oceanographic tripled in size. Navy contracts brought physicists and chemists to Woods Hole to work on science applied to naval warfare; many of these remained after the war, transformed by circumstance from defense scientists to oceanographers. Even when the postwar slump hit, contract research continued to bring bright, young, quantitatively trained scientists to the greatly expanded Oceanographic. In this environment the next steps in the development of Riley's plankton models were worked out. By the time he returned to Yale as an associate professor in the Bingham Oceanographic Laboratory in August 1948 he had radically modified his approach to plankton dynamics and was about to establish a technique of mathematically analyzing and modeling plankton dynamics that was quite unlike anything attempted before in biological oceanography.

After the war, before Riley returned to Yale, grazing was still an unresolved problem. If grazing animals played an important role in ending the spring bloom in the English Channel, as Harvey had suggested, they could be expected to be important on Georges Bank as well. Did grazers play the role that Harvey had supposed, and if so, how could it be quantified?

Harvey's hypothesis was far from universally accepted. His "ingenious theory"[2] that animals regulated the plant populations was only one way that reciprocal relationships between the abundances of phytoplankton and herbivores might come about. A. C. Hardy's animal exclusion hypothesis, according to which toxic plant products repelled copepods, suggested another reason for the patchy inverse distribution of plants and animals. But Hardy's mechanism could not be valid on Georges Bank, Riley and Bumpus reasoned, because mixing was so great there that the

---

1. Riley 1984, pp. 44–72.
2. Clarke 1939, p. 61.

zooplankton could not avoid plant cells. Despite this, as the season progressed the copepods developed a strikingly inverse relationship with the phytoplankton. Moreover, each species of zooplankter had a different pattern of correlation with the plant cells.[3] It appeared, just as Steemann Nielsen had suggested a decade before, that the inverse relationship had more to do with grazing and the lag between phytoplankton production and the zooplankton reproduction depending upon it than with toxic effects. Grazing was important on Georges Bank too; however, it had to be fitted in with the other factors controlling phytoplankton growth.

One way to evaluate grazing was to calculate it from the regression between phytoplankton and zooplankton abundances, which allowed the amount of plant material consumed by the animals to be calculated. The results showed that grazing was low during most months, but that it increased to more than 40% of the available supply at the end of the spring bloom in May. But even at that, Harvey and others had estimated far higher grazing rates in British waters. And the Georges Bank copepods grew more than would be expected from the modest decrease of phytoplankton. How could their correct grazing rate be calculated?[4]

In June 1939, after he had returned from a cruise on *Atlantis*, Riley had met Richard Fleming, a chemical oceanographer from Scripps Institution of Oceanography, at Woods Hole.[5] Fleming, apparently quite independently of the Lotka-Volterra equations, had developed a simple mathematical formulation of the control of the spring bloom by grazing at Plymouth. In it, the change of the phytoplankton population in time (t) could be expressed as

$$\frac{dP}{dt} = P[a - (b + ct)],$$

where P is the phytoplankton population, a is its division rate, b is the initial grazing rate of the herbivores, and c is the grazing rate as the copepod population grows. The result was a bell-shaped curve of the phytoplankton population which simulated quite closely the course of the spring bloom off Plymouth.[6] Riley's first reaction to Fleming's model

---

3. Riley and Bumpus 1946, p. 38.
4. Riley and Bumpus 1946, pp. 38–40.
5. Richard H. Fleming, b. 1909 in Victoria, British Columbia, completed a doctorate under T. W. Vaughan at Scripps in 1935 and remained on the faculty there until 1946. Later he worked in the U.S. Navy Hydrographic Office, and was on the oceanography faculty of the University of Washington from 1950 until retirement.
6. Fleming 1939, pp. 213–214.

was unfavorable—he found it unoriginal, biologically oversimplified, and too a priori to be useful.[7] But a few years later he saw that it could be used to estimate grazing by calculating b, the copepod grazing rate. Once calculated it proved to be similar to his previous calculation, thus still far too low to account for the growth of the animals. By ingeniously modifying Fleming's equation to account for turbulent mixing, Riley showed that the discrepancy arose because the animals were mixed less effectively than the phytoplankton, probably because of their diurnal migrations into deeper, less turbulent water.[8] So grazing proved recalcitrant, though there could be no denying that it played a role in changing the abundance of plankton on Georges Bank. Riley soon returned to the problem of grazing in a broader context, using Fleming's equation as the intellectual point of take-off for a wholly new formulation of plankton dynamics.

In 1946 the technique of multiple partial regressions had been Riley's principle technique for examining causal relationships in nature for eight years. Combined with experiment, especially the light and dark bottle technique, pigment extraction, nutrient analyses, and counts of zooplankton, it provided suggestive, though never fully definitive, evidence that spatial and temporal variations of plankton organisms could be accounted for by a few environmental variables and biological interactions. Riley introduced the technique to an audience of usually uninterested, occasionally hostile, ecologists whose reactions, at best, were never overwhelming. Riley described the reception of his early papers as "a dull thud."[9] Though he was far from indifferent to criticism, he persisted because he believed his techniques were the only quantitative ones available to penetrate what another mathematically inclined theorist called "blizzards of data" to achieve "honest and crucial tests for . . . theories."[10] When Riley did give up multiple correlations it was because he believed he had developed a better approach, not because of outside pressures.

By 1946 Riley found that the multiple correlation method, when used to test the significance of relationships inferred from nonquantitative theories, was seriously defective. His results were frequently hard to interpret or inconsistent from time to time and place to place. For example, during his work on the Dry Tortugas, phosphate was correlated with

7. Riley 1984, p. 37; and interview, October 20, 1980. Riley believed that Fleming did not know of the Lotka-Volterra equations in 1939.

8. Riley and Bumpus 1946, pp. 41–45. Riley (1976) reexamined this problem many years later, showing how hydrographic complexity and varying patterns of zooplankton vertical migration could give rise to zooplankton patches.

9. Riley, interview, October 20, 1980.

10. Haskell 1940, p. 5.

oxygen production, but it had no effect on phytoplankton growth when added to light and dark bottles. Also, in the western North Atlantic, oxygen production was positively correlated with plant pigments but not with phosphate, nitrate, or temperature, whereas in Long Island Sound oxygen was positively correlated with phosphate and temperature but negatively correlated with light and chlorophyll. The technique was still useful, but the generality of its results was becoming questionable. Riley summarized his problem by saying that "in each survey a reasonably high degree of correlation was found, but the empirical nature of the relationship was often confusing."[11] As long as this was the case, multiple correlations could not give definitive insights into the basic biological and physical processes at work in the sea.

Georges Bank was another case in point. The data from this, the most intensively studied of Riley's areas, readily yielded multiple regression equations expressing month by month the phytoplankton crop as a result of depth, temperature, nutrients, and grazing. And a general equation expressing seasonal variations as a function of the same variables closely simulated the changes of phytoplankton observed on the bank (Figure 11.1).[12] But to a mind inquiring into causes, too many complex positive and negative relationships were hidden in each component of the equation to be satisfactory. Riley concluded in 1946 that "the only way to avoid [the limitations of the statistical method] is by the opposite approach—that of developing the mathematical relationships on theoretical grounds and then testing them statistically by applying them to observed cases of growth in the natural environment."[13] This change in his thinking, certainly as important a step as Brandt's in using chemistry to explain the plankton cycle nearly fifty years before, was the result of Riley's recognition that multiple correlation had gone as far as it usefully could. The complexities hidden in coefficients of partial correlation were becoming more important to him than the general results they revealed. Statistics were not to be rejected, for "the statistical method is useful in determining whether a particular factor is significant but . . . the theoretical method carries on from there, discriminating between cause and effect and helping to establish certain quantitative relationships that are not likely to be derived empirically."[14]

The war had forced Riley to halt work, but he continued to think about his work on Georges Bank, incubating ideas until he had time to bring data and intuitions together again. Meanwhile he began to educate him-

11. Riley 1946, p. 55.
12. Riley 1946, pp. 55–60.
13. Riley 1946, p. 55.
14. Riley 1946, p. 70.

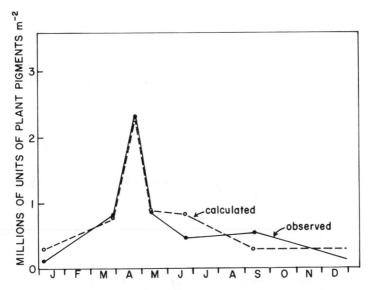

FIGURE 11.1 Changes of the phytoplankton population on Georges Bank compared with the population predicted by Riley's multiple regression analysis. Redrawn from Riley 1946, p. 60, with permission of Mrs. G. A. Riley, Halifax, Nova Scotia.

self in physical oceanography. When Sverdrup, Johnson, and Fleming's magisterial text *The Oceans* was published in 1942, he used it to learn the basic principles of mathematical physical oceanography. Then in 1945 he found a foil for his ideas; Henry Stommel had joined the Oceanographic the year before to study sound transmission, bringing a sharp inquiring mind trained in mathematics and physics. The two worked together at sea and talked over their ideas, laying the foundation for an important collaboration a few years later.[15] The physical approach, with its keen analytical tools became Riley's ideal.

In the 1940s it was natural that Riley considered his work a branch of population ecology, closely linked to the demographic ideas that were permeating American ecology.[16] The apparently clear-cut results of the Lotka-Volterra equations, specifying the outcome of predator-prey interactions, suggested a new approach to plankton dynamics. Fleming's equation could be made respectable by increasing its biological content, thereby increasing its explanatory power. Riley commented that since 1939, when he had met Fleming at Woods Hole, the equation "wasn't as much an inspiration as a burr under my saddle blanket, but that is an

15. Riley 1984, pp. 60–61.
16. Best expressed in Riley 1953b. See also Hutchinson 1979, pp. 21–25, 119–126, and Kingsland 1985 for historical accounts of this development.

inspiration of sorts."[17] Late in 1946 he began to grow the burr in soil provided by population ecology and mathematical physical oceanography.

The simple differential equation that was the basis of Riley's new work expressed phytoplankton growth in time (t) as the resultant of production and its removal by respiration and grazing, that is

$$\frac{dP}{dt} = P(Ph - R - G),$$

where P is the initial phytoplankton population, Ph is its photosynthetic rate, R its respiration, and G the grazing rate of the herbivores.[18] Each component had its own mathematical expression, determined in the field or in the laboratory, although the physiological information needed to develop the equation was limited in 1946. He elaborated the equation using a best-guess approach until more information became available, often developing his own equations, among them an expression for the mean photosynthetic rate in the water column, Ph, as a function of the light available to cells in a turbulent water column in which the nutrient supply was decreasing to a lower limit below which no growth could occur (Figure 11.2).[19]

The result was a predictive equation that expressed population changes as the result of only six variables: light intensity, the transparency of the water, nutrients, the depth of the mixed layer, surface temperature, and the abundance of grazers. In integral form it was too complex to solve, but the population during short periods of time could be calculated numerically using average values of the variables, yielding results that were quite similar to the annual phytoplankton cycle on Georges Bank (Figure 11.3). It could also be used to simulate the annual cycle in Vineyard Sound, where the hydrography and nutrients differed considerably from Georges Bank, and a three-year sequence of phytoplankton populations in a Korean harbor, without adding any new variables.[20]

17. Riley 1984, pp. 37–38.
18. Riley 1946, p. 60.
19. Riley 1946, pp. 61–66. Detailed information on many of the components of Riley's equation appeared rather slowly, mainly in the 1960s and 1970s, long after Riley had stopped active work on plankton modeling. His calculation of the mean photosynthetic rate in the euphotic zone (Riley 1946, pp. 62–63) was strikingly similar to the technique used by H. U. Sverdrup in 1953 to calculate the critical depth. Understandably, Riley was annoyed that Sverdrup did not mention his work (Riley 1984, pp. 90–91). The critical depth concept actually was rooted in Gran's ideas (cf. Gran 1915, 1931, 1932; Gran and Braarud 1935) as well as Riley's early work on Georges Bank (Riley 1941b, 1942), beginning earlier than I previously had thought (Mills 1982, p. 12).
20. Riley 1947a, b; Riley and Von Arx 1949.

$$\frac{dP}{dt} = P(Ph - R - G)$$

expanded into :

$$\frac{dP}{dt} = P\left[\underbrace{\frac{pIo}{kz_1}(1 - e^{-kz_1})}_{\text{effect of light}} \underbrace{(1-N)(1-V)}_{\substack{\text{nutrient} \\ \text{limitation}}} - \underbrace{Roe^{rT}}_{\text{respiration}} - \underbrace{gZ}_{\text{grazing}}\right]$$

FIGURE 11.2 Riley's development of his first differential equation, where P is plant population and t is time, expressing phytoplankton growth rate as a function of plant population, photosynthetic rate (Ph), respiration (R), and herbivore grazing (G). In the expansion, p is a photosynthetic constant, I_o the surface light intensity, Z_1 the depth of the euphotic zone, k the extinction coefficient of light, N the reduction of Ph due to nutrient depletion, V the reduction of Ph due to turbulent mixing, Ro the phytoplankton respiratory rate at $0°C$, r the temperature constant of respiration, T the temperature, g a grazing coefficient, and Z the amount of zooplankton. From Riley 1946, pp. 61–68.

His new analytical approach also gave Riley the ability to analyze fluctuations of the zooplankton, extending his modeling farther up the food chain. Just as the variations of phytoplankton could be expressed analytically, zooplankton fluctuated as the result of relatively few factors, among them their food supply, assimilation rate, respiration, and mortality (part of which was caused by predation). Developing the simple equation

$$\frac{dH}{dt} = H(A - R - C - D),$$

where H is the herbivore population, A their assimilation rate, R their respiration rate, C their rate of consumption by predators, and D their mortality rate apart from predation, he could readily evaluate the factors that shaped the copepod population and simulate its annual changes (Figure 11.4).[21] By developing the new theoretical approach he introduced in 1946, Riley began to work toward his early ideal, to provide a quantitative, causal framework, valid on theoretical grounds, for the broad-scale spatial and temporal variations of plankton populations in the sea, which were controlled not just by grazing but by the interaction of a few important biological and physical variables. It was the changing relative importance of these variables that was likely to account for

21. Riley 1947b, pp. 105–112; Riley 1984, p. 76.

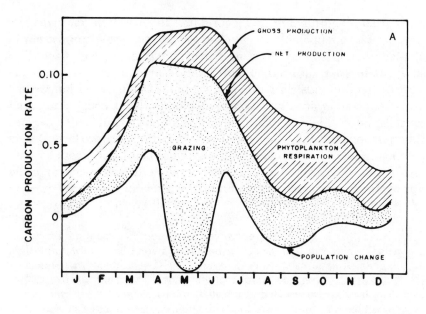

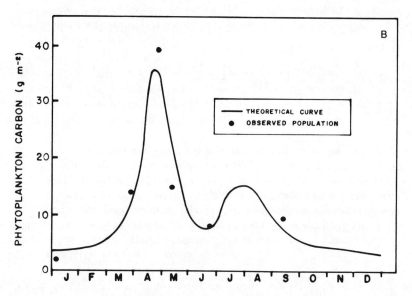

FIGURE 11.3 The results of Riley's first model of phytoplankton production using differential equations. *A*, The effect of respiration and grazing, yielding the bottom curve, which represents realized production. *B*, Seasonal changes in the standing crop of carbon as observed on Georges Bank (dots) and as calculated using Riley's first plankton model (solid line). Redrawn from Riley 1946, pp. 69–70, with permission of Mrs. G. A. Riley, Halifax, Nova Scotia.

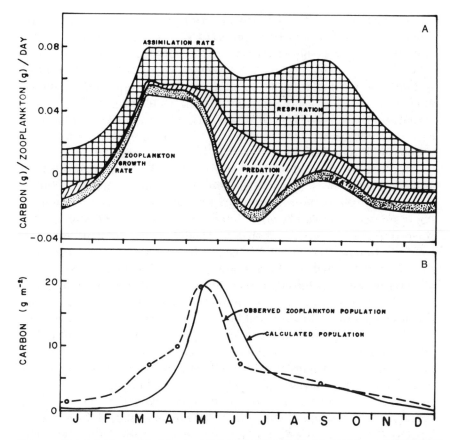

FIGURE 11.4 Riley's model of the copepod population on Georges Bank. *A*, Zooplankton growth and the major factors governing it (assimilation, respiration, and mortality). *B*, The population of zooplankton observed on Georges Bank compared with the population calculated using Riley's model of zooplankton growth. Redrawn from Riley 1947b, p. 111, with permission of Mrs. G. A. Riley, Halifax, Nova Scotia.

regional variations of plankton populations of the kind he had been observing since 1938.

The "synthetic approach," as Riley called it, of using differential equations to express the theoretical relationships involved in plankton dynamics, was based on his recognition that very few biological and physical factors interacted in nature. His innovation was the result of knowledge hard won between 1899 and 1946, but Riley's greatest debt was to the Plymouth group, who provided precise information on chemical factors, light, and grazing in the English Channel. To this was added the physical oceanographic approach he learned at Woods Hole, where his colleagues provided help with the mathematics and a growing data file on the physics of the Atlantic Ocean.

The simple basis of Riley's first mathematical modeling lay in the predator-prey equations, which allowed precise outcomes to be calculated when ecological relations were very simple. Their analytical simplicity was lost in the first Georges Bank model, but the predictive power of his first models encouraged Riley to think about how they could be applied to simple Eltonian food chains, a first approximation to complex planktonic food webs. In the five years following his first model in 1946, Riley introduced two new modeling studies unparalleled in their complexity and originality.

The first new study, published in 1949 in collaboration with Henry Stommel and Dean Bumpus, predicted the distribution of stable populations of phytoplankton and zooplankton as a result of changes in their physical environment. The second, published in 1951, examined biological activity in deep water, predicting the horizontal and vertical distribution of the nonconservative variables[22] oxygen, phosphate, and nitrate as a result of interacting biological and physical processes. Both were rooted in Riley's 1946 model, increased in sophistication and complexity by new data and by the extensive use of mathematical techniques to express the distribution and changes of properties in the water column. Riley described the 1949 paper as "a logical progression [from the 1946 paper] with addition of complicating details and culminating in an arithmetic tour de force."[23] The two more complicated papers are often cited as the apex of his plankton modeling, while the lonely originality of Riley's first models, which, he recalled, "required more hard cogitation than anything else I have done,"[24] remains unrecognized. The later models, however, did set the standard for a new generation of biological oceanographers through their use of physical techniques. Physical oceanographic analysis became firmly lodged within biological oceanography after 1949 as a result of Riley's work.

The simple differential equation that Riley introduced in 1946 could be used to predict seasonal and regional variations in plankton populations in environments as different as Georges Bank and a Korean harbor. But the early model had a number of flaws. First, although its simplicity was a virtue, the model was too simple to eliminate errors arising from the differences between dissimilar environments.[25] Next there was the prob-

---

22. Nonconservative variables in the ocean are those affected by or resulting from biological activity (e.g., oxygen and nutrients). Conservative variables (e.g., salinity and temperature) are changed only by physical processes such as incoming solar radiation, diffusion, and currents.
23. Riley 1984, p. 79.
24. Riley 1984, p. 62.
25. Riley et al. 1949, p. 7.

lem of getting absolute values from the model; calibrating it required rough-and-ready approximations that were unsatisfactory.[26] Most important of all was the way physical factors, which governed the over-all functioning of the planktonic system, had to be mixed with biological variables in an indiscriminate way that violated Riley's intuitive beliefs about their relationships. He summarized the theory that was the foundation of his new approach as follows:

> Any biological entity—an organism or an essential chemical element—is ecologically both a cause and an effect. The production of phytoplankton is essential for the production of zooplankton, but the latter, through its grazing, also affects phytoplankton. The chain of cause and effect ramifies through the whole biological system and returns to the starting point. Each interdependent biological factor is important ecologically, but is less significant from the standpoint of controlling the productivity of the ocean than are the independent physical factors. The latter, by controlling the rate of every physiological process in the community, determine the rate of passage of elements through the whole system.[27]

Practically speaking, the new approach, developed by trial and error with the assistance of Henry Stommel between 1946 and late 1948, involved developing a modeling technique that more clearly separated the governing environmental factors from their biological expression. The raw material was all the data that could be compiled on the physical circulation, nutrients, oxygen, and plankton populations of the western North Atlantic, so that areas that were believed to be quite different, such as Georges Bank, the coastal water, Gulf Stream, and Sargasso Sea were well represented. To allow all these areas to be compared Riley concluded that a period of relative equilibrium, May–June, would be analyzed; there were not enough data to model the populations present throughout the year. This had the considerable virtue of greatly simplifying the mathematical analysis. Only a few physical factors were included: light, temperature, vertical eddy diffusion, transparency of the water, and the amount of deep-water phosphate (which set the absolute levels of phytoplankton and surface phosphate calculated using the model). Currents were assumed to be negligible during the time represented by the model. Ecologically, too, the 1949 model was very simple: a linear food chain existed, involving phytoplankton, herbivorous zooplankton, their carnivores, and phosphate in the surface waters. These were in a steady

26. His technique was to set the average quantity of plankton predicted by the equation equal to an observable quantity, the average crop in nature (Riley 1952, p. 96).

27. Riley 1952, p. 91.

state, that is, their growth or increase was just balanced by consumption.[28]

The mathematical formulation of these ideas combined an equation like that used in 1946 with expressions developed by Sverdrup to analyze the vertical distribution of oxygen in the oceans.[29] In simple terms paraphrased from Sverdrup, this expressed the rate of change of a concentration (e.g., oxygen) as the result of diffusion minus advection plus biological effects. The trick was to solve the equation for biological effects; in a steady state (rate of change of concentration = 0), where current effects (advection) were negligible, they could be evaluated knowing only the rate of vertical eddy diffusion of the substance. While he and Stommel were developing their techniques, Riley experimented with the method in 1948 to calculate the phytoplankton production of the northern Sargasso Sea. In that warm, stable area his light and dark bottle experiments indicated that oxygen uptake (respiration) in the bottles was just about equal to its production, making calculations of the total production impossible at most times of the year. But phytoplankton pigments and oxygen were most abundant below the surface, representing a balance between photosynthetic production and its dissipation by vertical turbulence. This was just the problem that Sverdrup's equation was intended to solve. Using the measured oxygen gradient in the Sargasso Sea and estimating the coefficient of vertical eddy diffusion from the temperature gradient,[30] Riley calculated the biological oxygen production necessary to give the profile he had measured. From this it was a simple step to estimate primary production as carbon fixed by the phytoplankton.[31]

Using similar principles, Riley developed a series of equations for phytoplankton and phosphate concentrations in each of several depth intervals in the surface waters, along with single equations for the herbivores and primary carnivores. The carnivores, because they vertically migrated, were assumed to be present at each depth for equal amounts of time. The result was a set of equations in which the unknowns were the biological populations and the concentration of phosphate; they were controlled by the environmental factors light, temperature, vertical eddy diffusion, transparency (arbitrarily treated as a purely physical factor), and the amount of deep-water phosphate. Using computational tech-

28. For outlines of the model, see Riley 1952, pp. 97–98; Riley 1963c, pp. 444–447; Riley 1984, pp. 77–79.

29. Sverdrup 1938; more fully developed by Sverdrup et al. 1942, pp. 153–163.

30. The temperature gradient actually yields a coefficient of vertical eddy conductivity. For a discussion of early attempts to calculate eddy coefficients see Strickland 1960, pp. 126–127.

31. Riley and Gyorgy 1948, pp. 102–105.

niques unearthed by Stommel, the equations could be solved simultaneously once they were put in finite difference form and set equal to zero (representing a steady-state system). This was tedious—Riley estimated that in those precomputer days solving each pair of equations took 25 to 30 hours—but it yielded solutions giving the phytoplankton abundance, herbivore and carnivore populations, and phosphate levels in various parts of the Atlantic that matched the observed populations about 75% of the time. This was not more accurate than his earlier models, but it was the result of an approach that Riley thought had both greater generality and a better theoretical basis than his initial model of three years before.

The implications of the new model were actually greater than Riley, Stommel, and Bumpus's conclusion that

> the . . . theory states in effect that if in any one region one has numerical values for a few basic environmental factors—temperature, radiation, transparency, vertical eddy diffusivity, and the deep water nutrient concentration—it is possible to determine the concentrations of surface nutrients, phytoplankton, herbivore zooplankton, and its predators that will be supported by that region, at least . . . when they most nearly approximate the ideal conditions of a steady state.[32]

It could be extended to explain the annual plankton cycle by adding the dimension time to the matrix of equations used to predict spatial patterns. The solution of such a matrix would have been incredibly laborious in 1949, but the fact that it could be proposed shows how much Riley's quantitative approach had changed the largely qualitative study of the plankton cycle between 1938 and 1949.

With all its imperfections, Riley, Stommel, and Bumpus's "Quantitative Ecology of the Plankton of the Western North Atlantic" provided a fully oceanographic analysis of plankton distributions that might some day be applied to the full annual cycle in the surface waters. One enigmatic region it ignored was the deep water of the open oceans, where the ultimate oxidation of sinking plankton took place, and from which nutrients returned to the surface by upwelling, vertical diffusion, and advection at largely unknown rates. Brandt had recognized the significance of the abyss—its rich store of nutrients was the ultimate supply of the surface waters—when he saw the results from the expeditions on *Gauss*, *Planet*, and especially *Meteor* (see Chapters 3 and 5). But neither he nor anyone else had attempted to link deep-water nutrient supplies with surface production until Riley's work at Yale in the early 1950s,

---

32. Riley et al. 1949, p. 123.

resulting in the second of his great applications of physical oceanography to pelagic ecosystems.[33]

Once again, Riley's aim was to determine the unknown biological component in Sverdrup's theory of nonconservative properties, this time by estimating the magnitude of the physical processes governing the distribution of oxygen, phosphate, and nitrate in deep water. Once the biological component was known it could then be used with physical oceanographic variables to model the distribution of phosphate and nitrate in the Atlantic, allowing the validity of the original analysis to be tested against nature.

The original analysis, to determine biological rates of change, involved a complicated amalgam of the orthodox and unorthodox. Riley began by calculating the current patterns of the Atlantic from temperature and salinity distributions (a standard technique) using data averaged within squares 1000 km on a side, an innovation necessary to eliminate fluctuations of unknown origin in the small-scale pattern of currents. From the physical oceanographic calculations came rates of horizontal and vertical eddy diffusion that were necessary to calculate the effect of physical processes on oxygen and the nutrients. Then the only remaining unknown, the effects of organisms on the concentrations of oxygen, phosphate, and nitrate, could be calculated as the difference between observed concentrations and those to be expected as a result of the physical processes of diffusion and advection.

The results were intriguing biologically and yielded models of nutrient distribution that agreed moderately well with observational data. Certainly the most significant outcome was Riley's calculation that 90% of the organic matter produced by photosynthesis in the top 200 meters was oxidized there; the remaining 10% was oxidized in deep water, mainly by bacteria, resulting in the release of deep-water phosphate and nitrate. In general, biological activity decreased with depth (its index was the slow decline of oxygen with depth), but a few depths were especially interesting, for example, the mid-depth nitrate maximum and oxygen minimum. A subsidiary benefit of the analysis was Riley's demonstration that variations in the elemental ratios in seawater (Redfield ratios; see Chapter 9 near note 31) could be explained as a result of different patterns of biological activity, especially decomposition, assimilation, and excretion, modified by advection or diffusion.

The importance of this, the last of Riley's major contributions to the plankton dynamics of the open ocean, was not so much in the results (although their staying power has proved to be great) as in the originality

33. Riley 1951; 1984, pp. 88–92.

of his approach to a biological problem. His approach, using physical oceanographic techniques with chemical and biological data to solve intricate biological problems, provided the form that nearly all later analyses adhered to, at least in principle, if not in detail. The specific details of the analyses soon proved inadequate. In the late 1950s, using a new analysis of data from *Meteor*, Georg Wüst showed that currents in deep water were up to three times as great as Sverdrup and Riley had calculated. As a result, biological rates of change were probably significantly higher.[34] But the example of Riley's techniques, in both the 1949 and 1951 papers, survived the scientific inadequacies of the results. After 1951 quantitative biological oceanography that did not at least nod to physical oceanography became impossible.

In 1948, shortly before his work with Stommel and Bumpus was published, Riley returned to the Bingham Oceanographic Laboratory, making Long Island Sound once again the captive coastal sea of Yale oceanography. The Sound's accessibility made it an ideal testing area for ideas that had been brewing since his work there in 1938–1939. What factors governed the level of production from place to place? How efficiently was phytoplankton production transferred from one group of organisms to another? What was the importance of physical factors compared with biological ones in governing the level and pattern of production? And how could the amounts and fluctuations of fish production be accounted for ecologically? The questions—and their answers— were too complex, taken as a whole, to be considered using his synthetic modeling approach; what he called 'specialized studies" were needed to complement plankton modeling and to lay the groundwork for solutions to these "ultimate rather than immediate problems."[35]

At Yale this ideal could be realized with the help of graduate students and assistants, who made frequent field trips, and the help of the Milford Laboratory, where Riley's first work on the Sound had begun a decade earlier. In 1952, using his first major research grant, from the U.S. Office of Naval Research, he began sampling weekly in central Long Island Sound. Students and assistants collected phytoplankton, zooplankton, and benthic animals while Riley himself worked on the other biological, chemical, and physical variables. This program expanded to the whole Sound, sampled every two weeks between March 1954 and November 1955.[36] Not only could the results (which were published in 1956 and 1959) be compared with studies of other coastal areas, they showed how

34. Wüst 1955, 1958; Riley 1984, pp. 91–92.
35. Riley 1956a, pp. 9–10.
36. Riley 1984, pp. 101–106, 110–115.

quantitatively estimated physical processes affected the biological activity of a coastal ecosystem.[37]

In Long Island Sound, as on Georges Bank and in the Sargasso Sea, Riley took pains to separate the roles of physical and biological factors that were likely to control the abundance and production of organisms. Descriptive studies were not sufficient—quantitative dissection of the processes at work was a sine qua non of biological oceanographic analysis. This ideal lay behind his work on water exchange, the seasonal variation of plankton, and the budget of nutrients in the Sound. Taking this approach he added new dimensions to quantitative studies, such as H. W. Harvey's review of production in the English Channel, published in 1950, which concentrated entirely on estimating biological turnover but had nothing to say about how greatly the biology was affected by nonbiological variables such as eddy diffusion or advection.

By 1956, Riley had established that Long Island Sound was twice as productive as the English Channel—as long as only the phytoplankton was considered. But its zooplankton, benthos, and fish populations were scarcer and less productive in absolute terms, indicating that it was less efficient ecologically than the Channel. By using physical oceanographic analysis of changes in phosphate and oxygen, combined with the results of his early light and dark bottle experiments and new experimental work, he added new facets to Harvey's picture. Estimates of primary production, phytoplankton and zooplankton respiration, nutrient regeneration and uptake, details that Harvey could not deal with, were within reach by applying quantitative physical techniques to the data. Once the rates of change of oxygen and phosphate were divided between eddy diffusion (physical) and biological processes, the partitioning of biological activity among the phytoplankton, zooplankton, benthic animals, and bacteria was relatively straightforward.

Overall, the production of Long Island Sound proved to be limited by nitrogen supply from outside its boundaries, which itself was governed by a complex interplay between physical oceanographic factors and the sinking of the plankton.[38] Phytoplankton cells in Long Island Sound were demonstrably nitrogen limited, as Shirley Conover's work showed.[39] And Eugene Harris's work on the nitrogen cycle, before his untimely death in 1956, showed that at least half the nitrogen required by the phytoplankton was provided by ammonia excreted by the zooplankton; the remainder came from the benthos, bacterial regeneration, river runoff, and advection from outside the Sound. Harvey and Cooper, long

37. See especially Riley 1955, pp. 236–237; Riley 1956b; Riley 1956c, pp. 342–343.
38. Riley and Conover 1956, pp. 55–61.
39. Conover 1956, pp. 80–85.

before, had shown that phytoplankton used ammonia, but Harris's work quantified its importance.[40] It exemplified particularly clearly the kinds of results that Riley expected from his combined biological and physical analyses of marine ecosystems.

Riley's work in Long Island Sound was the natural outcome of his approach to scientific investigation. Patterns occurred in nature; the simpler ones could be analyzed or modeled mathematically, but more complex ones required detailed observational and experimental work that could take its place in, or be tested by, mathematical modeling. Although he never stated so explicitly, the Long Island Sound studies were the result of this long-evolved philosophical position about how scientific problems should be approached. His dissection and synthesis of the marine ecosystem in Long Island Sound were the culmination and logical outcome of his experimental, quantitative studies there at the beginning of his career, refined and amplified by his ability to use physical oceanographic techniques to help unravel complex biological problems. Riley never believed that plankton modeling was the only route to understanding. Detailed local studies were part of his plan to account for the different patterns of abundance and production evident in the sea.

The importance of Riley's Long Island Sound project was more than purely scientific. It also left a school of oceanographers. His ambitious scientific program required bodies and minds, many of them Yale undergraduates or graduate students from the zoology department, where G. E. Hutchinson was still teaching. Fortuitously, many of Riley's and Hutchinson's students, among them Georgianna B. Deevey, R. J. Conover, Shirley Conover, H. L. Sanders, A. G. Carey, and a number of others, got their degrees when oceanography in North America began to expand almost explosively in the late 1950s and early 1960s. As a result, oceanographers schooled in Riley's techniques and philosophical approach to oceanography, many of them deeply influenced by G. E. Hutchinson, just as Riley had been, moved on from Yale to the University of Rhode Island, Woods Hole Oceanographic Institution, Oregon State University, later Dalhousie University, the Bedford Institute of Oceanography, and many other institutions. While Harvey's influence had been realized through his scientific papers and books, and before him, Brandt had built a school of oceanography that eventually withered away, Riley had the satisfaction of seeing his scientific influence spread long after his work on Long Island Sound had ended.

The scientific and philosophical principles that governed Gordon Riley's work in biological oceanography are most evident in what he did,

---

40. On Harris and his work see Riley 1959. Harris's (1959) paper was written by Riley from notes and conversations.

rather than in what he said. Reticent in print, direct and to the point in conversation, he seldom gave easy access to his thoughts about principles. Instead they must be inferred from his experiences and his style of work. His statement that "the goal of quantitative ecological theory is to reduce the environmental factors to what we know are the real fundamentals—meteorological data and geographical facts—and use them to determine both the biological history of a local water mass and the modifying effects of animal migration and physical oceanographic dispersal"[41] is almost too explicit to be helpful. It hides a number of important intellectual threads that run through his work, uniting it with that of his colleagues, especially G. E. Hutchinson and H. W. Harvey.

The dominating motivation of Riley's work was his need to understand patterns in nature, especially the geographical and seasonal variations of plankton abundance. His quantitative approach to this problem, originally a reaction to descriptive, conceptually overloaded American ecology in the 1930s, became a well-thought-out theoretical position, based on parsimony, intended to get the maximum results from a minimum investment of scientific data and analytical effort. It was, after all, a struggle to get data at sea. It had to be used efficiently. Somewhat whimsically he stated in a symposium in 1952 that "ecologists are already swamped by the burden of routine observations, and every time we discover another important factor, the burden gets heavier. It is an attractive idea that we might get something for nothing, and theoretical analysis is practically free."[42] The time saved could be well spent, for example in detailed studies such as the Long Island Sound program.

To Riley, biological oceanography was marine population ecology,[43] a theme he made clearly explicit only in the 1950s after most of his theoretical work was finished. He began work on Georges Bank in 1939 to study its production, but the regularity of the seasonal cycle captured his attention. If environmental factors and physiological processes were correlated, the pattern of these marine populations could be analyzed and made comprehensible. Statistical analysis provided the answers for a time, but he found far greater satisfaction in taking over Fleming's equation, expressing the relations of phytoplankton and zooplankton as prey and predator, because of its similarity to the mathematical approach taken by Vito Volterra in his well-known paper of 1928.[44] Riley's first plankton model was deliberately based on the Volterra prey equation, in which ecological variables replaced the fixed coefficients, thus adapting

41. Riley 1953b, p. 222.
42. Riley 1953b, p. 222. The paper was delivered in Ottawa in January 1952.
43. Riley 1952, p. 79.
44. Volterra's work, dating from 1926, reached many marine biologists in 1928 as an English translation in the *Journal du Conseil International pour l'Exploration de la Mer*.

the equation to the patterns of geographical and seasonal change. He explained his approach succinctly. "In short, the prey equation for phytoplankton was rewritten in terms of physiological constants and six environmental factors: solar radiation, transparency, depth of the mixed layer, phosphate, temperature, and zooplankton."[45] The prey equation was an understandable starting point in the mid-1940s. Population ecology (nowhere more clearly understood than at Yale) was on the rise, beginning its pervasive influence on ecological thought. It was both scientifically fashionable and apparently rigorous. But, quite aware of what he was doing, Riley abandoned the analytical simplicity of the prey equation in 1946, preferring biological and physical reality to an a priori formulation.

His opportunistic conversion of the prey equation into a biological oceanographic model was fully consistent with principles Riley expressed when writing up his thesis research years before. Theories, based on the skill and knowledge of the scientist, always lay behind any investigation. Facts were then matched to theory—and if the match failed, theory could be adjusted, or the facts reexamined for accuracy. Statistical testing was an invaluable tool in this process of match and test, hypothesize and prove. But the most rigorous mathematical approach could hardly express the full complexity of nature; species were too many and their interrelationships too multifarious to be dealt with without shortcuts. Statistical analysis helped to sort the ecological wheat from the chaff of nature, but sophisticated guesses were necessary if any quantitative analyses of nature were to be possible.

> The relationships were too intricate to be understood by qualitative reasoning. The alternative was some method of quantitative ecological analysis, but it seemed impossibly difficult to apply quantitative methods to a population containing hundreds of species. It was expedient to simplify the analysis to a consideration of a few ecological groups which could be correlated with their environment by statistical or other methods. Such an analysis proceeds on the working hypothesis that the ecological group behaves as a unit, a sort of superspecies. . . . By such methods, it has been possible to derive general principles of population ecology that could not have been obtained in any other way.[46]

This pragmatic convenience removed many of the complexities of dealing with individual species.

In fact, the simplicity of Riley's approach was more apparent than real.

---

45. Riley 1953b, p. 218.
46. Riley 1952, pp. 85–86. See also Riley 1953b, p. 217. Riley was an unabashed group-selectionist throughout his life, although he deprecated many of the "holistic" concepts of American ecology prevalent in the 1930s.

Methodologically he viewed biological processes in the sea as the result of a few governing physical factors that affected the rates and limits of physiological processes such as photosynthesis, respiration, and grazing. But, as Evelyn Hutchinson had long taught, organisms should not be separated from the biogeochemical nexus of which they are only a limited part. It was a methodological necessity to assume that "a mere handful of meteorological and oceanographic factors will determine the major features of seasonal and regional plankton variations."[47] But while this might well be true of open-ocean ecosystems, the subjects of his first modeling studies, the analysis of real ecosystems with more than one or two trophic links was well beyond the limits of his technique in 1949. Even in 1963, when Riley reviewed and extended his modeling studies to include secondary carnivores—he called them "potential tuna"—he viewed the results more as an exploration of methods than as a simulation of ecological reality.[48] He knew that any ecological relationships beyond the complexity of a short Eltonian food chain lay well beyond the power of his methods.[49] Nonetheless, the power of Riley's analytical methods and experimental techniques to simulate nature, leading to new ideas about the balance of factors controlling plankton abundance, was impressive proof that his quantitative approach worked and was on the right track, even if the full complexity of nature was out of reach.

Riley's use of mathematical analyses was intended to bring together all the relevant information he could find from laboratory and field. He summarized his approach to the synthetic, that is modeling, technique as follows:

> The synthetic method brings a wide variety of information to bear on a particular ecological problem. It utilizes all available knowledge about physiological processes as determined from laboratory and field experiments. It combines this information with environmental data in such a way that the overall accuracy of the results is easily determined, although the reasons for the errors may remain obscure. Thus the application of the synthetic method fuses all the diverse ecological methods into a whole.[50]

He took pains to make it clear that his models were not intended to be predictive but were, as he said, "a way of thinking."[51] If they were kept as simple as possible, provided with realistic data from nature, and not extended beyond their limits, they led directly to further problems and

47. Riley 1953b, p. 222.
48. Riley 1963c, pp. 456–462.
49. Riley 1963a, pp. 230–231.
50. Riley 1952, p. 96.
51. Riley 1963a, p. 227.

new solutions. They were an important part of the whole scientific process, which he regarded as a form of artistic expression.[52] To critics, for instance, J. D. H. Strickland, who complained that models restated the obvious, Riley replied that the models clarified thought, but that they should be combined with the detailed study of biological processes to yield the best understanding of the complexity of biological relationships. To believe that his mathematical modeling expressed the self-evident was to misunderstand his approach to problem solving in science.[53]

In many ways Gordon Riley and H. W. Harvey were kindred spirits. Early in his work Riley consciously adopted the careful, analytical approach that was evident in the Plymouth group's work, and was inspired by Harvey's books. They corresponded extensively, especially about Riley's work in Long Island Sound and on Georges Bank; but despite their mutual regard, Harvey and Riley never met.[54] Their work shows many similarities in their approaches to scientific knowledge and some interesting differences.

The view that the natural world was closely integrated characterized both Harvey's and Riley's work. Each believed that nature was full of complex interdependencies that had to be approached using scientific simplification, despite its limitations. Mathematics and physics were the keys to this approach because of their problem-solving power, but they were only means to understanding the balance of biological interrelations. Riley, in particular, assumed, and probably fully believed, that nature was in a balanced state; his models frequently used the assumption that marine ecosystems were in equilibrium. It was a useful heuristic device, simplifying the solution of his models, but it was also an axiom of his understanding of nature. To Harvey also, balance in nature was real, but it was precarious, frequently disrupted by changes of the physical and biological environment. While Riley focused broadly on the seasonal regularities of the planktonic environment, Harvey frequently became fascinated by the biological consequences of imbalance, which were often outside the limits imposed by tractable mathematical models.[55]

Pattern in nature preoccupied Riley throughout his career. The fascinating differences of plankton standing crops that he observed from place to place were usually the subject of his work. He explored their

52. Riley 1984, p. 80.
53. See the discussion in Riley 1963a, pp. 227–229.
54. Riley 1984, p. 76.
55. Harvey 1955, pp. 114–120. It was a short step for Riley from his views on equilibrium to a belief that community evolution, which involved group selection not individual selective advantage, was an important process in the evolution of marine communities. See especially Riley 1963b, pp. 380–381, and Riley 1967, pp. 196–201. He never overcame this breach of Darwinian orthodoxy.

causes quantitatively, working single-mindedly on planktonic systems until he believed he had reached the limits of his methods; only then did he expand his work to study whole marine ecosystems such as Long Island Sound. Equally concerned with causes in nature, Harvey explored the control of plankton growth in the laboratory, choosing experiment rather than mathematical analysis. Little by little he began to develop broadly quantitative syntheses of the annual production in the English Channel which were far more general than Riley's models, but which lacked the rigor and predictive capability of the models. As late as 1950, Harvey believed that the application of Riley's kind of analysis to the English Channel would be premature.[56] He preferred to gain generality even though he had to sacrifice predictability. "There results," Harvey wrote, "a picture crude and in part indefinite, of the quantitative relations between the different ecological groups and of how change of circumstance may affect the populations."[57] For his part, only when the limits of his quantitative methods had been reached did Riley expand his work in Long Island Sound to a study of the whole ecosystem. Drawing explicit comparisons with Harvey's synthesis of knowledge about the English Channel, he left behind his plankton models but retained the mathematical tools of analysis that were so distinctive in his plankton studies. Harvey's and Riley's work converged in ecosystem studies in the mid-1950s despite their different starting points and approaches to the conditions of planktonic life.

Harvey's syntheses of research on the English Channel, emphasizing the fluctuating balance of forces leading from phytoplankton to fish, are typical of work at Plymouth in the Allen tradition. E. J. Allen's Plymouth Laboratory exerted no overt pressures on its members, but the ideal Allen established was to explain the production of the Channel so that the fluctuations of the fishery would become explicable. Food-chain modeling of Harvey's rough, moderately quantitative kind was an important step in this direction and a logical outcome of the expectations that Allen, and later F. S. Russell, had of work at the laboratory. Harvey responded to the intellectual and practical demands made of him; by the 1950s his work was emblematic of the Plymouth style. By contrast, Gordon Riley's work developed almost entirely within an academic framework. During the 1930s and 1940s at Yale and Woods Hole Oceanographic Institution, Riley experienced few external pressures. His zeal to reform ecology, which he shared with Evelyn Hutchinson, shaped by mathematical analysis and theoretical ecology, grew into a highly individual approach

56. Harvey 1950, p. 111.
57. Harvey 1950, p. 97.

to solving problems of rather limited scope in marine ecology. His most distinctive work, the quantitative models he devised between 1946 and 1951, were attempts to test the implications of biological and physical interactions and to sort out the causal from the noncausal factors governing plankton abundance. As a result, Riley's work was more mathematical, more predictive, but of necessity more narrowly aimed than Harvey's. Although they shared many of the same ideals, Harvey and Riley took distinctively different routes toward the convergence of their work in the study of large-scale marine ecosystems that took place in the mid-1950s.

Mathematical modeling of plankton dynamics and the introduction of physical oceanographic techniques into biological oceanography were undoubtedly Gordon Riley's greatest scientific achievements. But, ironically, Riley's work has had little impact on pure ecology, which he consciously set out to reform. Even modern accounts of the background of ecology do not acknowledge his role as a marine population ecologist, nor as one of the most creative minds to enter the field in the past five decades.[58] Riley often felt isolated and misunderstood. Even after the triumph of his modeling efforts in the 1950s he was never elected to the National Academy of Sciences.[59] Although he and Harvey rank as the two most influential biological oceanographers of their era, his work has had surprisingly little impact outside oceanography and is often still underappreciated within it.

The mixed success of Riley's work is due to both personal and scientific factors. His greatest originality lay in fusing biology and physics in a way that lay outside the developing disciplinary structure of oceanography. During the 1930s and 1940s very few biologists could handle the mathematics Riley used, and very few physical oceanographers were interested in biological problems. For a few fortunate years at Woods Hole Oceanographic Institution, under Bigelow and Iselin, he worked in a small, informal group in which unorthodox ideas could be discussed and developed. Once Riley returned to Yale he found himself alone; even had he stayed at Woods Hole, however, he would have been isolated by the boundaries that rapidly developed in oceanography during the 1950s and 1960s. Admittedly, too, his personality contributed to Riley's isolation. He was basically a loner; when he sought company it was frequently individualistic friends such as F. C. Fuglister, Henry Stommel, and R. B. Montgomery, not empire builders or group leaders. Riley was never a popularizer nor an entrepreneur. Even as a university administrator later

58. Notably Egerton 1983, 1985, and McIntosh 1985.
59. Riley 1984, p. 1.

in his life he kept up individual work that he discussed with only a small circle of friends.[60] Had he promoted his work more vigorously his material success would certainly have been greater.

Nearly all of Riley's publications during the formative years of his career were in publications that were not likely to promote revolutionary developments in biological oceanography to ecologists. While they were certainly respectable journals, Yale's *Journal of Marine Research* and *Bulletin of the Bingham Oceanographic Collection* did not have the prestige of, say, MIT's *Papers in Physical Oceanography and Meteorology*, not to biologists of the journal *Ecology*. They were too much like house journals. Presumably Riley chose them for convenience because they would publish his lengthy, often highly mathematical, monographs. As a result, ecologists, who should have appreciated his papers the most, either could not understand them or did not see them.

Finally, I believe that Riley's exposition of his work, combined with the other factors, was partly responsible for the limited recognition it received. His prose has a deceptive lucidity, but the results of his work are not necessarily transparent. Frequently his analyses are hard to follow; logical leaps occur that are not fully explained. Detailed calculations are often necessary to follow his arguments—a situation that is familiar to mathematicians and physicists but that has never been welcomed by biologists. He seldom popularized, although when he did so, the logic, clarity, and beauty of his scientific ideas are evident.[61] His audience was nearly always other numerate scientists, few though they were, who cared about quantifying biological oceanographic processes, not the ecologist who dabbled with a few statistical analyses of data.

Despite the limited recognition that Riley's work received outside biological oceanography, within the field it was quickly accepted as a new standard of quality. The unprecedented originality of his publications between 1946 and 1951 lay mainly in their mathematical approach and their sophistication. Once he leaped beyond the use of statistical generalizations, Riley applied the power of mathematical physical oceanography to problems that were still resolutely qualitative, even in the hands of the Plymouth group. Riley's approach to data also distinguished his work. Never afraid to work with simple data, he used the limited information available in innovative and ingenious ways to test hypotheses. He never regarded his successes or his failures as ultimate; they were merely stepping stones to the conceptual changes he expected when data, analysis,

60. Riley told me once that during university committee meetings he wrote scientific papers to escape the tedium.
61. See esp. Riley 1952, 1953b, 1972.

hypotheses, and interpretation improved. Finally, although it would have been easy for him to be sidetracked into recondite mathematical modeling, Riley always kept his analyses close to the data and to simply conceived biological problems. Linear food chains of the Eltonian kind were far too simple to represent natural complexity, but they were likely to result in solutions. To bog down in biological complexity was no progress at all.

It is easy to discern in the development of Riley's work up to 1951 a hidden agenda of opposition to the nonmathematical tradition in American ecology. But by the 1950s that battle had been won. What persisted in his work was an interest in the patterns of nature. During the 1960s American ecologists' enthusiasm for energy diverted attention from Riley's kind of analyses to ones involving rates of flow and the partitioning of energy. But, even in his analyses of production, Riley always betrayed his underlying interest in the factors governing the patterns of standing crop, production, and biological interaction. By 1951 this, not the battle to establish quantitative methods, was the feature that distinguished his always individualistic philosophy of scientific activity. Even though Harvey's descriptive approach rather than Riley's analytical one established the pattern for many later analyses of marine ecosystems,[62] it was Riley who accomplished the difficult task, envisioned first by Nathansohn and Gran, of uniting the physics of the sea with its biology.

---

62. Beginning with the highly influential book by Steele (1974), which drew a great deal of its information from Harvey's summaries of work in the English Channel.

Conclusion:
The End of One Tale
and the Beginning of Another

The story that began with Victor Hensen's wish to calculate the production of the seas has ended with my account of how mathematical physical oceanography came to be used in a new branch of science, biological oceanography. Little by little, more by punctuated equilibrium than by the gradual accumulation of variations or by orthogenesis, certainly not by any linear progression, the nitrogen cycle and chemical control, vertical circulation and light, algal growth, and the stability of the surface waters began to play roles in explanations of the abundance and production of plant cells in the sea. Victor Hensen's ideal, an estimate of marine production that would allow the abundance of fishes to be calculated or predicted, was still elusive in 1960, as it is now. What is clear is that by 1960 a standard view of how that search should be conducted and the kind of scientific form it should take had evolved, based mostly on the work of the Kiel school, Nathansohn, Gran and Braarud, the Plymouth group, and Gordon Riley.

Riley's work after 1957 and the effect that his mathematical modeling had on the marine science community provide excellent evidence that a new, stable view of how biological systems operate in the ocean had been reached by 1960. In Chapter 11, I described how Riley returned to detailed studies of Long Island Sound once he had reached the limits of his modeling attempts. This was not merely a strategic redeployment of forces. It also reflected Riley's belief that the accepted scheme of how the plankton cycle was controlled, reasonable though it was, required new empirical information before it could be made more realistic. He quickly became involved in a wholly new, rather distantly related line of research, the ecological role of nonliving particulate organic matter, that took up most of the rest of his research time.[1] Occasionally, though, Riley returned briefly to plankton modeling, for example in 1957 when he defended his estimates of production in the Sargasso Sea against Einar Steemann Nielsen's measurements using a new method, uptake of the radioisotope [14]C.[2] Once again he returned to plankton modeling in 1965, showing how a simple model of regional variations in plankton could be made more realistic by allowing the depth of the euphotic zone to vary inversely with the abundance of phytoplankton, as always occurred in nature. This time, however, his work was an extension of a new model, based on Riley, Stommel; and Bumpus's "Quantitative Ecology" of 1949, that had been devised by the Scottish mathematical biologist John Steele (b. 1926) during the 1950s.[3]

Steele's work drew its inspiration directly from Riley's most advanced modeling and was founded indirectly on the same body of theory and experimental results that had been brought together by Riley, Stommel, and Bumpus, that is, on the widely accepted theory of plankton dynamics which traced its roots to Brandt, Gran, and the Plymouth group. It grew directly out of work by the Aberdeen Marine Laboratory on herring surveys beginning in 1951, which by the mid-1950s was attempting, in Steele's words, to provide "connexions between plankton and fishing theory" on the Fladen Ground, a fishing bank in the northern North Sea.[4] First, by accounting for nutrient regeneration and using a Sverdrup-type approach (as Riley, Stommel, and Bumpus had done) to calculate biological changes, Steele estimated the annual production on and around the Fladen Ground. Then he calculated the seasonal changes of phytoplankton in the water column in terms of their populations, nutrients (he used phosphate), vertical turbulence, and grazing, adding a temporal component to Riley, Stommel, and Bumpus's steady-state model. Two years later Steele applied this work to the whole North Sea, intending, just as Riley had done, "to see to what extent there is consistency between the various modes of approach and to use the inconsistencies to suggest further investigations."[5] This time he used his two-layered model to simulate and explore the consequences of mixing across a thermocline at 40 meters depth.[6] The result was a mathematical model of the seasonal

1. Summarized in Riley 1970.
2. Riley 1953a, 1957; Steemann Nielsen 1952, 1954. For an outline of the controversy see Peterson 1980.
3. See Steele 1956, 1958, 1959.
4. Steele 1956, p. 1.
5. Steele 1958, p. 3.
6. The model was based on the idea that production occurred above the thermocline, and nutrient regeneration (but no primary production) below it.

cycle showing that phytoplankton production and populations in the North Sea were not controlled directly by nutrient exhaustion but by grazing and mixing. Steele arbitrarily but fruitfully chose to simplify the plankton system in his models so that the effects of single factors could be explored, bringing problems into focus and highlighting the importance of variables such as tidal mixing, the effect of currents, and the effect of differences in zooplankton behavior. It was this adaptation and development of his work that brought Riley back, although briefly, to plankton modeling in 1965, when he explored the effects of varying the depth of the euphotic zone in a model based directly on Steele's work.

Steele and Riley held the same philosophy of investigation. Steele expressed this clearly in 1958.

> To set up a mathematical model of the process of production, it is necessary to put some of these loosely defined relations [such as the interplay of light, nutrients, mixing, and grazing] into a rigid structure and to neglect the others completely. Such a procedure appears arbitrary, but it is only in this way that the logical consequences of these relations can be explored.[7]

The scientist had to choose intelligently the factors to be investigated.

> There is a basic choice about what is important and what is unimportant. In this paper [1958], the bias is towards a very simple set of postulates which can be fitted into a mathematical scheme and used to explain events in a chosen area. . . . It is certain that the simplicity described by the postulates will not in fact be realised; the use of this approach is to discuss merely the extent to which it holds. It can point to the conditions when it breaks down and suggest the methods and alternative hypotheses for further investigation.[8]

Thus, using just the kind of brush-clearing logic of investigation that Riley had begun to expound in 1938, Steele carried Riley, Stommel, and Bumpus's 1949 approach into new areas and in new directions. Although the areas and directions were new, Steele's mathematical techniques and conceptual approach were based firmly on the foundation that Riley had established between 1938 and 1949, itself derived from the Kiel and Plymouth work between 1897 and 1935. Even though some of the details could be troublesome, the theoretical structure of plankton dynamics seems to have solidified by the late 1950s, if we judge by Steele's development of Riley's ideas and by the way other biologists took up both the

7. Steele 1958, p. 3.
8. Steele 1958, p. 34. Similar ideas are expressed in Steele 1961, pp. 524–525, 534–535.

presuppositions and the mathematical approaches introduced by Riley.[9] Enriched detail, not radical conceptual change, was the concern of plankton modelers by and after 1960.

Further evidence that a stable theory of the plankton cycle had been accepted by 1960 appears in Steele's review of the factors governing primary production which was published in 1959.[10] Despite the theory's successes, mathematical and qualitative, there were many unknown or problematic factors in 1959. The relationship between carbon and chlorophyll in phytoplankton cells was so variable that it was difficult to find a common currency for biomass or production estimates. Very little was known about the response of most phytoplankton cells to light, so that it was unclear if a single or many photosynthesis: light intensity (PI) curves would have to be determined. What was the level of plant respiration, and could it be considered constant in modeling or metabolic calculations? And how could one best express the uptake of nutrients by plant cells, including the complications caused by inhibiting levels and the previous history of the cells? The environmental setting of the cells also posed many unsolved problems, many of which had been overcome by unrealistic expedients in Riley's and Steele's models. Nutrient regeneration was such a problem; where, by what mechanisms, and at what rates did it take place during the seasonal plankton cycle? How could vertical mixing best be estimated and how did it affect nutrient supply to the euphotic zone? What environmental factors governed the varying sinking rates of the cells, thus controlling their nutrient and light relations? Grazing, accepted as a major controlling factor since at least the mid-1930s, still could not be measured reliably. Even primary production, the most basic of measurements, had become uncertain because of the discrepancy between estimates based on ^{14}C uptake and those on the light and dark bottle technique. How could this be overcome?[11] Even the

9. Including David Cushing (1955; 1959a, b) of the English Lowestoft Laboratory, who used elaborations of Riley's first model (1946) to simulate phytoplankton production, grazing, and the seasonal cycle. Cushing often hid variables such as photosynthesis, respiration, and turbulence in the coefficients of his equations (e.g., Cushing 1955, p. 74), just the reverse of Riley's dissection of causal factors out of the nexus of processes governing plankton crops. Unlike Riley and Steele, Cushing was more concerned with simulating changes in the plankton than in dissecting out causal factors.

10. Equally convincing is his 1961 review, the text of a talk at the First International Congress of Oceanography in New York in 1959.

11. Some of the answers were provided in another landmark review by J. D. H. Strickland (1920–1970) while he was working for the Fisheries Research Board of Canada in 1960. Ostensibly a review of primary production, it was actually a personal essay, incisive and idiosyncratic, that brought together biological oceanographic work, plant physiology, cell-growth kinetics, and a chemist's careful analysis of techniques. Strickland formalized much of

successful models, which deliberately evaded or ignored these problems, were very simple; they seldom could be extended beyond the grazing animals, or at best beyond the primary carnivores, so that models encompassing a whole food web from phytoplankton to fish seemed to be beyond reach.

All of these problems, which read like a prescription for much of the research in biological oceanography that occurred between 1960 and the early 1980s, were serious impediments to a truly realistic theory of plankton dynamics. Even the postulate that the sea was uniform enough to sample easily had to be questioned, for as Steele showed, the patchiness of plankton and of physical properties in the sea made it vital to select the appropriate areas and appropriate time and space scales of sampling so the results would not be seriously in error.[12] Similarly, the reasonable postulate that the time delays occurring between, say, the growth of phytoplankton and the growth of the herbivores feeding on them could be ignored in simple models was unrealistic. Steele showed that simple correlations were likely to seriously misrepresent biological relationships because of time lags and nonlinearities that had not been accounted for (indeed were often not even recognized) in simple models.[13] Clearly both a great deal of new oceanographic information and a theoretically well-informed approach would be needed if the shortcomings of mathematical plankton dynamics were to be overcome in the 1960s.

Yet even with these reservations, it is evident that the question by 1960 was not *what* was to be included in the theory of quantitative plankton dynamics, but *how* (and with what level of sophistication) it was to be done. The slow accumulation of information and the brief periods of synthesis between Victor Hensen's first studies and the theoretical models of the 1940s and 1950s had yielded a consistent, conceptually simple picture. Steele could have been speaking for his predecessors when he said in 1959 that

> it has been shown that the models are extremely simple mathematically; that the observables, phosphate, chlorophyll, dry weight, are only very rough indices of the variables nutrient, plant carbon, herbivore carbon; that the coefficients photosynthetic rate, sinking rate, grazing rate, have very wide limits, especially in experimental studies in the laboratory. Yet

the terminology used in primary production studies in this review. And as he did later (in discussion, Riley 1963a, pp. 227–229), he cast a baleful eye on plankton modeling, which he believed was data-poor and hence premature.

12. Steele 1961, pp. 525–535; see also Riley 1963c, p. 452.
13. Steele 1961, pp. 524–527.

the fact that these bases are sufficient for adequate descriptions of the main features of the changing plant life in the sea implies that the factors controlling this life must be simple in character and few in number.[14]

By design or by accident rather few of the features of plankton models used by Riley and Steele and suggested by Steele's review in 1959 have been the focus of work on plankton biology since the early 1960s. The brief description that follows is my view of the currents of change in biological oceanography that have followed their work. First, the photosynthetic characteristics (PI curves) of both laboratory and natural populations of phytoplankton have been intensively investigated, showing, among other features, that significant changes in the reactions of the photosynthesizing cells can occur quickly in response to changes in illumination and nutrient abundance. The cells' reactions to nutrients have been elucidated; not only do phytoplankton species compete for specific nutrients such as phosphate, silicate, nitrate, or ammonia, their rates and amounts of uptake may govern the types of species found during the seasonal succession. A radically new approach to the nutrient relations of the cells and to the whole subject of photosynthetic primary production was introduced in the late 1970s when it was recognized that production based on nutrients that enter the euphotic zone from deep water (e.g., nitrate) could be distinguished from that based on rapidly recycled nutrients such as ammonia released directly into the euphotic zone by zooplankton.

Since Riley's last modeling attempts, herbivore grazing has been studied with increasing sophistication, mainly in laboratory cultures. The filtering characteristic of herbivorous zooplankton proved to be easily characterizable as curvilinear relationships attributable to a mechanically operating filter. Even more recently, grazing has been shown to operate (at least for small animals) in the world of small-scale viscosity, so that most small zooplankton cannot truly filter water but must find ways to overcome its drag. Even without knowing the mechanisms of zooplankton feeding the problems of grazing that Harvey and Riley battled with have been solved to the extent that they are no longer particularly contentious.

Undoubtedly the most controversial topic between the 1960s and the 1980s in biological oceanography has been the use of ^{14}C-labeled compounds to estimate primary production, a technique that, it was hoped, would eliminate the tedious and error-prone light and dark bottle oxygen-change incubation method described in Chapter 5. Although [14]C

14. Steele 1959, p. 148.

compounds, products of nuclear reactors, were first used by plant physiologists and biochemists to investigate photosynthesis during the 1940s,
it took the ingenuity of the Danish pharmacologist E. Steemann Nielsen
to use ^{14}C uptake to measure primary production in the oceans. In the
early 1950s, notably on the Danish *Galathea* expedition of 1950–1952,
he made a number of estimates of phytoplankton primary production
that were significantly lower than those using the light and dark bottle
oxygen method. This controversy has still not been completely resolved,
but by attempting to reconcile the results obtained using the two methods, biological oceanographers have come to realize that methods using
different time scales—minutes to hours in the case of ^{14}C uptake, hours
to days in the case of the light and dark bottle methods—may not be
readily compared because they represent different scales of variation in
nature. The equilibrium populations that Riley postulated and the mosaic of changes and perturbations that Harvey believed existed in the
water column are the extremes of a continuum of variation in time and
space in the productive upper layers of the oceans. This was made even
more evident during the late 1960s and 1970s when rapid methods of
determining photosynthetic pigments such as chlorophyll were developed using first spectrophotometry, then fluorometry. Phytoplankton
biomass, measured as photosynthetic pigments (as Harvey had done
using his plant pigment units) proved to be just as patchy and variable as
the primary production measurements made in the surface waters. The
first plankton models of Riley and Steele provided a framework of investigation, but one that did not readily take into account the variation and
variability of nature.

Finally, and most recently, it is now clear that much of the primary
production in the oceans, especially the tropics, is carried out by very
small cells, the smallest of Lohmann's nanoplankton, now called picoplankton. These tiny cells, 2 μm or less in diameter, are surprisingly
abundant and may account for significant fractions of the total primary
production even at high latitudes. Students of phytoplankton primary
production, who in the past often concentrated almost entirely on large
cells such as diatoms and dinoflagellates, have had to adapt rapidly to the
fact that very small cells, practically unknown to them before the 1970s,
are of great ecological importance.

This brief summary of some of the changes in the study of phytoplankton dynamics since 1960 suggests that biological oceanography
has changed dramatically during recent years. At first glance this conclusion might be borne out by my experiences in teaching biological oceanography between 1971 and the late 1980s: the specific content of my
course has changed so much in less than twenty years that the 1971

course is virtually unrecognizable. But the differences in content are superficial rather than profound; biological oceanography still orients its thinking around the concepts developed by Hensen, Brandt, and the others, and attacks many of the same research problems that Riley and Steele outlined. What seems to have changed significantly in biological oceanography, as in all of oceanography, is the way science is done, and especially the way it is funded.

As I have tried to show, by 1960, biology, chemistry, physics, and mathematics had been diffidently, then confidently brought together. In fact, so many plankton models had been born and were proliferating by the late 1960s that they could be arranged in a taxonomic scheme.[15] In the decade after 1960, government funding transformed a great deal of science at sea from small-scale cruises to bandwagon, multiship operations in which the work was matched to and modified by computer-generated sampling schemes.[16] Group projects (accompanied by group strife as well as solidarity) and group funding by government agencies became common. "Relevance," the catchword of elected officials and bureaucrats, became more and more a criterion for funding and doing research. The lone oceanographer adrift on a sea of thought became an endangered species.

But this vignette, or perhaps caricature, of the current scene in biological oceanography should be tempered by a more realistic, historically based view of the conditions under which earlier biological oceanographers worked. Relevance is not new, just as environmental problems, including fishery failures, are not new. How and to what extent did the marine scientists at Kiel, in Plymouth, and in the United States during the development of plankton dynamics contend with, adapt to, or ignore the practical demands of their times?

As Chapter 1 shows, a strong stimulus for Victor Hensen's interest in marine production was the depressed economy of his native Schleswig-Holstein. As an influential Professor ordinarius in a German university he could use his position to influence the kind of research that was done, directing state funds into research on fisheries and on more basic topics through the Kiel Commission. Karl Brandt, Hensen's intellectual successor in these endeavors, became even more influential through his university post, which included control over the Zoological Institute and Museum, and through his membership in the German commission carrying out work for ICES. But Brandt found himself in an enviable position,

15. Patten (1968) describes the main plankton models, arranging them in sequences that lack much historical reality (see esp. figure 1).
16. Walsh (1972) outlines this new brand of research.

318   *Conclusion*

empowered to carry out basic marine research without direct attention to practical problems, mainly because Germany's work for ICES on fisheries problems took place elsewhere, in Helgoland, where the Königliche Biologische Anstalt Helgoland was established under Friedrich Heincke in 1892. Thereby Brandt and his group were freed to do basic research on the control of the plankton cycle without the need to justify their work in purely practical terms. Brandt and his associate Otto Krümmel, who was in charge of Germany's hydrographic work for ICES, frequently found their work limited by insufficient funds, but, viewed objectively, they were in an unusual and enviable position, empowered and enabled to do a great deal of pure research without any pressing need to justify it, thanks to Helgoland and its active program of fisheries research.

The situation in England was quite different. E. J. Allen found himself the director of a very small private laboratory that was dependent for life on a yearly government grant that only barely allowed its existence. Allen had well-defined and clearly articulated views of what marine research should do.

> Slowly . . . we are building up a true science of the fisheries. The direction and velocity of the currents, the differences in temperature and salinity of the water, and the variations in these factors from season to season and from year to year upon which the fluctuations of fish must very largely depend, are being gradually worked out and understood. The effects of wind and weather and of the varying amount of sunshine falling on the water in different years are questions which are being studied. Then again, the natural history of the fishes themselves is the subject of much research.

So it was in 1917. At issue in Allen's mind was "the question of how the primary or fundamental food supply of the sea is built up";[17] the sea was "a self-contained whole"[18] that required study on a broad front by scientists trained in quantitative disciplines. How could a small poorly funded laboratory carry out such an ambitious program?

Allen combined scientific vision, political skill, and a large element of good luck. The Plymouth Laboratory's early life was ensured when Great Britain became involved in ICES in 1902, for only the Fishery Board for Scotland and the Plymouth Laboratory had the ability or the credentials to do the requisite scientific work in the ICES area. Rather than continuing basic fisheries research only at Plymouth, Allen ensured that it would be unified, and noticed, at Lowestoft, where the laboratory's research on North Sea demersal fish, directed by Walter Garstang, was centered

17. Allen 1917, p. 383.
18. Allen 1917, p. 397.

between 1902 and 1910.[19] Lowestoft rapidly became known as the English laboratory for fisheries research, freeing the Plymouth scientists for more basic research.

At Plymouth, Atkins and Harvey shared the view that the sea could be "regarded as a pasture," its productivity dependent on light, nutrients, and unknown or partially known physical factors. But Atkins's early belief that phosphate uptake and fisheries production were directly linked[20] and Harvey's view that "we are dealing with *living organisms* which are subject to the resultant of so many physical laws that their behaviour is extremely complex"[21] were not likely to cut much ice with officials of the treasury or with the administrators of the Board of Agriculture and Fisheries. Although it might well be true, as Harvey claimed, that it was necessary to study "total populations" in the sea "because fluctuations in the total population are probably linked up with the economically important fluctuations of particular fish,"[22] Allen would have had trouble justifying such an approach on its own grounds. Practical results were a necessity.

Viewed from the 1980s, the history of Plymouth Laboratory in the 1920s is distinguished by the work of W. R. G. Atkins, H. W. Harvey, and the others whose work is described in Chapter 8. But from the viewpoint of an English fisheries administrator circa 1928, the situation was different. J. A. Borley, outlining Great Britain's contribution to ICES during the 1920s, conceded the growth in England of "a more lively acceptance of the need of fundamental researches, both physical and biological, and a fuller realisation of their ultimate utility,"[23] but he ignored entirely Atkins's and Harvey's remarkable early work. To Borley, it was the Lowestoft laboratory's work on plaice that distinguished the Plymouth Laboratory's contribution to marine science.

We must give credit to Allen for being able to maintain his laboratory under difficult political and financial conditions. His skill at eliciting basic research in Plymouth, fisheries research at Lowestoft, and political support (which meant money) in London was essential to the survival of the Plymouth Laboratory. He was the linchpin of its day-to-day and long-term operation, under conditions that would be recognizable to laboratory administrators in the 1980s—the need to conduct basic research in an indifferent or hostile financial and intellectual climate. Allen's ability,

19. The Lowestoft Laboratory was taken over by the English Board of Agriculture and Fisheries in 1910; see Chapter 7 at n. 39.
20. Atkins 1925a, p. 164.
21. Harvey 1928a, p. 164.
22. Harvey 1928a, p. 181.
23. Borley 1928, p. 153.

his scientific skill, and his plausibility in the world of affairs meant the difference between the success of the Plymouth Laboratory and its total failure.

At Yale, early in his career, Gordon Riley was probably more free of outside, utilitarian pressures, than any of his predecessors. As a graduate student under G. E. Hutchinson, then as a postdoctoral researcher, he was free to follow his own ideas about the controls on plankton abundance. Yale, particularly through its Bingham Oceanographic Laboratory, played a large part in Riley's independence, as did Woods Hole Oceanographic Institution later.

Harry Payne Bingham (1887–1955), industrialist and businessman, had been a Yale undergraduate. In middle age he became interested in marine biology, especially fishes, and began to develop a collection that soon developed into a private museum. Hiring specialists to document his collection, Bingham supported publication of the results in what became the *Bulletin of the Bingham Oceanographic Collection*. The collection was transferred, with an endowment, to Yale in 1930, to be directed by Bingham's friend, the Norwegian-American zoologist A. E. Parr (b. 1900).[24] It was Parr's and the Bingham Oceanographic Laboratory's research expeditions in the Gulf of Mexico that gave Riley his first opportunity to take his limnology to sea. The result was profound; Riley took research on plankton production in the sea as his life's work. A little later, the Bingham laboratory housed and supported his work on Long Island Sound, the beginnings of his research on Georges Bank, and published his monographs on plankton dynamics.

Woods Hole Oceanographic Institution provided a similar haven for Riley during the 1940s. The Oceanographic originated in Frank R. Lillie's belief in the mid-1920s that the United States needed a major oceanographic laboratory to match or surpass European oceanographic work.[25] The U.S. National Academy of Science's Committee on Oceanography, founded in 1927, chaired by Lillie, gave support and plausibility to this view, and its report, written by the committee's secretary, Henry Bryant Bigelow of Harvard, provided a blueprint for U.S. oceanographic work in the 1930s. The General Education Board, linked to the Rockefeller Foundation, provided the money, allowing Woods Hole Oceanographic Institution, directed by Bigelow, to become a reality in 1930. A year later the Oceanographic had a ship, the large steel ketch

24. Merriman 1956.
25. On the history of Woods Hole Oceanographic Institution, see Schlee 1973, pp. 273–280; Schlee 1980, pp. 49–50; and Burstyn 1980.

Atlantis, enabling scientists at the Oceanographic to do full-scale research on the open ocean.[26]

Bigelow had begun research on the oceanography of the Gulf of Maine in 1912, linking his work, of necessity, with the U.S. Fish Commission (later named the U.S. Bureau of Commercial Fisheries).[27] He had always found it difficult to get enough ship time; now work could be renewed on the Gulf, and on Georges Bank, with its important, highly fluctuating stocks of commercially exploited haddock. The Oceanographic's program of research on Georges Bank, cut short by World War II, provided Riley with regular cruises and the intellectual setting in which some of his most significant early work took place.

The Oceanographic, in its early years, as a private institution, was under no particular pressure to justify its programs of research. The presence of its neighbor, the laboratory of the U.S. Bureau of Commercial Fisheries, a few steps away down Water Street in Woods Hole meant that the pressure to do fisheries research in Bigelow's Oceanographic was insignificant; the work could be done by choice, not by necessity. Scientific curiosity appears to have been the main motivation of the scientists who staffed the new institution during its first busy summers.[28]

The war brought important changes to the oceanographic community and significantly changed the way that marine research was supported and conducted in the United States and overseas. Both Woods Hole Oceanographic Institution and the Scripps Institution of Oceanography rapidly became active centers of defense-related research, much of it supported by the United States Navy. In Washington, oceanographers were appointed to U.S. Navy agencies responsible for initiating and funding oceanographic research.[29] The outcome after the war was significant: an oceanographic community had been born that for a time had nothing to do.[30] Into the gap stepped the U.S. Navy's Office of Naval Research (ONR), established in 1946 and staffed by several young oceanographers who had been involved in directing naval oceanographic research during the war. ONR grants after the war allowed Woods Hole Oceanographic Institution and Scripps Institution of Oceanography to expand, and, because many of the officials of the granting agency were

26. On *Atlantis*, see Schlee 1973, pp. 275–280; Schlee 1978, 1980.
27. See esp. Brosco 1985 on Bigelow's work. The bureau is now the National Marine Fisheries Service.
28. Until war research began in the 1940s nearly all of the scientific staff of Woods Hole Oceanographic Institution returned to their universities at the end of each summer.
29. See esp. Revelle 1969, 1980, 1985, and 1987.
30. Schlee 1973, pp. 311–313.

oceanographers, the whole science of oceanography, not just its defense-linked branches, received support.[31] Gordon Riley's first large research grant, for work on the marine ecosystem of Long Island Sound, came from the ONR in 1952. By the 1960s regular grant support, much of it from agencies such as the ONR and the U.S. National Science Foundation (founded in 1950) was the norm, not an exception. Science in general, not just oceanography, was becoming firmly emplaced in, and highly dependent upon, state support. Private laboratories, individual financially unsupported research, and research dependent solely upon university resources, the kind familiar to Brandt and Allen (and to Riley in his early career), gave way to big science and the one-track strategy of the research grant.

These recent developments, since 1945, have not yet found their historian, though they have been profoundly important in the development of recent science, including oceanography. My aim, in conclusion, has been to give a sketch of how the world of affairs and the world of biological oceanography interacted from the time of Victor Hensen until the days when oceanography, including biological oceanography, was about to become big science. The core study of biological oceanography, plankton dynamics, still has elements that each of the pioneers would recognize as his own contribution despite the technological, economic, and social transformations of marine science that have taken place from the end of World War II to the 1980s. These modern transformations, sketched above from a personal viewpoint, must be another story. Mine ends here, having outlined how the changes of marine plankton in space and time were isolated from the welter of observable events in nature and interpreted as natural phenomena governed by relatively simple biological, chemical, and physical processes. Haltingly at first, then with rapid accelerations and decelerations, a quantitative theory of plankton dynamics reached stable maturity 70 years after Victor Hensen's great monograph *Über die Bestimmung des Planktons* first claimed that the sea, like the human body and its fluids, could be studied quantitatively.

31. Cochrane 1978, pp. 462–463; Pfeiffer 1949.

REFERENCES

Allee, W.C., and T. Park. 1939. Concerning ecological principles. Science 89: 166–169.

Allen, E.J. 1899. On the fauna and bottom deposits near the thirty-fathom line from the Eddystone grounds to Start Point. Journal of the Marine Biological Association of the United Kingdom 5: 365–542.

Allen, E.J. 1903. Director's report. *In* Report of the Council, 1902–1903. Journal of the Marine Biological Association of the United Kingdom 6: 639–656.

Allen, E.J. 1905. First report of the Council of the Marine Biological Association of the United Kingdom on work carried out in connection with the International Fishery Investigations. Journal of the Marine Biological Association of the United Kingdom 7: 383–390.

Allen, E.J. 1909. Mackerel and sunshine. Journal of the Marine Biological Association of the United Kingdom 8: 394–406.

Allen, E.J. 1914. On the culture of the plankton diatom *Thalassiosira gravida* Cleve, in artificial sea-water. Journal of the Marine Biological Association of the United Kingdom 10: 417–439.

Allen, E.J. 1917. Food from the sea. Journal of the Marine Biological Association of the United Kingdom 11:380–398.

Allen, E.J. 1922. The progression of life in the sea. Address to Section D (Zoology). Report of the British Association for the Advancement of Science 90: 79–93.

Allen, E.J. 1930. The origin of adaptations. The Hooker Lecture. Proceedings of the Linnean Society of London 141: 119–138.

Allen, E.J. 1938. Edward Thomas Browne, 1866–1937. Journal of the Marine Biological Association of the United Kingdom 22: 405–408.

Allen, E.J., and E.W. Nelson. 1910. On the artificial culture of marine plankton organisms. Journal of the Marine Biological Association of the United Kingdom 8:421–474.

Alter, P. 1982. *Wissenschaft, Staat, Mäzene: Anfänge moderner Wissenschaftspolitik in Grossbritannien, 1850–1920*. Stuttgart: Klett-Kotta. 327 pp.

Alter, P. 1987. *The reluctant patron. Science and the state in Britain, 1850–1920*. New York: Berg. 292 pp.

Andersson, K.A. 1949. Johan Hjort, 1869–1948. Journal du Conseil International pour l'Exploration de la Mer 16: 3–8.

Anonymous. 1874–1895. *Ergebnisse der Beobachtungsstationen an den deutschen Küsten über die physikalischen Eigenschaften der Ostsee und Nordsee und die Fischerei. Veröffentlicht von ,der Ministerial-Kommission zur Untersuchung der deutschen Meere in Kiel, Bd 1–20, Jahrgang 1873–1893*. Berlin.

Anonymous. 1884a. A plea for a national marine zoological survey. Nature 30: 25–26.

Anonymous. 1884b. Marine Biological Association of the United Kingdom. Nature 30: 40.

Anonymous. 1884c. The Marine Biological Association. Nature 30: 350–351.

Anonymous. 1884d. *Report of the meeting held in the rooms of the Royal Society, on March 31st, 1884, for the purpose of forming a society for the biological investigation of the coasts of the United Kingdom*. London: Hepburn and Co. 20 pp.

Anonymous. 1886. Second annual report of the Council of the Marine Biological Association of the United Kingdom. Nature 34: 177–181.

Anonymous. 1887. The history of the foundation of the Marine Biological Association of the United Kingdom, Old Series 1: 17–21.

Anonymous. 1888a. Opening of the Marine Biological Laboratory. Journal of the Marine Biological Association of the United Kingdom, Old Series 1: 125–141.

Anonymous. 1888b. The opening of the Marine Biological Laboratory at Plymouth. Nature 38: 236–237.

Anonymous. 1899. Conférence internationale pour l'exploration de la mer, réunie à Stockholm. Stockholm: Imprimerie K.L. Beckman. lvi + 28 pp., map.

Anonymous. 1905. General review. Rapports et Procès-Verbaux des Réunions, Conseil International pour l'Exploration de la Mer 3: 8–11.

Anonymous. 1928. La Grande Bretagne. 2. L'Ecosse. *In* H.G. Maurice (ed.). Rapport jubilaire (1902–1927). Rapports et Procès-Verbaux des Réunions, Conseil Intertional pour l'Exploration de la Mer 47(III): 171–182.

Anonymous. 1941. Cunningham, Joseph Thomas. P. 317 in *Who Was Who (1929–1940)*, vol. 3.

Apstein, C. 1896. *Das Süsswasserplankton. Methode und Resultate der quantitativen Untersuchungen*. Kiel and Leipzig: Lipsius und Tischer. 200 pp., 5 Tabellen.

Apstein, C. 1905. Die Schätzungsmethode in der Planktonforschung. Wissenschaftliche Meeresuntersuchungen, Abteilung Kiel, Neue Folge 8: 103–123.

Apstein, C. 1908. Übersicht über das Plankton, 1902–1907. Pp. 41–54 in *Die Beteiligung Deutschlands an der internationalen Meeresforschung. IV/V Bericht. I. Abteilung Keil*. Berlin: Verlag von Otto Salle.

Ardoynaud, A. 1875. Recherches sur l'ammoniaque contenu dans les eaux marines et dans celles des marais salants du voisinage de Montpellier. Comptes Rendus de l'Académie des Sciences de Paris 81: 619–621.

Armstrong, F.A.J., and H.W. Harvey. 1950. The cycle of phosphorus in the waters of the English Channel. Journal of the Marine Biological Association of the United Kingdom 29: 145–162.

Atkins, W.R.G. 1922a. The hydrogen ion concentration of sea water in its biological relations. Journal of the Marine Biological Association of the United Kingdom 12: 717–771.

Atkins, W.R.G. 1922b. The respirable organic matter of sea water. Journal of the Marine Biological Association of the United Kingdom 12: 772–780.

Atkins, W.R.G. 1923a. The hydrogen ion concentration of sea water in its relation to photosynthetic changes. Part II. Journal of the Marine Biological Association of the United Kingdom 13: 93–118.

Atkins, W.R.G. 1923b. Note on the oxidisable organic matter of sea water. Journal of the Marine Biological Association of the United Kingdom 13: 160–163.

Atkins, W.R.G. 1923c. The phosphate content of fresh and salt waters in its relationship to the growth of the algal plankton. Journal of the Marine Biological Association of the United Kingdom 13: 119–150.

Atkins, W.R.G. 1924. On the vertical mixing of sea-water and its importance for the algal plankton. Journal of the Marine Biological Association of the United Kingdom 13: 319–324.

Atkins, W.R.G. 1925a. The ocean regarded as a pasture. Marine Observer 2: 162–164.

Atkins, W.R.G. 1925b. On the thermal stratification of sea water and its importance for the algal plankton. Journal of the Marine Biological Association of the United Kingdom 13:693–699.

Atkins, W.R.G. 1925c. Seasonal changes in the phosphate content of sea water in relation to the growth of the algal plankton during 1923 and 1924. Journal of the Marine Biological Association of the United Kingdom 13: 700–720.

Atkins, W.R.G. 1926a. The phosphate content of sea water in relation to the growth of the algal plankton. Part III. Journal of the Marine Biological Association of the United Kingdom 14: 447–467.

Atkins, W.R.G. 1926b. A quantitative consideration of some factors concerned in plant growth in water. Part I. Some physical factors. Journal du Conseil International pour l'Exploration de la Mer 1: 99–126.

Atkins, W.R.G. 1926c. A quantitative consideration of some factors concerned in plant growth in water. Part II. Some chemical factors. Journal du Conseil International pour l'Exploration de la Mer 1: 197–226.

Atkins, W.R.G. 1928. Seasonal variations in the phosphate and silicate content of sea water during 1926 and 1927 in relation to the phytoplankton crop. Journal of the Marine Biological Association of the United Kingdom 15: 191–205.

Atkins, W.R.G. 1930a. Seasonal changes in the nitrite content of sea-water. Journal of the Marine Biological Association of the United Kingdom 16: 515–518.

Atkins, W.R.G. 1930b. Seasonal variations in the phosphate and silicate content of sea-water in relation to the phytoplankton crop. Part V. November 1927 to April 1929, compared with earlier years from 1923. Journal of the Marine Biological Association of the United Kingdom 16: 821–852.

Atkins, W.R.G. 1932a. The chemistry of sea-water in relation to the productivity of the sea. Science Progress no. 106, pp. 298–312.

Atkins, W.R.G. 1932b. Nitrate in sea-water and its estimation by means of diphenylbenzidine. Journal of the Marine Biological Association of the United Kingdom 18: 167–192.

Atkins, W.R.G. 1932c. Solar radiation and its transmission through air and water. Journal du Conseil International pour l'Exploration de la Mer 7: 171–211.

Atkins, W.R.G. 1945. Conditions for the vernal increase in the phytoplankton and a supposed lag in the process. Nature 156: 599.

Atkins, W.R.G., G.L. Clarke, H. Pettersson, H.H. Poole, C.L. Utterback, and A. Angström. 1938. Measurement of submarine daylight. Journal du Conseil International pour l'Exploration de la Mer 13: 37–57.

Atkins, W.R.G., and H.W. Harvey. 1925. The variation with depth of certain salts utilized in plant growth in the sea. Nature 116: 784–785.

Atkins, W.R.G., and H.H. Poole. 1930a. Methods for the photo-electric and photochemical measurement of daylight. Biological Reviews of the Cambridge Philosophical Society 5: 91–113.

Atkins, W.R.G., and H.H. Poole. 1930b. The photo-chemical and photo-electric measurement of submarine illumination. Journal of the Marine Biological Association of the United Kingdom 16: 509–514.

Atkins, W.R.G., and H.H. Poole. 1933a. The photo-electric measurement of the penetration of light of various wave-lengths into the sea and the physiological bearing of the results. Philosophical Transactions of the Royal Society of London B 222: 129–164.

Atkins, W.R.G., and H.H. Poole. 1933b. The use of cuprous oxide and other rectifier photocells in submarine photometry. Journal of the Marine Biological Association of the United Kingdom 19: 67–72.

Atkins, W.R.G., and H.H. Poole 1936. The photo-electric measurement of the diurnal and seasonal variations in daylight and a globe integrating photometer. Philosophical Transactions of the Royal Society of London A 235: 245–272.

Atkins, W.R.G., and E.G. Wilson. 1926. The colorimetric estimation of minute amounts of compounds of silicon, of phosphorus, and of arsenic. Biochemical Journal 20: 1223–1228.

Atkins, W.R.G., and E.G. Wilson. 1927. The phosphorus and arsenic compounds of sea water. Journal of the Marine Biological Association of the United Kingdom 14: 609–614.

Aulie, R.P. 1970. Boussingault and the nitrogen cycle. Proceedings of the American Philosophical Society 114: 435–479.

Aurivillius, C.W.S. 1896. Das Plankton des baltischen Meeres. Kungliga Svenska Vetenskapsakadamiens Handlingar 21, N(8), 78 pp., 2 Tabellen.

Aurivillius, C.W.S. 1898. Vergleichende thiergeographische Untersuchungen über die Plankton-Fauna des Skageracks in den Jahren 1893–1897. Kungliga Svenska Vetenskapsakadamiens Handlingar B, 30(3), 427 pp.

Austerfield, P. 1979. Swashbuckling zoologist. New Scientist 83: 529–531.

Backus, R.H. (ed.). 1987. *Georges Bank*. Cambridge, Mass.: MIT Press. x + 595 pp.

Baker, J.R., and C.M. Pitman. 1949. Bourne, Gilbert Charles. Pp. 93–94 in *Dictionary of National Biography: 1931–1940*.

Baker, R.A., and R.A. Bayliss. 1984. Walter Garstang (1868–1949): zoological pioneer and poet. Naturalist 109: 41–53.

Barnes, B. 1982. *T.S. Kuhn and social science*. London: Macmillan. xix + 135 pp.

Barnes, H. 1966. Foreword. P. 7 in H. Barnes (ed.). *Some contemporary studies in marine science*. London: George Allen and Unwin.

Bartrip, P. 1985. Food for the body and food for the mind: The regulation of freshwater fisheries in the 1870's. Victorian Studies 28: 285–304.

Baur, E. 1902. Ueber zwei denitrificierenden Bakterien aus der Ostsee. Wissenschaftliche Meeresuntersuchungen, Abteilung Kiel, Neue Folge 6: 9–21.

Beijerinck, M. 1890. Over lichtvoedsel en plastisch voedsel van lichtbakterien. Verslagen en Mededeelingen der Koninklijke Akademie der Wetenschnapen, ser. 3, 7: 237–238.

Beijerinck, M. 1901. Ueber oligonitrophile Mikroben. Centralblatt für Bakteriologie, Abteilung II, 7: 561–582.

Ben-David, J. 1971. *The scientist's role in society. A comparative study*. Englewood Cliffs, N.J.: Prentice Hall. xxvi + 209 pp.

Ben-David, J., and A. Zloczower. 1972. Universities and academic systems in modern societies. Archives Européennes de Sociologie 3: 45–84.

Benecke, W., and J. Keutner. 1903. Über stickstoffbindende Bakterien aus der Ostsee. Bericht der Deutschen Botanischen Gesellschaft 21: 333–345.

Bergmann, L. 1931. Über eine neue Selen-Sperrschicht-Photozelle. Physikalische Zeitschrift 32: 286–288.

Berkeley, C. 1919. A study of the marine bacteria, Straits of Georgia, B.C. Transactions of the Royal Society of Canada 13 (V): 15–43.

Bernard, F. 1937. Résultats d'une année de recherches quantitatives sur le phytoplancton de Monaco. Comparison avec les mers voisines. Rapports et Procès-Verbaux des Réunions, Conseil International pour l'Exploration de la Mer 105(III): 38–29.

Bernard, F. 1938. Cycle annuel du nannoplankton à Monaco et Banyuls. Paris, Annales de l'Institut Océanographique 17: 349–406.

Bernard, F. 1939. Etude sur les variations de fertilité des eaux méditerranéennes. Climat et nanoplancton à Monaco en 1937–38. Journal du Conseil International pour l'Exploration de la Mer 14: 228–241.

Bidder, G.P. 1935. Joseph Thomas Cunningham (1859–1935). Journal du Conseil International pour l'Exploration de la Mer 10: 245–248.

Bidder, G.P. 1943. Edgar Johnson Allen 1866–1942. Journal of the Marine Biological Association of the United Kingdom 25: 671–684.

Bigelow, H.B. 1926a. Plankton of the offshore waters of the Gulf of Maine. Bulletin of the United States Bureau of Fisheries 40(2): 1–509.

Bigelow, H.B. 1926b. Physical oceanography of the Gulf of Maine. Bulletin of the United States Bureau of Fisheries 40(2): 511–1027.

Bigelow, H.B. 1931. *Oceanography, its scope, problems, and economic importance*. Boston: Houghton Mifflin. vi + 263 pp.

Bigelow, H.B., and W.W. Welsh. 1924. Fishes of the Gulf of Maine. Bulletin of the United States Bureau of Fisheries 40: 1–56.

Birge, E.A., and C. Juday. 1922. The inland lakes of Wisconsin. Wisconsin Geological and Natural History Survey Bulletin 22. 259 pp.

Birge, E.A., and C. Juday. 1911. The inland lakes of Wisconsin. The plankton. I. Its quantity and chemical composition. Wisconsin Geological and Natural History Survey Bulletin 64, Scientific ser. no. 13. 222 pp.

Bjerknes, V. 1898. Ueber einen hydrodynamischen Fundamentalsatz, und seine Anwendung besonders auf die Mechanik der Atmosphäre und des Weltmeeres. Stockholm, Kungliga Svenska Vetenskapsakadamiens Handlingar 31(4): 1–35.

Blaxter, J.H.S., A.J. Southward, and C.M. Yonge. 1984. Sir Frederick Russell, 1897–1984. Advances in Marine Biology 21: vii–viii.

Blohm, G. 1968. Landwirtschaftwissenschaften. Pp. 230–247 in K. Jordan (ed.). *Geschichte der Christian-Albrechts-Universität Kiel, 1665–1965.* Band 6. *Geschichte der Mathematik, der Naturwissenschaften und der Landwirtschaftwissenschaften.* Neumünster: Karl Wachholtz Verlag.

Borchardt, K. 1973. The industrial revolution in Germany, 1700–1914. Pp. 76–160 in C.M. Cipolla (ed.). *The Fontana economic history of Europe. The emergence of industrial societies—1.* Glasgow: Fontana/Collins.

Borley, J.O. 1928. La Grande Bretagne. 1. L'Angleterre, In H.G. Maurice (ed.). Rapport jubilaire (1902–1927). Rapports et Procès-Verbaux des Réunions, Conseil International pour l'Exploration de la Mer 47(III): 152–170.

Borriss, H. 1956. Die Entwicklung der Botanik und der botanischen Einrichtungen an der Universität Greifswald. Festschrift zur 500-Jahr-Feier 1956. Wissenschaftliche Zeitschrift der Ernst-Moritz-Arndt-Universität Greifswald. 515 pp.

Böttiger, J., and A.G. Nathorst. 1928. A la memoire du roi Oscar II. *In* H.G. Maurice (ed.). Rapport jubilaire (1902–1927). Rapports et Procès-Verbaux des Réunions, Conseil International pour l'Exploration de la Mer 47(II): 1–13.

Bourne, G.C. [G.C.B.]. 1888. The opening of the Marine Biological Laboratory at Plymouth. Nature 38: 198–201.

Bourne, G.C. 1930. Edwin Ray Lankester, 1847–1929. Journal of the Marine Biological Association of the United Kingdom 16: 365–371.

Braarud, T. 1935. The Øst-expedition to the Denmark Strait, 1929. II. The phytoplankton and its condition of growth. (Including some qualitative data from the Arctic in 1930.) Hvalradets Skrifter 10: 1–173.

Braarud, T. 1956a. Haaken Haasberg Gran, 1870–1955. Journal du Conseil International pour l'Exploration de la Mer 21: 121–124.

Braarud, T. 1956b. In memoriam. Haakon Hasberg Gran, 1870–1955. Deep-sea Research 3: 232–233.

Braarud, T., and B.R. Føyn. 1931. Beiträge zur Kenntnis des Stoffwechsels im Meere. Avhandlinger utgitt av det Norske Videnskaps-Akademi I Oslo, 1930. 1. Matematisk-Naturvidenskapelig Klasse 14: 1–24.

Braarud, T., and A. Klem. 1931. Hydrographical and chemical investigations in the sea off Møre and in the Romsdalsfjord. Skrifter utgitt av Hvalraadet ved Universitetets Biologiske Laboratorium, Oslo. Hvalradets Skrifter 1: 1–88.

Brandt, K. 1885. *Die coloniebildenden Radiolarien (Sphaerotsen) des Golfes von Neapel.* Fauna und Flora des Golfes von Neapel 13: viii + 276 pp.

Brandt, K. 1887. Über *Actinosphaerium eichhorni.* Inaugural-Dissertation zur Erlangung der philosophischen Doctorwürde welche mit Genehmigung der Philosophischen Facultät der vereinigten Friedrichs-Universität Halle-Wittenberg. Halle. 54 pp.

Brandt, K. 1891. Haeckels Ansichten über die Plankton-Expedition. Schriften d. Naturwissenschaftliche Vereins für Schleswig-Holstein 8: 199–213.

Brandt, K. 1896. Ueber das Stettiner Haff. Wissenschaftliche Meeresuntersuchungen, Abteilung Kiel, Neue Folge 1: 105–141.

Brandt, K. 1897. Die Fauna der Ostsee, ins besondere die der Kieler Bucht. Verhandlungen der Deutschen Zoologischen Gesellschaft 1897: 10–34.

Brandt, K. 1898. Beiträge zur Kenntnis der chemischen Zusammensetzung des Planktons. Wissenschaftliche Meeresuntersuchungen, Abteilung Kiel, Neue Folge 3: 43–90.

Brandt, K. 1899a. Ueber den Stoffwechsel im Meere. Rede beim Antritt des Rektorates der Königlichen Christian-Albrechts-Universität zu Kiel am 6. März 1899. Kiel: Universitäts-Buchhandlung. 36 pp.

Brandt, K. 1899b. Ueber den Stoffwechsel im Meere. Wissenschaftliche Meeresuntersuchungen, Abteilung Kiel, Neue Folge 4: 215–230.

Brandt, K. 1901. Life in the ocean. Pp. 493–506 in *Annual Report of the Smithsonian Institution for 1900*. Washington, D.C.: Smithsonian Institution.

Brandt, K. 1902a. Ueber die demnächst beginnenden internationalen Untersuchungen der nordischen Meere. Pp. 290–295 in *Verhandlungen des V. Internationalen Zoölogischen-Congresses zu Berlin, 12–16 August 1901*. Jena: Verlag von Gustav Fischer.

Brandt, K. 1902b. Ueber den Stoffwechsel im Meere. 2. Abhandlung. Wissenschaftliche Meeresuntersuchungen, Abteilung Kiel, Neue Folge 6: 23–79.

Brandt, K. 1904. Ueber die Bedeutung der Stickstoffverbindungen für die Produktion im Meere. Beihefte zum Botanischen Zentralblatt 16: 383–402.

Brandt, K. 1905a. Bericht über allgemeine biologische Meeresuntersuchungen. Pp. 7–17 in *Die Beteiligung Deutschlands an der internationalen Meeresforschung*. I. Bericht. I. Abteilung Kiel. Berlin: Verlag von Otto Salle.

Brandt, K. 1905b. On the production and conditions of production in the sea. Rapports et Procès-Verbaux des Réunions, Conseil International pour l'Exploration de la Mer 3: appendix D, 12 pp.

Brandt, K. 1908. Bericht über allgemeine biologische Meeresuntersuchungen. Pp. 17–40 in *Die Beteiligung Deutschlands an der internationalen Meeresforschung*. IV/V Jahresbericht. I. Abteilung Kiel. Berlin: Verlag von Otto Salle.

Brandt, K. 1911. Die wichtigsten Ergebnisse der im Kieler Meereslaboratorium ausgeführten quantitativen Untersuchungen über die anorganischen Stickstoffverbindungen des Meerwassers. Rapports et Procès-Verbaux des Réunions, Conseil International pour l'Exploration de la Mer 13: 77–79.

Brandt, K. 1915. Über den Nitratgehalt des Ozeanwassers und seine biologische Bedeutung. Abhandlungen der Kaiser Leopoldinisch-Carolinischen Deutschen Akademie der Naturforscher, Halle (Nova Acta Leopoldina) 100(4): 1–56.

Brandt, K. 1920. Ueber den Stoffwechsel im Meere. 3. Abhandlung. Wissenschaftliche Meeresuntersuchungen, Abteilung Kiel, Neue Folge 18: 185–429.

Brandt, K. 1921a. Die zoologischen Arbeiten der Kieler Kommission, 1870–1920. Pp. 76–194 in *Festschrift der Preussischen Kommission zur wissenschaftlichen Untersuchung der deutschen Meer zu Kiel aus Anlass ihres 50 jährigen Bestehens*. Kiel and Leipzig: Lipsius und Tischer.

Brandt, K. 1921b. Ueber Menge und Bedeutung der wichtigsten Pflanzennährstoffen im Meere. Pp. 7–38 in *Festschrift der Preussischen Kommission zur wissenschaftlichen Untersuchung der deutschen Meer zu Kiel aus Anlass ihres 50 jährigen Bestehens*. Kiel and Leipzig: Lipsius und Tischer.

Brandt, K. 1925a. Bericht über allgemeine biologische Meeresuntersuchungen. Berichte der Deutschen wissenschaftlichen Kommission für Meeresforschung, Neue Folge 1: 7–9.

Brandt, K. 1925b. Die Produktion in den heimischen Meeren und das Wirkungsgesetz der Wachstumsfaktoren. Berichte der Deutschen wissenschaftlichen Kommission für Meeresforschung, Neue Folge 1: 67–102.

Brandt, K. 1925c. Geheimrat Dr. V. Hensen †. Berichte der Deutschen wissenschaftlichen Kommission für Meeresforschung, Neue Folge 1: vii–x.

Brandt, K. 1925d. Victor Hensen und die Meeresforschung. Wissenschaftliche Meeresuntersuchungen, Abteilung Kiel, Neue Folge 20: 49–103.

Brandt, K. 1927. Stickstoffverbindungen im Meere. I. Wissenschaftliche Meeresuntersuchungen, Abteilung Kiel, Neue Folge 20: 201–292.

Brandt, K. 1928. Die beiden Meereslaboratorien in Kiel. Rapports et Procès-Verbaux des Réunions, Conseil International pour l'Exploration de la Mer 47: 3–16.

Brandt, K. 1929. Phosphate und Stickstoffverbindungen als Minimumstoffe für die Produktion im Meere. Rapports et Procès-Verbaux des Réunions, Conseil International pour l'Exploration de la Mer 53: 5–35.

Brannigan, A. 1982. *The social basis of scientific discoveries.* Cambridge: Cambridge Univ. Press. xi + 212 pp.

Brattström, H. 1967. The biological stations of the Bergens Museum and the University of Bergen, 1892–1967. Sarsia 29: 7–80.

Brennecke, W. 1909. *Forschungsreise S.M.S. "Planet" 1906/07 herausgegeben vom Reichs-Marine-Amt*, Band 3:*Ozeanographie*. Berlin.

Broch, H. 1954. Sars, Georg Ossian, 1837–1927. Pp. 260–264 in *Norsk biografisk lexicon*, vol. 12.

Brosco, J.P. 1985. Henry Bryant Bigelow and the transformation of American oceanography. B.A. thesis, Univ. of Pennsylvania. 80 pp.

Browne, C.A. 1944. *A source book of agricultural chemistry.* Waltham, Mass.: Chronica Botanica.

Buch, K. 1920. Ammoniakstudien an Meer- und Hafenwasserproben. Helsingfors, Merentutkimuslaitoksen julkaisu, Havforskningsinstitutets Skrifter no. 2.

Buch, K., H.W. Harvey, H. Wattenberg, and S. Gripenberg. 1932. Über das Kohlensäuresystem im Meerwasser. Rapports et Procès-Verbaux des Réunions, Conseil International pour l'Exploration de la Mer 79: 1–70.

Buchanan-Wollaston, H.J. 1933. Some modern statistical methods: Their application to the solution of herring race problems. With a foreword by R.A. Fisher. Journal du Conseil International pour l'Exploration de la Mer 8:7–47.

Buchanan-Wollaston, H.J. 1935. The philosophic basis of statistical analysis. Journal du Conseil International pour l'Exploration de la Mer 10: 249–265.

Buchanan-Wollaston, H.J. 1936. The philosophic basis of statistical analysis. Journal du Conseil International pour l'Exploration de la Mer 11: 7–26.

Bulloch, W. [1938] 1960. *The history of bacteriology.* London: Oxford Univ. Press. xii + 422 pp.

Burstyn, H.L. 1980. Reviving American oceanography. Frank Lillie, Wickliffe Rose, and the founding of Woods Hole Oceanographic Institution. Pp. 57–66 in M. Sears and D. Merriman (eds.). *Oceanography: The past.* New York: Springer-Verlag.

Busch, A. 1962. The vicissitudes of the Privatdozent: Breakdown and adaptation in the recruitment of the German university teacher. Minerva 1: 319–341.

Busch, W. 1916. Ueber das Plankton der Kieler Föhrde im Jahre 1912/13. I. Teil. (Vorwort von K. Brandt). Wissenschaftliche Meeresuntersuchungen, Abteilung Kiel, Neue Folge 18: 25–142.

Büse, T. 1915. Quantitative Untersuchungen von Planktonfangen des Feuerschiffes "Fehmarnbelt" vom April 1910 bis März 1911. Wissenschaftliche Meeresuntersuchungen, Abteilung Kiel, Neue Folge 17: 229–279.

Calkins, G.N. 1893. The seasonal distribution of microscopical organisms in surface waters. Annual Report of the Massachusetts State Board of Health 24: 381–390.

Calman, W.T. 1927. Georg Ossian Sars. Proceedings of the Linnean Society of London, 139th Session, pp. 98–100.

Capone, D.G., and E.J. Carpenter 1982. Nitrogen fixation in the marine environment. Science 217: 1140–1142.

Cardwell, D.S.L. 1972. *The organization of science in England.* London: Heinemann. xii + 268 pp.

Carey, C.L. 1938. The occurrence and distribution of nitrifying bacteria in the sea. Journal of Marine Research 1: 291–304.

Carpenter, E.J., and D.G. Capone (eds.). 1983. *Nitrogen in the marine environment.* New York: Academic Press. xvii + 900 pp.

Carruthers, J.N. 1956. Donald John Matthews, 1873–1956. Journal du Conseil International pour l'Exploration de la Mer 22:3–8.

Chumley, J. 1918. *The fauna of the Clyde Sea area.* Glasgow: The University Press. vii + 200 pp.

Cittadino, E. 1981. Plant adaptation and natural selection after Darwin: Physiological plant ecology in the German Empire, 1880–1900. Ph.D. diss., Univ. of Wisconsin, Madison. iii + 260 pp.

Clark, A.M. 1952. William Leadbetter Calderwood, I.S.O. P. 78 in *Royal Society of Edinburgh Yearbook for 1951.*

Clarke, G.L. 1939. The relation between diatoms and copepods as a factor in the productivity of the sea. Quarterly Review of Biology 14: 60–64.

Clarke, G.L. 1942. Productivity of Georges Bank. Pp. 19–20 in *The Woods Hole Oceanographic Institution Report for the Year 1940.* Woods Hole Oceanographic Institution Collected Reprints 1941.

Clarke, G.L. 1946. Dynamics of production in a marine area. Ecological Monographs 16: 322–335.

Clements, F.E., and V.E. Shelford. 1939. *Bio-ecology.* New York: John Wiley and Sons. vi + 425 pp.

Cleve, P.T. 1896a. Microscopic marine organisms in the service of hydrography. Nature 55: 89–90.

Cleve, P.T. 1896b. Planktonundersökningar: vegetabilisk plankton. Kungliga Svenska Vetenskapsakadamiens Handlingar 22(III)(5): 1–33.

Cleve, P.T. 1897. *A treatise on the phytoplankton of the Atlantic and its tributaries and on the periodical changes of the plankton of Skagerak.* Uppsala. 28 pp., 15 tables, 4 pls.

Cleve, P.T. 1900a. The plankton of the North Sea, the English Channel and the Skagerak in 1898 and 1899. Kungliga Svenska Vetenskapsakadamiens Handlingar 32(VIII): 1–53.

Cleve, P.T. 1900b. *The seasonal distribution of Atlantic plankton organisms.* Göteborg: Bonniers Tryckeri Aktiebolag. 369 pp.

Cleve, P.T. 1902. *Additional notes on the seasonal distribution of Atlantic plankton organisms.* Göteborg: Bonniers Tryckeri Aktiebolag. 51 pp.

Cleve-Euler, A. 1928. Per Theodor Cleve. *In* H.G. Maurice (ed.). Rapport jubilaire (1902–1927). Rapports et Procès-Verbaux des Réunions, Conseil International pour l'Exploration de la Mer 47(II): 35–38.

Cochrane, R.C. 1978. *The National Academy of Sciences. The first hundred years, 1863–1963.* Washington, D.C.: National Academy of Sciences. xv + 694 pp.

Collard, P. 1976. *The development of microbiology.* London: Cambridge Univ. Press. vii + 201 pp.

Collins, H. 1981. The place of the 'core-set' in modern science: Social contingency with methodological propriety in science. History of Science 19: 6–19.

Committee on Fishery Investigations. 1908. *Report of the committee appointed to inquire into the scientific and statistical investigations now being carried on in relation to the fishing industry of the United Kingdom.* Parliamentary Papers 1908. XIII (Cd. 4628). London: HMSO. xx + 571 pp.

Committee on Ichthyological Research. 1902. *Report of the committee appointed to inquire and report as to the best means by which the state or local authorities can assist scientific research as applied to problems affecting the fisheries of Great Britain and Ireland, together with minutes of evidence, appendix, and index.* Parliamentary Papers 1902. XV (Cd. 1312). London: HMSO. xxv + 168 pp.

Conover, S.A.M. 1956. Oceanography of Long Island Sound, 1952–1954. IV. Phytoplankton. Bulletin of the Bingham Oceanographic Collection 15: 62–112.

Cook, R.E. 1977. Raymond Lindeman and the trophic-dynamic concept in ecology. Science 198: 22–26.

Cooper, L.H.N. 1933a. Chemical constituents of biological importance in the English Channel. Part I. Phosphate, silicate, nitrate, nitrite, ammonia. Journal of the Marine Biological Association of the United Kingdom 18: 677–728.

Cooper, L.H.N. 1933b. Chemical constituents of biological importance in the English Channel, November 1930 to January 1932. Part II. Hydrogen ion concentration, excess base, carbon dioxide, and oxygen. Journal of the Marine Biological Association of the United Kingdom 18: 729–753.

Cooper, L.H.N. 1933c. Chemical constituents of biological importance in the English Channel. Part III. June–December, 1932. Phosphate, silicate, nitrite, hydrogen ion concentration, with a comparison with wind records. Journal of the Marine Biological Association of the United Kingdom 19: 55–62.

Cooper, L.H.N. 1933d. A system of rational units for reporting nutrient salts in sea water. Journal du Conseil International pour l'Exploration de la Mer 8: 331–334.

Cooper, L.H.N. 1934a. The determination of phosphorus and nitrogen in plankton. Journal of the Marine Biological Association of the United Kingdom 19: 755–760.

Cooper, L.H.N. 1934b. The variation of excess base with depth in the English Channel with reference to the seasonal consumption of calcium by plankton. Journal of the Marine Biological Association of the United Kingdom 19: 747–754.

Cooper, L.H.N. 1935a. Iron in the sea and in marine plankton. Proceedings of the Royal Society of London B 118: 419–438.

Cooper, L.H.N. 1935b. The rate of liberation of phosphate in sea water by the breakdown of plankton organisms. Journal of the Marine Biological Association of the United Kingdom 20: 197–200.

Cooper, L.H.N. 1937a. The nitrogen cycle in the sea. Journal of the Marine Biological Association of the United Kingdom 22: 183–204.

Cooper, L.H.N. 1937b. On the ratio of nitrogen to phosphorus in the sea. Journal of the Marine Biological Association of the United Kingdom 22: 177–182.

Cooper, L.H.N. 1937c. Some conditions governing the solubility of iron. Proceedings of the Royal Society of London B 124: 299–307.

Cooper, L.H.N. 1938a. Phosphate in the English Channel, 1933–38, with a comparison with earlier years, 1916 and 1923–32. Journal of the Marine Biological Association of the United Kingdom 23: 181–195.

Cooper, L.H.N. 1938b. Redefinition of the anomaly of the nitrate-phosphate ratio. Journal of the Marine Biological Association of the United Kingdom 23: 179.

Cooper, L.H.N. 1938c. Salt error in determinations of phosphate in sea water. Journal of the Marine Biological Association of the United Kingdom 23: 171–178.

Cooper, L.H.N. 1948a. The distribution of iron in the waters of the western English Channel. Journal of the Marine Biological Association of the United Kingdom 27: 279–313.

Cooper, L.H.N. 1948b. Phosphate and fisheries. Journal of the Marine Biological Association of the United Kingdom 27: 326–336.

Cooper, L.H.N. 1948c. Some chemical considerations on the distribution of iron in the sea. Journal of the Marine Biological Association of the United Kingdom 27: 314–321.

Cooper, L.H.N. 1952. Processes of enrichment of surface water with nutrients due to strong winds blowing onto a continental slope. Journal of the Marine Biological Association of the United Kingdom 30: 453–464.

Cooper, L.H.N. 1955a. Deep water movements in the North Atlantic as a link between climatic changes around Iceland and biological productivity of the English Channel and Celtic Sea. Journal of Marine Research 14: 347–362.

Cooper, L.H.N. 1955b. Hypotheses connecting fluctuations in Arctic climate with biological productivity of the English Channel. Deep-sea Research 3(Supplement): 212–223.

Cooper, L.H.N. 1960. Obituary. W.R.G. Atkins, C.B.E., O.B.E. (Mil.), Sc.D., F.R.I.C., F. Inst. P., F.R.S., 1884–1959. Journal of the Marine Biological Association of the United Kingdom 39: 153–154, portrait.

Cooper, L.H.N. 1972a. Hildebrand Wolfe Harvey, 1887–1970. Biographical Memoirs of Fellows of the Royal Society 18: 331–347.

Cooper, L.H.N. 1972b. Obituary. Hildebrand Wolfe Harvey. Journal of the Marine Biological Association of the United Kingdom 52: 773–775.

Cooper, L.H.N., and D. Vaux. 1949. Cascading over the continental slope of water from the Celtic Sea. Journal of the Marine Biological Association of the United Kingdom 28: 719–750.

Cox, D.L. 1979. Charles Elton and the emergence of modern ecology. Ph.D. diss., Washington Univ. viii + 232 pp.

Cranefield, P.F. 1957. The organic physics of 1847 and the biophysics of today. Journal of the History of Medicine and Allied Sciences 12: 407–423.

Cranefield, P.F. 1966. The philosophical and cultural interests of the biophysics movement of 1847. Journal of the History of Medicine and Allied Sciences 21: 1–7.

Culotta, C. 1975. German biophysics, objective knowledge, and romanticism. Historical Studies in the Physical Sciences 4: 3–38.

Cunningham, J.T. 1896. Physical and biological conditions in the North Sea. Journal of the Marine Biological Association of the United Kingdom 4: 233–263.

Cushing, D.H. 1955. Production and a pelagic fishery. Ministry of Agriculture, Fisheries and Food of the United Kingdom, Fishery Investigations, ser. 2, 18(7): i–vi, 1–104.

Cushing, D.H. 1959a. The seasonal variation in oceanic production as a problem in population dynamics. Journal du Conseil International pour l'Exploration de la Mer 24: 455–464.

Cushing, D.H. 1959b. On the nature of production in the sea. Ministry of Agriculture, Fisheries and Food of the United Kingdom, Fishery Investigations 22: 1–40.

Cushing, D.H. 1966. *The arctic cod.* Oxford: Pergamon Press. 93 pp.

Cushing, D.H. 1975. *Marine ecology and fisheries.* Cambridge: Cambridge Univ. Press. xiv + 278 pp.

Cushing, D.H. 1982. *Climate and fisheries.* London: Academic Press. xiv + 373 pp.

Dakin, W.J. 1908a. Methods of plankton research. Proceedings and Transactions of the Liverpool Biological Society 22: 500–553.

Dakin, W.J. 1908b. Notes on the alimentary canal and food of the Copepoda. Internationale Revue der gesamten Hydrobiologie 1: 772–782.

Dakin, W.J. 1909. The filtration coefficient of plankton nets. Proceedings and Transactions of the Liverpool Biological Society, Session 1908–1909, 23: 228–242.

Damkaer, D., and R. Mrozek-Dahl. 1980. The Plankton Expedition and the copepod studies of Friedrich and Maria Dahl. Pp. 462–473 in M. Sears and D. Merriman (eds.). *Oceanography: The past.* New York: Springer-Verlag.

Davy, H. 1814. *Elements of agricultural chemistry* (2nd ed.). London: Longman.

Deacon, M. 1971. *Scientists and the sea, 1650–1900. A study of marine science.* London: Academic Press. xvi + 445 pp.

Deacon, M.B. 1984. G. Herbert Fowler (1861–1940): The forgotten oceanographer. Notes and Records of the Royal Society of London 38: 261–296.

DeBeer, G. 1973. Lankester, Edwin Ray. Pages 26–27 in *Dictionary of Scientific Biography*, vol. 8.

Dehérain, P.P., and L. Maquenne. 1882. Sur la réduction des nitrates dans la terre arable. Comptes Rendus de l'Académie des Sciences de Paris 95: 691–693, 732–734, 854–856.

Delwiche, C.C. 1970. The nitrogen cycle. Scientific American 223(3): 136–146.

Delwiche, C.C. 1981. The nitrogen cycle and nitrous oxide. Pp. 1–5 in C.C. Delwiche (ed.). *Denitrification, nitrification, and atmospheric nitrous oxide.* New York, Wiley Interscience.

Denton, E.J., and A.J. Southward. 1986. Frederick Stratten Russell, 1897–1984. Elected F.R.S. 1938. Biographical Memoirs of Fellows of the Royal Society 32: 463–493.

Dittmer, R. 1902. *Die Deutsche Hochsee-, See- und Küstenfischerei im 19. Jahrhundert und bis zum Jahre 1902. Bearbeitet im Auftrage des Deutschen Seefischereivereins.* Hannover u. Leipzig: Hahnsche Buchhandlung. 70 pp.

Drew, G.H. 1911. The action of some denitrifying bacteria in tropical and temperate seas, and the bacterial precipitation of calcium carbonate in the sea. Journal of the Marine Biological Association of the United Kingdom 9: 142–155.

Drew, G.H. 1913. On the precipitation of calcium carbonate in the sea by marine bacteria, and on the action of denitrifying bacteria in tropical and temperate seas. Journal of the Marine Biological Association of the United Kingdom 9: 479–524.

Drygalski, E. von. 1897. *Grönland-Expedition der Gesellschaft für Erdkunde zu Berlin, 1891–1893, unter Leitung von Erich von Drygalski.* 2 vols. Berlin: W.H. Kuhl.

Du Bois-Reymond, E. 1882. The seven world-problems. Popular Science Monthly 20: 433–447.

Du Bois-Reymond, E. 1886. Die sieben Welträthsel. Pp. 381–417 in *Reden von Emil Du Bois-Reymond. Erste Folge. Litteratur. Philosophie. Zeitgeschichte.* Leipzig: Verlag von Veit and Co. 550 pp.

Du Bois-Reymond, E. 1890. Humboldt-Stiftung: Bericht von Herrn E. Du Bois-Reymond. Pp. 82–87 in *Sitzungsberichte der Königlichen Preussischen Akademie der Wissenschaften zu Berlin, Jahrgang 1890, Erster Halb.*

Egerton, F.N. 1983. The history of ecology: Achievements and opportunities. Journal of the History of Biology 16:259–310.

Egerton, F.N. 1985. The history of ecology: Achievements and opportunities, part two. Journal of the History of Biology 18: 103–143.

Ekman, F.L. 1875a. Notice sur les movements de l'eau de la mer, dans le voisinage de l'embouchure des fleuves, pour servir à la connaissance de la nature des courants marins. Archives des Sciences Physiques et Naturelles 54: 62–71.

Ekman, F.L. 1875b. Om de strömmar som uppsto vid podymynnigar, samt om hafströmmarnes allmänna orsakei. Stockholm, Ofversight af Kungliga Vetenskapsakadamiens Förhandlingar 1875, no. 7.

Ekman, F.L. 1876. On the general causes of the ocean currents. Uppsala. Nova Acta Regiae Societas Scientiarum Upsaliensis, ser. III, 10(6): 1–52.

Ekman, F.L., and O. Pettersson. 1893. Den Svenska hydrografiska expeditionen ar 1877 under ledning af F.L. Ekman. Kungliga Svenska Vetenskapsakadamiens Handlingar 25(1): 1–163, 14 Taf.

Ekman, V.W. 1905. On the influence of the earth's rotation on ocean currents. Arkiv für Matematik, Astronomi, och Fysik 2(11): 1–52.

Eliassen, A. 1982. Vilhelm Bjerknes and his students. Annual Review of Fluid Mechanics 14: 1–11.

Elmhirst, R. 1939. Marine biology in the Firth of Clyde. Scottish Naturalist no. 238, pp. 89–93.

Elton, C. 1927. *Animal ecology.* New York: Macmillan. xxi + 207 pp.

Elton, C. 1940a. American ecology. Journal of Animal Ecology 9: 148–149.

Elton, C. 1940b. Scholasticism in ecology. Journal of Animal Ecology 9: 151–152.

Farrar, W.V. 1976. Science and the German university system, 1790–1850. Pp. 179–192 in M.P. Crosland (ed.). *The emergence of science in western Europe.* New York: Science History Publications.

Feitel, R. 1903. Beiträge zur Kenntnis denitrificierender Meeresbakterien. Wissenschaftliche Meeresuntersuchungen, Abteilung Kiel, Neue Folge 7: 89–109.

Fischer, B. 1894a. Die Bakterien des Meeres nach den Untersuchungen der Plankton-Expedition unter gleichzeitiger Berücksichtigung einiger alterer und neuerer Untersuchungen. Centralblatt für Bakteriologie 15: 657–666.

Fischer, B. 1894b. Die Bakterien des Meeres nach den Untersuchungen der Plankton-

Expedition unter gleichzeitiger Berücksichtigung einiger alterer und neuerer Untersuchungen. Ergebnisse der Plankton-Expedition der Humboldt-Stiftung 4: 1–83.

Fisher, R.A. 1925. *Statistical methods for research workers.* Edinburgh: Oliver and Boyd. viii + 239 pp.

Fjeldstad, J.E. 1958. Harald Ulrik Sverdrup, 1888–1957. Journal du Conseil International pour l'Exploration de la Mer 23: 147–150.

Fleming, R.H. 1939. The control of diatom populations by grazing. Journal du Conseil International pour l'Exploration de la Mer 14: 210–227.

Føyn, B.R. 1929. Quantitative examination of the phytoplankton at Lofoten March–April 1922–1927. Skrifter utgitt av det Norske Videnskaps-Akademi i Oslo. I. Matematisk-Naturvidenskapelig Klasse 1928, no. 10, pp. 1–71.

Frey, D.G. 1963. Wisconsin: the Birge-Juday era. Pp. 3–54 in D.G. Frey (ed.). *Limnology in North America.* Madison: Univ. of Wisconsin Press.

Fulton, T.W. 1889. The scientific work of the Fishery Board for Scotland. Journal of the Marine Biological Association of the United Kingdom 1: 75–91.

Gaarder, T. 1915. Surstoffet i Fjordene (de vestlandske fjordes hydrografi I). Bergens Museums Aarbok 1915–1916. Naturvidenskapelig raekke no. 2.

Gaarder, T. 1917. Die Hydroxylzahl des Meerwassers (Die Hydrographie der Fjorde des westlichen Norwegens no. 2). Bergens Museums Aarbok 1916–1917.

Gaarder, T., and H.H. Gran. 1927. Investigations of the production of plankton in the Oslo Fjord. Rapports et Procès-Verbaux des Réunions, Conseil International pour l'Exploration de la Mer 42: 1–48.

Gall, D.C., and W.R.G. Atkins. 1931. Apparatus for the photo-electric measurement of submarine illumination assembled for the U.S.A. research ship *Atlantis.* Journal of the Marine Biological Association of the United Kingdom 17: 1017–1028.

Gardiner, J.S. 1920. The physico-chemical conditions underlying life in the sea. Rapports et Procès-Verbaux des Réunions, Conseil International pour l'Exploration de la Mer 26: 71–73.

Garstang, W. 1900. The impoverishment of the sea. A critical summary of the experimental and statistical evidence bearing upon the alleged depletion of the fishing grounds. Journal of the Marine Biological Association of the United Kingdom 6: 1–69.

Garstang, W. 1919. Sea-fishery investigations and the balance of life. Nature 104: 48–49.

Gayon, U., and G. Dupetit. 1882a. Sur la fermentation des nitrates. Comptes Rendus de l'Académie des Sciences de Paris 95: 644–646.

Gayon, U., and G. Dupetit. 1882b. Sur la transformation des nitrates en nitrites. Comptes Rendus de l'Académie des Sciences de Paris 95: 1365–1367.

Gayon, U., and G. Dupetit. 1886. Recherches sur la réduction des nitrates par les infiniments petits. Mémoires de la Société de Médicine et de Chirurgie de Bordeaux, ser. 3, 2: 201–307.

Gazert, H. 1903. Bakteriologischer Bericht. Pp. 154–160 in E. von Drygalski et al. *Die Deutsche Südpolar-Expedition auf dem Schiff "Gauss" unter Leitung von Erich von Drygalski. Bericht über die wissenschaftlichen Arbeiten seit der Abfahrt von Kerguelen bis zur Rückkehr nach Kapstadt 31. Januar 1902 bis 9. Juni 1903 und die Thätigkeit auf der Kerguelen-Station vom 1. April 1902 bis 1 April 1903.* Veröffentlichungen des Instituts für Meereskunde und des Geographischen Instituts an der Universität Berlin. Heft 5, 181 pp., 3 Taf.

Gazert, H. 1909. I. Untersuchungen über Denitrifikation und Nitrifikation im Meere während der Reise des "Gauss," sowie Sammlung und Aufbewahrung von Wasserproben für die quantitative Stickstoffbestimmung in der Heimat. Pp. 120–125 in J. Gebbing, *Chemische Untersuchungen von Meeresboden-, Meerwasser- und Luft-Proben der Deutschen Südpolar-Expedition, 1901–1903*. Pp. 75–234 in E. von Drygalski, *Deutsche Südpolar-Expedition, 1901–1903*, Band 7(2).

Gazert, H. 1912. Untersuchungen über Meeresbakterien und ihren Einfluss auf den Stoffwechsel im Meere. Pp. 235–296 in E. von Drygalski, *Deutsche Südpolar-Expedition, 1901–1903*, Band 7(3).

Gebbing, J. 1909. Chemische Untersuchungen von Meeresboden-, Meerwasser- und Luft-Proben der Deutschen Südpolar-Expedition 1901–1903. Pp. 75–234 in E. von Drygalski, *Deutsche Südpolar-Expedition, 1901–1903*, Band 7(2).

Gebbing, J. 1910. Über den Gehalt des Meeres an Stickstoffnährsalzen. Untersuchungsergebnisse der von der Deutschen Südpolar-Expedition (1901–1903) gesammelten Meerwasserproben. Internationale Revue der gesamten Hydrobiologie 3: 50–66.

Geison, G.L. 1981. Scientific change, emerging specialties, and research schools. History of Science 19: 20–40.

Gilson, H.C. 1937. Chemical and physical investigations. The nitrogen cycle. The John Murray Expedition, 1933–34. Scientific Reports 2(2): 21–81.

Goodrich, E.S. 1930. Edwin Ray Lankester. Proceedings of the Royal Society of London B 105: x–xv.

Goppelsröder, F. 1862. Beiträge zum Studium der Salpeterbildungen. Annalen der Physik und Chemie 115: 125–137.

Gräf, Dr. 1909. *Forschungsreise S.M.S. "Planet" 1906/07 herausgegeben von Reichs-Marine-Amt*, Band 4: *Biologie*. 198 pp.

Gran, H.H. 1902a. Das Plankton des norwegischen Nordmeeres von biologischen und hydrographischen Gesichtspunkten behandelt. *Report on Norwegian Fishery and Marine Investigations* 2, part III(5). viii + 222 pp.

Gran, H.H. 1902b. Studien über Meeresbakterien. I. Reduktion von Nitraten und Nitriten. Bergens Museums Aarbok 1901, no. 10, pp. 1–23.

Gran, H.H. 1903. Havets bakterier og deres stofskifte. Naturen 27: 33–40, 72–84.

Gran, H.H. 1912a. Pelagic plant life. Pp. 307–386 in J. Murray and J. Hjort (eds.). *The depths of the ocean*. London: Macmillan.

Gran, H.H. 1912b. Preservation of samples and quantitative determination of the plankton. Conseil International pour Exploration de la Mer, Publication de Circonstance, no. 62.

Gran, H.H. 1915. The plankton production in the North European waters in the spring of 1912. Conseil International pour l'Exploration de la Mer, Bulletin Planctonique pour l'année 1912, pp. 1–142.

Gran, H.H. 1918. Kulturforsøk med planktonalger. Skandinaviske Naturforskeres 16de møte Kristiania den 10–15. Juli 1916, Forhandlinger, p. 391.

Gran, H.H. 1919. Quantitative investigations as to phytoplankton and pelagic Protozoa in the Gulf of St. Lawrence and outside the same. Pp. 489–501 in *Canadian Fisheries Expedition, 1914–1915. Investigations in the Gulf of St. Lawrence and Atlantic waters of Canada*. Ottawa: Dept. of Naval Service.

Gran, H.H. 1923. Snesmeltningen som hovedaarsak til den rike produktion i vort kysthav om vaaren. Samtiden 34.

Gran, H.H. 1927. The production of plankton in the coastal waters off Bergen. March–April 1922. *Report on Norwegian Fishery and Marine Investigations* 3(8): 1–74.

Gran, H.H. 1928. La Norvège. 2. The conditions of life for plankton in the coastal waters of northern Europe. *In* H.G. Maurice (ed.). Rapport jubilaire (1902–1927). Rapports et Procès-Verbaux des Réunions, Conseil International pour l'Exploration de la Mer 47(III): 196–204.

Gran, H.H. 1929a. Investigation of the production of plankton outside the Romsdalsfjord, 1926–1927. Rapports et Procès-Verbaux des Réunions, Conseil International pour l'Exploration de la Mer 56: 1–112.

Gran, H.H. 1929b. Quantitative plankton investigations carried out during the expedition with the "Michael Sars," July–Sept. 1924. Rapports et Procès-Verbaux des Réunions, Conseil International pour l'Exploration de la Mer 56: 1–50.

Gran, H.H. 1930. The spring growth of the plankton at Møre in 1928–29 and at Lofoten in 1929 in relation to its limiting factors. Skrifter utgitt av det Norske Videnskaps-Akademi i Oslo. I. Matematisk-Naturvidenskapelig Klasse, no. 5, pp. 1–77.

Gran, H.H. 1931. On the conditions for the production of plankton in the sea. Rapports et Procès-Verbaux des Réunions, Conseil International pour l'Exploration de la Mer 75:37–46.

Gran, H.H. 1932. Phytoplankton. Methods and problems. Journal du Conseil International pour l'Exploration de la Mer 7: 343–358.

Gran, H.H. 1933a. Norway. *In* Reports of area committees. Plankton Committee 1932. Rapports et Procès-Verbaux des Réunions, Conseil International pour l'Exploration de la Mer 85(II): 68–70.

Gran, H.H. 1933b. Studies on the biology and chemistry of the Gulf of Maine. II. Distribution of phytoplankton in August, 1932. Biological Bulletin 64: 159–182.

Gran, H.H., and T. Braarud. 1935. A quantitative study of the phytoplankton in the Bay of Fundy and the Gulf of Maine (including observations on hydrography, chemistry and turbidity). Journal of the Biological Board of Canada 1: 279–433.

Gran, H.H., and T. Gaarder. 1918. Über den Einfluss der atmosphärischen Veränderung Nordeuropas auf die hydrographischen Verhältnisse des Kristianiafjords bei Drøbak in März 1916. Conseil International pour l'Exploration de la Mer, Publication de Circonstance, no. 71. 29 pp.

Gran, H.H., and T.G. Thompson. 1930. The diatoms and the physical and chemical conditions of the sea water of the San Juan Archipelago. Publications of the Puget Sound Biological Station 7: 169–204.

Grisebach, A.H.R. 1872. *Die Vegetation der Erde nach ihrer klimatischen Anordnung.* Leipzig: Engelmann.

Günther, N. 1970. Abbe, Ernst. P. 609 in *Dictionary of Scientific Biography,* vol. 1.

Gutina, V. 1976. Vinogradsky, Sergey Nikolaevich. Pp. 36–38 in *Dictionary of Scientific Biography,* vol. 14. 36–38.

Haeckel, E. 1890a. *Plankton-Studien.* Jena: Gustav Fischer. 104 pp.

Haeckel, E. 1890b. Plankton-Studien—Vergleichende Untersuchungen über die Bedeutung und Zusammensetzung der pelagischen Fauna und Flora. Jenaische Zeitschrift für Naturwissenschaft 25: 232–336.

Haeckel, E. 1893. Plankton studies: A comparative investigation of the importance and constitution of the pelagic fauna and flora. (Trans. George W. Field). Report of the United States Fish Commission, 1889–1891 16: 565–641.

Hagstrom, W.O. 1965. *The scientific community*. New York: Basic Books. x + 340 pp.

Haines, G. 1958. German influence upon scientific instruction in England, 1867–1887. Victorian Studies 1: 215–244.

Haines, G. 1969. *Essays on German influence upon English education and science, 1850–1919*. Hamden, Conn.: Archon Books. x + 188 pp.

Hardy, A.C. 1946. Stanley Wells Kemp, 1882–1945. Journal of the Marine Biological Association of the United Kingdom 26: 219–234.

Hardy, A.C. 1950a. Johan Hjort, 1869–1948. Obituary Notices of Fellows of the Royal Society of London 7: 167–181.

Hardy, A.C. 1950b. Walter Garstang. Proceedings of the Linnean Society of London 162: 99–105.

Hardy, A.C. 1950c. Walter Garstang. Journal du Conseil International pour l'Exploration de la Mer 17: 7–12.

Hardy, A.C. 1951. Walter Garstang, 1868–1949. Journal of the Marine Biological Association of the United Kingdom 29: 561–566.

Hardy, A.C. 1968. Charles Elton's influence in ecology. Journal of Animal Ecology 37: 3–8.

Hardy, A.C., and E.R. Gunther. 1935. The plankton of the South Georgia whaling grounds and adjacent waters, 1926–27. Discovery Reports 11: 1–456.

Harmer, S.J. 1933. Gilbert Bourne, F.R.S. Obituary Notices of Fellows of the Royal Society of London 2: 126–130.

Harris, E. 1959. The nitrogen cycle in Long Island Sound. Bulletin of the Bingham Oceanographic Collection 17: 31–65.

Harvey, H.W. 1923. Hydrographic features of the water in the neighbourhood of Plymouth during the years 1921 and 1922. Journal of the Marine Biological Association of the United Kingdom 13: 225–235.

Harvey, H.W. 1925a. Evaporation and temperature changes in the English Channel. Journal of the Marine Biological Association of the United Kingdom 13: 678–692.

Harvey, H.W. 1925b. Hydrography of the English Channel. Rapports et Procès-Verbaux des Réunions, Conseil International pour l'Exploration de la Mer 37: 59–89.

Harvey, H.W. 1925c. Water movement and sea temperature in the English Channel. Journal of the Marine Biological Association of the United Kingdom 13: 659–664.

Harvey, H.W. 1926. Nitrate in the sea. Journal of the Marine Biological Association of the United Kingdom 14: 71–88.

Harvey, H.W. 1928a. *Biological chemistry and physics of sea water*. Cambridge: Cambridge Univ. Press. x + 194 pp.

Harvey, H.W. 1928b. Nitrate in the sea. II. Journal of the Marine Biological Association of the United Kingdom 15: 183–190.

Harvey, H.W. 1929. Hydrodynamics of the waters south-east of Ireland. Journal du Conseil International pour l'Exploration de la Mer 4: 80–92.

Harvey, H.W. 1930. Hydrography of the mouth of the English Channel, 1925–1928. Journal of the Marine Biological Association of the United Kingdom 16: 791–820.

Harvey, H.W. 1933. On the rate of diatom growth. Journal of the Marine Biological Association of the United Kingdom 19: 253–275.

Harvey, H.W. 1934a. Annual variation of planktonic vegetation, 1933. Journal of the Marine Biological Association of the United Kingdom 19: 775–792.

Harvey, H.W. 1934b. Hydrography of the mouth of the English Channel, 1929–

1932. Journal of the Marine Biological Association of the United Kingdom 19: 737–746.

Harvey, H.W. 1934c. Measurement of phytoplankton population. Journal of the Marine Biological Association of the United Kingdom 19: 761–773.

Harvey, H.W. 1935. Note concerning a measuring plankton-net. Journal du Conseil International pour l'Exploration de la Mer 10: 179–184.

Harvey, H.W. 1937a. Note on colloidal ferric hydroxide in sea water. Journal of the Marine Biological Association of the United Kingdom 22: 221–225.

Harvey, H.W. 1937b. The supply of iron to diatoms. Journal of the Marine Biological Association of the United Kingdom 22: 205–219.

Harvey, H.W. 1939. Substances controlling the growth of a diatom. Journal of the Marine Biological Association of the United Kingdom 23: 499–520.

Harvey, H.W. 1940. Nitrogen and phosphorus required for the growth of phytoplankton. Journal of the Marine Biological Association of the United Kingdom 24: 115–123.

Harvey, H.W. 1942. Production of life in the sea. Biological Reviews 17: 221–246.

Harvey, H.W. 1945. *Recent advances in the chemistry and biology of sea water.* Cambridge: Cambridge Univ. Press. vii + 164 pp.

Harvey, H.W. 1947. Manganese and the growth of phytoplankton. Journal of the Marine Biological Association of the United Kingdom 26: 562–579.

Harvey, H.W. 1949. On manganese in sea and fresh waters. Journal of the Marine Biological Association of the United Kingdom 29: 97–137.

Harvey, H.W. 1950. On the production of living matter in the sea off Plymouth. Journal of the Marine Biological Association of the United Kingdom 29: 97–137.

Harvey, H.W. 1953. Note on the absorption of organic phosphorus compounds by *Nitzschia closterium* in the dark. Journal of the Marine Biological Association of the United Kingdom 31: 475–476.

Harvey, H.W. 1955. *The chemistry and fertility of sea waters.* Cambridge: Cambridge Univ. Press. viii + 234 pp.

Harvey, H.W. 1957. Bio-assay of nitrogen available to two species of phytoplankton in an off-shore water. Journal of the Marine Biological Association of the United Kingdom 36: 157–160.

Harvey, H.W., L.H.N. Cooper, M.V. Lebour, and F.S. Russell. 1935. Plankton production and its control. Journal of the Marine Biological Association of the United Kingdom 20: 407–441.

Haskell, E.F. 1940. Mathematical systematization of "environment," "organism" and "habitat." Ecology 21: 1–16.

Hattori, A. 1982. The nitrogen cycle in the sea with special reference to biogeochemical processes. Journal of the Oceanographical Society of Japan 38: 245–265.

Heape, W. 1887. Description of the laboratory of the Marine Biological Association at Plymouth. Journal of the Marine Biological Association of the United Kingdom, Old Series 1: 96–104.

Heincke, F. 1889. Die Untersuchungen von Hensen über die Produktion des Meeres an belebter Substanz. Deutscher Fischerei-Verein, Mittheilungen der Sektion für Küsten-und Hochsee-Fischerei 15(3–5): 35–58.

Helland-Hansen, B., and F. Nansen. 1909. The Norwegian Sea. Its physical oceanography based upon the Norwegian researches, 1900–1904. *Report on Norwegian*

Fishery and Marine Investigations 2(2): i–xx, 1–390; Supplement I–XII, pls. I–XXV.

Hellriegel, J., and H. Wilfarth. 1889. Erfolgt die Assimilation des freien Stickstoffs durch die Leguminosen unter Mitwirkung niederer Organismen? Mittheilung einiger neuerer Culturversuche. Berichte der Deutschen Botanischen Gesellschaft 7: 138–143.

Henking, H. 1910. *25 Jahre im Dienste der deutschen Seefischerei. Ein Rückblick auf die Tätigkeit des Deutschen Seefischerei-Vereins.* Abhandlungen des Deutschen Seefischerei-Vereins, Band 11, pp. 1–92.

Hensen, V. 1857. Ueber die Zuckerbildung in der Leber. Virchows Archiv für pathologische Anatomie und Physiologie 11: 395–398.

Hensen, V. 1863. Zur Morphologie der Schnecke des Menschen und der Säugethiere. Zeitschrift für wissenschaftliche Zoologie 13: 481–512, Taf. XXXI–XXXIV.

Hensen, V. 1869. Ueber ein neues Strukturverhältnis der quergestreiften Muskelfaser. P. 172 ff. in *Arbeiten aus dem Kieler physiologischen Institut 1868*, herausgegeben von Victor Hensen. Kiel.

Hensen, V. 1873. Betreffend den Fischfang auf der Expedition. In *Die Expedition zur physikalisch-chemischen und biologischen Untersuchung der Ostsee im Sommer 1871 aus S.M. Avisodampfer Pommerania. Jahresbericht der Commission zur wissenschaftlichen Untersuchung der deutschen Meere in Kiel für 1871. I Jahrgang*, pp. 155–159.

Hensen, V. 1875. *Ueber die Befischung der deutschen Küsten. Jahresbericht der Commission zur wissenschaftlichen Untersuchung der deutschen Meere in Kiel für 1872, 1873. I & II Jahrgang*, pp. 341–380, Taf. I–IX.

Hensen, V. 1876. Beobachtungen über Befruchtung und Entwicklung des Kaninchens und Meerschweinchens. Zeitschrift für anatomische Entwicklungsgeschichte 1: 213–273, 353–423, Taf. VIII–XII.

Hensen, V. 1878. *Resultate der statistischen Beobachtungen über die Fischerei an den deutschen Küsten. III Jahresbericht der Commission zur wissenschaftlichen Untersuchung der deutschen Meere in Kiel für die Jahre 1874–1876. IV–VI Jahrgang*, pp. 133–171.

Hensen, V. 1884. *Über das Vorkommen und die Menge der Eier einiger Ostseefische ins besondere der Scholle* (Platessa platessa), *der Flunder* (Platessa vulgaris) *und des Dorsches* (Gadus morrhua). *Vierter Bericht der Commission zur wissenschaftlichen Untersuchung der deutschen Meere in Kiel für die Jahre 1877 bis 1881. VII bis XI Jahrgang*, pp. 299–313.

Hensen, V. 1887a. Die Naturwissenschaft in Universitätsverband. Rede beim Antritt des Rektorats der Königlichen Christian-Albrechts-Universität zu Kiel am 5. März 1884. Kiel: Universitäts-Buchhandlung. 16 pp.

Hensen, V. 1887b. *Über die Bestimmung des Planktons oder des im Meere treibenden Materials an Pflanzen und Thieren. Kommission zur wissenschaftlichen Untersuchung der deutschen Meere in Kiel, 1882–1886. V. Bericht, Jahrgang 12–16*, pp. 1–107.

Hensen, V. 1890. *Einige Ergebnisse der Plankton-Expedition der Humboldt-Stiftung. Sitzungsberichte der Königlichen Preussischen Akademie der Wissenschaften zu Berlin, Jahrgang 1890, I Halbband*, pp. 243–253.

Hensen, V. 1891. *Die Plankton-Expedition und Haeckel's Darwinismus. Ueber einige Aufgaben und Ziele der beschreibenden Naturwissenschaften.* Kiel and Leipzig: Lipsius und Tischer. 87 pp.

Hensen, V. 1892a. Einige Ergebnisse der Expedition. *Ergebnisse der Plankton-Expedition der Humboldt-Stiftung,* Band 1.A., pp. 18–46.

Hensen, V. 1892b. Entwicklung des Reiseplans. *Ergebnisse der Plankton-Expedition der Humboldt-Stiftung,* Band 1.A., pp. 1–17.

Hensen, V. 1895. Methodik der Untersuchungen. *Ergebnisse der Plankton-Expedition der Humboldt-Stiftung,* Band 1.B., pp. 1–200, 12 pls.

Hensen, V. 1897. Ueber die Fruchtbarkeit des Wassers. Pp. 79–91 in V. Hensen and C. Apstein 1897. Ueber die Eimenge der im Winter laichenden Fische. *In* Die Nordsee-Expedition 1895 des Deutschen Seefischerei-Vereins. Wissenschaftliche Meeresuntersuchungen, Abteilung Kiel, Neue Folge 2(2): 1–98.

Hensen, V. 1906. Die Biologie des Meeres. Rede am Stiftungsfest des Naturwissenschaftlichen Vereins in Kiel. Archiv für Hydrobiologie 1: 360–377.

Hensen, V. 1911. Das Leben im Ozean nach Zählung seiner Bewohner. Übersicht und Resultate der quantitativen Untersuchungen. *Ergebnisse der Plankton-Expedition der Humboldt-Stiftung,* Band V.O., pp. 1–406, tables, figs.

Hensen, V. 1912. Zur Feststellung der Unregelmässigkeiten in der Verteilung der Planktonten mit besonderer Berücksichtigung der Schlauchfänge. Wissenschaftliche Meeresuntersuchungen, Abteilung Kiel, Neue Folge 14: 191–203.

Hensen, V., and C. Apstein. 1897. Die Nordsee-Expedition 1895 des Deutschen Seefischerei-Vereins. Ueber die Eimenge der im Winter laichenden Fische. Wissenschaftliche Meeresuntersuchungen, Abteilung Kiel, Neue Folge 2(2): 1–98.

Hentschel, E. 1926. Bericht über die biologischen Arbeiten. Die Deutsche Atlantische Expedition auf dem Forschungs- und Vermessungsschiff "Meteor." Zeitschrift der Gesellschaft für Erdkunde zu Berlin, Jahrgang 1926, no. 1.

Hentschel, E. 1927. Bericht über die biologischen Arbeiten. Die Deutsche Atlantische Expedition auf dem Forschungs- und Vermessungsschiff "Meteor." Zeitschrift der Gesellschaft für Erdkunde zu Berlin, Jahrgang 1927 no. 5–6.

Hentschel, E. 1933. Hans Lohmann zum 70. Geburtstage. Forschungen und Fortschritte 9(27): 403.

Herdman, W.A. 1893a. Report on the investigations carried on in 1892 in connection with the Lancashire Sea-Fisheries Laboratory at University College, Liverpool. Proceedings and Transactions of the Liverpool Biological Society, Session 1892–93, 7: 100–147.

Herdman, W.A. 1893b. Sixth annual report of the Liverpool Marine Biology Committee, and their biological station at Port Erin. Proceedings and Transactions of the Liverpool Biological Society, Session 1892–93, 7: 45–97.

Herdman, W.A. 1895. Eighth annual report of the Liverpool Marine Biology Committee and their biological station at Port Erin. Proceedings and Transactions of the Liverpool Biological Society, Session 1894–95, 9: 26–75.

Herdman, W.A. 1903. The new biological station at Port Erin. Being the sixteenth annual report of the Liverpool Marine Biology Committee. Proceedings and Transactions of the Liverpool Biological Society, Session 1902–03, 17: 15–80.

Herdman, W.A. 1907. Sea-fisheries research in England. (A statement bearing on the present situation.) Proceedings and Transactions of the Liverpool Biological Society, Session 1906–1907, 21: 109–128.

Herdman, W.A. (ed.). 1913. Report on the investigations carried on during 1912 in connection with the Lancashire Sea-Fisheries Laboratory at the University of Liverpool, and the sea-fish hatchery at Piel, near Barrow. Proceedings and Transactions of the Liverpool Biological Society, Session 1912–1913, 27: 177–494.

Herdman, W.A. 1920. Oceanography and the sea-fisheries. Report of the British Association for the Advancement of Science, Cardiff 88: 1–33.

Herdman, W.A. 1922. Spolia Runiana. V. Summary of the results of continuous investigation of the plankton of the Irish Sea during fifteen years. Journal of the Linnean Society, Botany 46: 141–170, pl. 7.

Herdman, W.A., and A. Scott. 1908. An intensive study of the marine plankton around the south end of the Isle of Man. Proceedings and Transactions of the Liverpool Biological Society, Session 1907–1908, 22: 186–289.

Herdman, W.A., A. Scott, and J. Johnstone. 1912. Report on the investigations carried on during 1911 in connection with the Lancashire Sea-Fisheries Laboratory at the University of Liverpool, and the sea-fish hatchery at Piel, near Barrow. Proceedings and Transactions of the Liverpool Biological Society, Session 1911–1912, 26: 71–367.

Hesselberg, T., and H.U. Sverdrup. 1915. Die Stabilitätsverhältnisse des Seewassers bei vertikalen Verschiebungen. Bergens Museums Aarbok 1914–1915, no. 15, pp. 1–16.

Hickson, S.J. 1925. Sir William A. Herdman—1858–1924. Proceedings of the Royal Society of London B 98: x–xiv.

Hjort, J. 1901. Die erste Nordmeerfahrt des norwegischen Fischereidampfers "Michael Sars" im Jahre 1900 unter Leitung von Johan Hjort. Petermann's geographische Mitteilungen 4: 73–83, 97–106.

Hjort, J. 1927. North-eastern area 1925–1926. Rapports et Procès-Verbaux des Réunions, Conseil International pour l'Exploration de la Mer 41: 107–119.

Hjort, J. 1928. La Norvège. 1. The first cruises with the steamship "Michael Sars." In H.G. Maurice (ed.). Rapport jubilaire (1902–1927). Rapports et Procès-Verbaux des Réunions, Conseil International pour l'Exploration de la Mer 47(III): 188–195.

Hjort, J. 1945. International exploration of the sea. Rapports et Procès-Verbaux des Réunions, Conseil International pour l'Exploration de la Mer 105: 3–19.

Hoek, P.P.C. 1903. Report of administration. Rapports et Procès-Verbaux des Réunions, Conseil International pour l'Exploration de la Mer 1: i–xxxix.

Holmboe, J. 1921. Gran, Haaken Hasberg. Pp. 558–560 in Norsk biografisk Leksikon.

Holme, N.A. 1953. The biomass of the bottom fauna in the English Channel off Plymouth. Journal of the Marine Biological Association of the United Kingdom 32: 1–49.

Holmes, F.L. 1973. Liebig, Justus. Pp. 329–350 in Dictionary of Scientific Biography, Vol. 8.

Holmes, R.W. 1957. Solar radiation, submarine daylight, and photosynthesis. Pp. 109–128 in J.W. Hedgpeth (ed.). Treatise on marine ecology and paleoecology. Vol. 1: Ecology. Geological Society of America, Memoir 67.

Holt, E.W.L. 1895. An examination of the present state of the Grimsby trawl fishery, with especial reference to the destruction of immature fish. Journal of the Marine Biological Association of the United Kingdom 3: 339–448.

Hopkins, F.G. 1926. Dr. Edward J. Bles. Nature 118: 90–91.

Hoppe, B. 1966. Meeresbiologie in den ersten hundert Jahren der Versammlungen Deutscher Naturforscher und Ärzte. Medizinhistorische Journal 1: 31–42.

Hutchinson, G.E. 1936. *The clear mirror. A pattern of life in Goa and in Indian Tibet.* Cambridge: Cambridge Univ. Press. xi + 171 pp.

Hutchinson, G.E. 1940. Bio-ecology. [Review of F.E. Clements and V.E. Shelford. 1939.] Ecology 21: 267–268.

Hutchinson, G.E. 1941. Limnological studies in Connecticut. IV. Mechanisms of intermediary metabolism in stratified lakes. Ecological Monographs 11: 21–60.

Hutchinson, G.E. 1979. *The kindly fruits of the earth. Recollections of an embryo ecologist.* New Haven: Yale Univ. Press. xiii + 264 pp.

Hutchinson, G.E. 1982. Reminiscences and notes on some otherwise undiscussed papers. Pp. vii–ix in J.S. Wroblewski (ed.). *Selected works of Gordon A. Riley.* Halifax, Nova Scotia: Dalhousie Univ.

Illing, R. 1923a. Die Entwicklung der Seefischerei an der Nordseeküste Schleswig-Holsteins. Zeitschrift der Gesellschaft für Schleswig-Holsteinische Geschichte 52: 1–71.

Illing, R. 1923b. Die Entwicklung der Seefischerei an der Nordseeküste Schleswig-Holsteins. Zweiter Teil. Die Seefischerei unter der preussischen Herrschaft von 1867 bis 1914. Zeitschrift der Gesellschaft für Schleswig-Holsteinische Geschichte 53: 135–188.

Iselin, C.O'D. 1942. Eleventh annual report of the director for the year 1940. Woods Hole Oceanographic Institution Collected Reprints 1941, pp. 9–16.

Issatchenko, B.L. 1908. Zur Frage von der Nitrifikation des Meeres. Centralblatt für Bakteriologie II 21: 430.

Issatchenko, B.L. 1914. *Recherches sur les microbes de l'océan glaciale arctique. L'Expédition scientifique pour l'exploration des pêcheries de la Côte Mourmane.* Petrograd: V.F. Kirshbaum. vii + 297 pp.

Issatchenko, B.L. 1926. Sur le nitrification dans les mers. Comptes Rendus de l'Académie des Sciences de Paris 182: 185–186.

Jarausch, K.H. 1982. *Students, society, and politics in Imperial Germany. The rise of academic illiberalism.* Princeton, N.J.: Princeton Univ. Press. xvi + 448 pp.

Jarausch, K.H. 1983a. Higher education and social change: Some comparative perspectives. Pp. 9–36 in K.H. Jarausch (ed.). *The transformation of higher learning, 1860–1930.* Stuttgart: Klett-Cotta.

Jarausch, K.H. (ed.). 1983b. *The transformation of higher learning, 1860–1930: Expansion, diversification, social opening, and professionalization in England, Germany, Russia, and the United States.* Stuttgart: Klett-Cotta. 375 pp.

Jeffreys, H. 1931. *Scientific inference.* Cambridge: Cambridge Univ. Press. vii + 272 pp.

Jenkin, P.M. 1937. Oxygen production by the diatom *Coscinodiscus excentricus* Ehr. in relation to submarine illumination in the English Channel. Journal of the Marine Biological Association of the United Kingdom 22: 301–342.

Jenkins, J.T. 1901. The methods and results of the German plankton investigations, with special reference to the Hensen nets. Proceedings and Transactions of the Liverpool Biological Society 15: 279–341.

Jenkins, J.T. 1920. *The sea fisheries.* London: Constable and Co. xxxi + 299 pp.

Jenkins, J.T. 1921. *A textbook of oceanography*. London: Constable and Co. x + 206 pp.

Jensen, J.V. 1970. The X Club: Fraternity of Victorian scientists. British Journal for the History of Science 5: 63–72.

Johnstone, J. 1905. *British fisheries. Their administration and their problems*. London: Williams and Norgate. xxxi + 350 pp.

Johnstone, J. 1908. *Conditions of life in the sea*. Cambridge: Cambridge Univ. Press. xiv + 332 pp.

Johnstone, J. 1923. Professor Benjamin Moore. Proceedings and Transactions of the Liverpool Biological Society, Session 1922–1923, 37: 12–14.

Johnstone, J. 1925. William Abbott Herdman, 1858–1924. Proceedings and Transactions of the Liverpool Biological Society, Session 1924–1924, 39: 19–28, 65–69.

Jörberg, L. 1973. The Nordic countries, 1850–1914. Transl. P.B. Austin. Pp. 375–485 in C.M. Cipolla (ed.). *The Fontana economic history of Europe. The emergence of industrial societies*. Part two. Glasgow: Fontana/Collins.

Just, T. (ed.). 1939. Plant and animal communities. American Midland Naturalist 21: 1–255.

Karsten, G. 1887. *Die Beobachtungen an den Küstenstationen. Kommission zur wissenschaftlichen Meeresuntersuchung der deutschen Meere in Kiel, 1882–1886, Jahrgang 12 bis 16*, 5 Berichte, pp. 135–157.

Keding, M. 1906. Weitere Untersuchungen über stickstoffbindende Bakterien. Wissenschaftliche Meeresuntersuchungen, Abteilung Kiel, Neue Folge 9: 273–308.

Kemp, S., and A.V. Hill. 1943. Edgar Johnson Allen, 1866–1942. Obituary Notices of Fellows of the Royal Society 4: 357–367.

Kerr, J.G. 1949. The Scottish Marine Biological Association. Notes and Records of the Royal Society of London 7: 81–96.

Keutner, J. 1905. Über das Vorkommen und die Verbreitung stickstoffbindender Bakterien im Meere. Wissenschaftliche Meeresuntersuchungen, Abteilung Kiel, Neue Folge 8: 27–55.

Kingsland, S.E. 1985. *Modelling nature. Episodes in the history of population ecology*. Chicago: Univ. Chicago Press. ix + 267 pp.

Kingsland, S.E. 1986. Mathematical figments, biological facts: Population ecology in the thirties. Journal of the History of Biology 19: 235–256.

Klatt, B. 1935. Hans Lohmann.†Hamburg. Mitteilungen der Zoologischen Staatsinstitut und Zoologischen Museum 45: i–x, portrait.

Kofoid, C.A. 1897. On some important sources of error in the plankton method. Science 6 (N.S.): 829–832.

Kofoid, C.A. 1910. The biological stations of Europe. United States Bureau of Education Bulletin no. 4, whole no. 440. Washington, D.C.: Government Printing Office. xiii + 360 pp.

Kohler, R.E. 1982. *From medical chemistry to biochemistry. The making of a biomedical discipline*. Cambridge: Cambridge Univ. Press. ix + 399 pp.

Koller, G. 1958. *Das Leben des Biologen Johannes Müllers, 1801–1858*. Stuttgart: Wissenschaftliche Verlagsgesellschaft M.B.H.

König, R. 1981. Karl Möbius, eine kurze Biographie. Mitteilungen aus dem Zoologischen Museum der Universität Kiel 1(7): 5–15.

Kramer, E. 1890. *Die Bakteriologie in ihren Beziehungen zur Landwirtschaft und den landwirtschaftlichen-technischen Gewerben.* I. Theil. Wien.

Kreps, E. 1934. Organic catalysts or enzymes in seawater. Pp. 193–202 in R.V. Daniel (ed.). *James Johnstone memorial volume.* Liverpool: Liverpool Univ. Press.

Kreps, E., and N. Verjbinskaya. 1930. Seasonal changes in the phosphate and nitrate content and in hydrogen ion concentration in the Barents Sea. Journal du Conseil International pour l'Exploration de la Mer 5: 326–346.

Krohn, W., and W. Schäfer. 1976. The origins and structure of agricultural chemistry. Pp. 27–52 in G. Lemaine et al. (eds.). *Perspectives on the emergence of scientific disciplines.* The Hague: Mouton; Chicago: Aldine.

Krümmel, O. 1892. Reisebeschreibung der Plankton-Expedition. *Ergebnisse der Plankton-Expedition der Humboldt-Stiftung,* Band 1.A., pp. 47–69, 80–104, 113–134, 150–167, 185–203, 210–231, 315–330.

Krümmel, O. 1893. Geophysikalische Beobachtungen. *Ergebnisse der Plankton-Expedition der Humboldt-Stiftung,* Band 1.C., pp. 1–118.

Krümmel, O. 1895. Zur Physik der Ostsee. Petermann's Geographische Mitteilungen 41 (4): 81–86, 111–118.

Kutzbach, G. 1979. *The thermal theory of cyclones. A history of meteorological thought in the nineteenth century.* American Meteorological Society, Historical Monograph Series. Boston: American Meteorological Society. xiv + 255 pp.

Lamb, H.H. 1982. *Climate, history and the modern world.* London: Methuen. xix + 387 pp.

Lankester, E.R. 1884. The Marine Biological Association. Nature 30: 123.

Lechevalier, H.A., and M. Solotorovsky. 1965. *Three centuries of microbiology.* New York: McGraw-Hill. 536 pp.

Lenoir, T. 1981. Teleology without regrets. The transformation of physiology in Germany: 1790–1847. Studies in the History and Philosophy of Science 12: 293–354.

Lenoir, T. 1982. *The strategy of life. Teleology and mechanics in nineteenth century German biology.* Dordrecht: D. Reidel. xii + 314 pp.

Lenz, J. 1981. Phytoplankton standing stock and primary production in the western Baltic. Kieler Meeresforschungen, Sonderheft 5: 29–40.

Liebert, F. 1915. Über mikrobiologische Nitrit- und Nitratbildung im Meere. Rapporten en Verhandelingen uitgegeven door het Rijksinstitut voor visscherijonderzoek 1(3).

Liebig, J. von. 1843. *Organic chemistry in its relation to agriculture and physiology.* (3rd ed.) London: Taylor and Walton.

Lipman, C.B. 1922. Does nitrification occur in sea water? Science 56: 501–503.

Lloyd, B. 1931. Bacterial denitrification: An historical and critical survey. Journal of the Royal Technical College, Glasgow 2: 530–550.

Lohff, B. 1977. *Johannes Müller (1801–1858) als akademischer Lehrer.* Dissertation zur Erlangung des Doktorgrades des Fachbereichs Mathematik der Universität Hamburg. vi + 227 pp.

Lohmann, H. 1896. Die Appendicularien der Plankton-Expedition. *Ergebnisse der Plankton-Expedition der Humboldt-Stiftung,* Band 2.E.c, pp. 1–148, 24 Taf.

Lohmann, H. 1899. Untersuchungen über den Auftrieb der Strasse von Messina mit besonderer Berücksichtigung der Appendicularien und Challengerien. Sitzungsberichte der Königlichen Preussischen Akademie der Wissenschaften in Berlin, Jahrgang 1899(1): 384–400.

Lohmann, H. 1901. Ueber das Fischen mit Netzen aus Müllergaze Nr 20 zu dem Zwecke quantitativen Untersuchungen des Auftriebs. Wissenschaftliche Meeresuntersuchungen, Abteilung Kiel, Neue Folge 5:46–66.

Lohmann, H. 1903a. Neue Untersuchungen über den Reichtum des Meeres an Plankton und über die Brauchbarkeit der verschiedenen Fangmethoden. Zugleich auch ein Beitrag zur Kenntnis des Mittelmeerauftriebs. Wissenschaftliche Meeresuntersuchungen, Abteilung Kiel, Neue Folge 7: 1–86, 4 pls.

Lohmann, H. 1903b. Untersuchungen über die Tier und Pflanzenwelt sowie über die Boden-sedimente des Nordatlantischen Ozeans zwischen dem 38. und 50. Grade nördl. Breite. Sitzungsberichte der Königlichen Preussischen Akademie der Wissenschaften, Physikalische-Mathematische Klasse 26: 560–583, Taf. I.

Lohmann, H. 1908a. Neues aus dem Gebiete der Planktonforschung. Wissenschaftliche Wochenschrift, Neue Folge 7(51): 801–810.

Lohmann, H. 1908b. Untersuchungen zur Feststellung des vollständigen Gehaltes des Meeres an Plankton. Wissenschaftliche Meeresuntersuchungen, Abteilung Kiel, Neue Folge 10: 129–370.

Lohmann, H. 1909a. Die Gehäuse und Gallertblasen der Appendicularien und ihre Bedeutung für die Erforschung des Lebens im Meer. Verhandlungen der Deutschen Zoologischen Gesellschaft 19: 200–239.

Lohmann, H. 1909b. Die Strömungen in der Strasse von Messina und die Verteilung des Planktons in derselben. Internationale Revue der gesamten Hydrobiologie 2: 505–556.

Lohmann, H. 1912. Die Probleme der modernen Planktonforschung. Verhandlungen der Deutschen Zoologischen Gesellschaft 22: 16–109.

Lohmann, H. 1918. Ernst Vanhöffen. Mitteilungen aus der zoologischen Museum in Berlin 9(1): 72–90.

Lohmann, H. 1920. Die Bevölkerung des Ozeans mit Plankton nach den Ergebnissen der Zentrifugenfänge während der Ausreise der "Deutschland" 1911. Zugleich ein Beitrag zur Biologie des Atlantischen Ozeans. Archiv für Biontologie 4(3). Teil I, pp. 1–470; Teil II, pp. 471–617, 16 Taf.

Lovelock, J.E. 1979. *Gaia, a new look at life on earth.* New York: Oxford Univ. Press. xi + 157 pp.

Lubbert, Korvetten-Kapitan. 1909. Reisebeschreibung. Forschungsreise S.M.S. "Planet" 1906/07. I. Band. Berlin: Verlag von Karl Siegismund. xvii + 104 pp.

Lucas, C.E. 1938. Some aspects of integration in plankton communities. Journal du Conseil International pour l'Exploration de la Mer 13: 309–322.

Lucas, C.E. 1956. The Scottish Home Department's marine laboratory in Aberdeen. Proceedings of the Royal Society of Edinburgh 66B: 222–234.

Lundgreen, P. 1980. The organization of science and technology in France: A German perspective. Pp. 311–332 in R. Fox and G. Weisz (eds.). *The organization of science and technology in France, 1808–1914.* Cambridge: Cambridge Univ. Press.

Lussenhop, J. 1974. Victor Hensen and the development of sampling methods in ecology. Journal of the History of Biology 7:319–337.

McClelland, C.E. 1980. *State, society and university in Germany, 1700–1914*. Cambridge: Cambridge Univ. Press. ix + 381 pp.

McConnell, A. 1982. *No sea too deep. The history of oceanographic instruments*. Bristol: Adam Hilger. xii + 162 pp.

McCormmach, R. 1974. On academic scientists in Wilhelmian Germany. Daedalus 103: 157–171.

McCormmach, R. 1982. *Night thoughts of a classical physicist*. Cambridge, Mass.: Harvard Univ. Press. 219 pp.

McCosh, F.W.J. 1984. *Boussingault. Chemist and agriculturalist*. Dordrecht: D. Reidel. vxiii + 280 pp.

McIntosh, R.P. 1976. Ecology since 1900. Pp. 353–372 in B.J. Taylor and T.J. White (eds.). *Issues and ideas in America*. Norman: Univ. of Oklahoma Press.

McIntosh, R.P. 1980. The background and some current problems of theoretical ecology. Pp. 1–61 in E. Saarinen (ed.). *Conceptual issues in ecology*. Boston: D. Reidel.

McIntosh, R.P. 1985. *The background of ecology. Concept and theory*. Cambridge: Cambridge Univ. Press. xiii + 383 pp.

M'Intosh, W.C. 1896. *The Gatty Marine Laboratory, and the steps which led to its foundation in the University of St. Andrews*. Dundee: John Lang and Co. 90 pp.

M'Intosh, W.C. 1899. *The resources of the sea*. Cambridge: Cambridge Univ. Press.

M'Intosh, W.C. 1907. Scientific work in the sea-fisheries. Zoologist, 4th Series, 11: 201–220, 247–266.

M'Intosh, W.C. 1919. The fisheries and the International Council. Nature 103: 355–358, 376–378.

M'Intosh, W.C. 1921. *The resources of the sea* (2nd ed.). Cambridge: Cambridge Univ. Press. xvi + 352 pp.

MacLeod, R.M. 1968. Government and resource conservation: The Salmon Acts Administration, 1860–1886. Journal of British Studies 7: 114–150.

MacLeod, R.M. 1970. The X-Club: A social network of science in late-Victorian England. Notes and Records of the Royal Society of London 24: 305–322.

MacLeod, R.M. 1971. The support of Victorian science: The endowment of research movement in Great Britain, 1868–1900. Minerva 9: 197–230.

MacLeod, R.M. 1976. Science and the treasury: Principles, personalities and policies, 1870–1885. Pp. 115–172 in G.L'E. Turner (ed.). *The patronage of science in the nineteenth century*. Leyden: Noordhoff International.

Marine Biological Association of the United Kingdom. 1903. The International Fisheries Investigations. Pp. 646–652 in Report of the Council, 1902–1903. Journal of the Marine Biological Association of the United Kingdom 6: 639–656.

Marine Biological Association of the United Kingdom. 1911. Report of the Council, 1910–1911. Journal of the Marine Biological Association of the United Kingdom 9: 248–271.

Marine Biological Association of the United Kingdom. 1912. Report of the Council, 1911–1912. Journal of the Marine Biological Association of the United Kingdom 9: 578–598.

Marine Biological Association of the United Kingdom. 1915. Report of the Council, 1914. Journal of the Marine Biological Association of the United Kingdom 10: 647–653.

Marine Biological Association of the United Kingdom. 1921. Report of the Council,

1920. Journal of the Marine Biological Association of the United Kingdom 12: 562–568.

Marine Biological Association of the United Kingdom. 1922. Report of the Council, 1921. Journal of the Marine Biological Association of the United Kingdom 12: 835–848.

Marine Biological Association of the United Kingdom. 1923. Report of the Council, 1922. Journal of the Marine Biological Association of the United Kingdom 13: 286–318.

Marine Biological Association of the United Kingdom. 1931. Report of the Council, 1930. Journal of the Marine Biological Association of the United Kingdom 17: 587–615.

Marine Biological Association of the United Kingdom. 1933. Report of the Council for 1932. Journal of the Marine Biological Association of the United Kingdom 19: 447–465.

Marine Biological Association of the United Kingdom. 1934. Report of the Council for 1933. Journal of the Marine Biological Association of the United Kingdom 19: 951–983.

Marine Biological Association of the United Kingdom. 1935. Report of the Council for 1934. Journal of the Marine Biological Association of the United Kingdom 20: 463–491.

Marine Biological Association of the United Kingdom. 1936. Report of the Council for 1935. Journal of the Marine Biological Association of the United Kingdom 21: 445–475.

Marine Biological Association of the United Kingdom. 1938. Report of the Council for 1937. Journal of the Marine Biological Association of the United Kingdom 23: 253–277.

Marine Biological Association of the United Kingdom. 1940. Report of the Council for 1938. Journal of the Marine Biological Association of the United Kingdom 24: 435–459.

Marine Biological Association of the United Kingdom. 1941. Report of the Council for 1940–41. Journal of the Marine Biological Association of the United Kingdom 25: 423–439.

Marine Biological Association of the United Kingdom. 1947. Report of the Council for 1945–46. Journal of the Marine Biological Association of the United Kingdom 26: 664–680.

Marine Biological Association of the United Kingdom. 1955. Report of the Council for 1954–55. Journal of the Marine Biological Association of the United Kingdom 34: 655–678.

Marine Biological Association of the United Kingdom. 1956. Report of the Council for 1955–56. Journal of the Marine Biological Association of the United Kingdom 34: 647–698.

Marshall, F.H.A. [F.H.A.M.]. 1930. Walter Heape—1885–1929. Proceedings of the Royal Society of London B 105: xv–xviii.

Marshall, S.M. 1987. *The Marine Station at Millport*. Millport, Scotland: University Marine Biological Station. v + 133 pp.

Marshall, S.M. 1963. Andrew Picken Orr, M.A., D.Sc., F.R.S.E. Pp. 4–7 in Scottish Marine Biological Association, Annual Report 1962–63.

Marshall, S.M., A.G. Nicholls, and A.P. Orr. 1934. On the biology of *Calanus finmarchicus*. Seasonal distribution, size, weight, and chemical composition in Loch Striven in 1933, and their relation to the phytoplankton. Journal of the Marine Biological Association of the United Kingdom 19: 793–827.

Marshall, S.M., and A.P. Orr. 1927. The relation of the plankton to some chemical and physical factors in the Clyde Sea area. Journal of the Marine Biological Association of the United Kingdom 14: 837–868.

Marshall, S.M., and A.P. Orr. 1928. The photosynthesis of diatom cultures in the sea. Journal of the Marine Biological Association of the United Kingdom 15: 321–360.

Marshall, S.M., and A.P. Orr. 1930. A study of the spring diatom increase in Loch Striven. Journal of the Marine Biological Association of the United Kingdom 16: 853–878.

Matthews, D.J. 1916. On the amount of phosphoric acid in the sea water off Plymouth Sound. Journal of the Marine Biological Association of the United Kingdom 11: 122–130, 251–275.

Maurice, H.G. (ed.). 1928. Rapport jubilaire (1902–1927). Rapports et Procès-Verbaux des Réunions, Conseil International pour l'Exploration de la Mer 47(I): 1–41; (II): 1–13; (III): 1–271.

Maurice, H.G. 1948. Prof. Johan Hjort, For. Mem. R.S. Nature 162: 765–766.

Mendelsohn, E. 1965. Physical models and physiological concepts: explanation in 19th century biology. Pp. 127–150 in R.S. Cohen and M.W. Wartofsky (eds.). *Essays in honour of Philipp Frank*. Boston Studies in Philosophy of Science 2.

Merriman, D. 1956. Harry Payne Bingham, 1887–1955. Bulletin of the Bingham Oceanographic Collection 15: 5–8.

Merriman, D. 1982. The history of Georges Bank. Pp. 11–30 in G.C. McLeod and J.H. Prescott (eds.). *Georges Bank. Past, present, and future of a marine environment*. Boulder, Colo.: Westview.

Meyer, H.A. 1871. *Untersuchungen über physikalische Verhältnisse des westlichen Theiles der Ostsee—ein Beitrag zur Physik des Meeres*. Kiel: Schwers'che Buchhandlung.

Meyer, H.A. 1875. *Zur Physik des Meeres. Beobachtungen über Meeresströmungen, Temperatur und specifisches Gewicht des Meerwassers. II–III Jahresbericht, Kommission zur wissenschaftlichen Untersuchung der deutschen Meere in Kiel, Jahre 1872, 1873, II und III Jahrgang*, pp. 1–41.

Meyer, H.A., K. Möbius, G. Karsten, and V. Hensen. 1873. *I Vorbericht der Commission. Jahresbericht der Kommission zur wissenschaftlichen Untersuchung der deutschen Meere in Kiel, 1871. I. Jahrgang*, pp. v–xi.

Mills, E.L. 1973. H.M.S. *Challenger* and the controversy about how the oceans circulate. Pp. 54–73 in E.L. Mills (ed.). *One hundred years of oceanography. Essays commemorating the visit of H.M.S. Challenger to Halifax, May 9–19, 1873*. Halifax, N.S.: Dalhousie Univ.

Mills, E.L. 1980. Alexander Agassiz, Carl Chun and the problem of the intermediate fauna. Pp. 360–372 in M. Sears and D. Merriman (eds.). *Oceanography: The Past*. New York: Springer-Verlag.

Mills, E.L. 1982. Saint-Simon and the oceanographers: patterns of change in the study of plankton dynamics 1897–1946. Pp. 3–18 in J.S. Wroblewski (ed.). *Selected works of Gordon A. Riley*. Halifax, N.S.: Dalhousie Univ.

Mills, E.L. 1983. Problems of deep-sea biology: an historical perspective. Pp. 1–79 in

G.T. Rowe (ed.). *The sea*. Vol. 8: *Deep-sea biology*. New York: John Wiley and Sons.

Mohn, H. 1885. Die Strömungen des europäischen Nordmeeres. Petermann's Geographische Mitteilungen. Erganzungsband 17, Nr 79, pp. 1–20, 4 Taf.

Mohn, H. 1887. The North Ocean, its depth, temperature and circulation. In *The Norwegian North-Atlantic Expedition, 1876–1878*. Christiania: Grøndahl and Son. 212 pp., 48 pls., 3 woodcuts.

Moncrieff, C.S., and D.W. Thompson. 1903. Report of the British delegates attending the meeting of the International Council held at Copenhagen in July 1902. Pp. 93–118 in *North Sea fishery investigations. Reports of the British delegates attending the international conferences, held at Stockholm, Christiania, and Copenhagen, with respect to fishery and hydrographical investigations in the North Sea, and correspondence related thereto*. Parliamentary Papers 1902. XV (Cd. 1313). London: HMSO. iii + 118 pp.

Moore, B., E.B.R. Prideaux, and G.A. Herdman. 1915. Studies of certain photosynthetic phenomena in sea-water. I. Seasonal variations in the reaction of sea-water in relation to the activities of vegetable and animal plankton II. The limitations of photosynthesis by algae in sea-water. Proceedings and Transactions of the Liverpool Biological Society, Session 1914–1915, 29: 233–264.

Moore, B., E. Whitley, and T.A. Webster. 1921. Studies of photosynthesis in marine algae. Proceedings of the Royal Society of London B 92: 51–60.

Mörth, W. 1892. I. Die Ausrüstung S.M. Schiffes "Pola" für Tiefsee-Untersuchungen. Berichte der Commission für Erforschung des östlichen Mittelmeeres. Erste Reihe. Denkschriften der Kaiserlichen Akademie der Wissenschaften, Wien, 59: 1–16, Taf. I–IX.

Mortimer, C.H. 1956. An explorer of lakes. Pp. 165–211 in G.S. Sellery (ed.). *E.A. Birge. A memoir*. Madison: Univ. of Wisconsin Press.

Mulkay, M.J. 1975. Three models of scientific development. Sociological Review 23: 509–526, 535–537.

Mulkay, M.J. 1979. *Science and the sociology of knowledge*. London: George Allen and Unwin. ix + 132 pp.

Mulkay, M.J., G.N. Gilbert, and S. Woolgar. 1975. Problem areas and research networks in science. Sociology 9: 187–203.

Murray, J., and J. Hjort. 1912. *The depths of the ocean. A general account of the modern science of oceanography based largely on the scientific researches of the Norwegian steamer Michael Sars in the North Atlantic*. London: Macmillan. 821 pp.

Murray, J., and J. Hjort. 1914–1956. *Report on the scientific results of the "Michael Sars" North Atlantic deep-sea expedition 1910*. 5 vols. Bergen: Bergens Museum.

Nansen, F. 1902. The oceanography of the North Polar Basin. Pp. 1–427 in F. Nansen (ed.). *The Norwegian North Polar Expedition, 1893–1896*. Scientific Results III. New York: Longmans, Green and Co.

Nathansohn, A. 1900. Physiologische Untersuchungen über amitotische Kerntheilung. Inaugural-Dissertation der hohen philosophischen Facultät der Universität Leipzig zur Erlangung der Doktorwürde. Jahrbuch für wissenschaftliche Botanik 35: 48–79.

Nathansohn, A. 1902. Über Regulationserscheinungen im Stoffaustausch. Jahrbuch für wissenschaftliche Botanik 38: 241–290.

Nathansohn, A. 1906a. Sur l'influence de la circulation verticale des eaux sur la production du plancton marin. Bulletin de la Musée Océanographique de Monaco, no. 62, pp. 1–12.

Nathansohn, A. 1906b. Über die Bedeutung vertikaler Wasserbewegung für die Produktion des Planktons. Abhandlungen der Königlichen Sächsischen Gesellschaft der Wissenschaften zu Leipzig, Mathematische-Physische Klasse 29: 355–441.

Nathansohn, A. 1906c. Vertikale Wasserbewegung und quantitative Verteilung des Planktons in Meere. Annalen der Hydrographie 34: 66–72.

Nathansohn, A. 1908. Über die allgemeinen Produktionsbedingungen im Meere. Beiträge zur Biologie des Planktons, von H.H. Gran and Alexander Nathansohn. Internationale Revue der gesamten Hydrobiologie 1: 38–72.

Nathansohn, A. 1909a. Sur les relations qui existent entre les changements du plankton végétal et les phénomènes hydrographiques, d'après les recherches faites à bord de l'*Eider*, au large de Monaco, en 1907–1908. Bulletin de l'Institut Océanographique, no. 140, pp. 1–93, 9 pls.

Nathansohn, A. 1909b. Beiträge zur Biologie des Planktons, von H.H. Gran und Alexander Nathansohn. II. Vertikalzirkulation und Planktonmaxima im Mittelmeer. Internationale Revue der gesamten Hydrobiologie 2: 580–632, Taf. 20–29.

Nathansohn, A. 1910. Etudes hydrobiologiques après les recherches faites à bord de l'"Eider" au large de Monaco de janvier a juillet 1909. Paris. Annales de l'Institut Océanographique 1(5). 27 pp.

Natterer, K. 1892a. Chemische Untersuchungen im östlichen Mittelmeer. Berichte der Commission für Erforschung des östlichen Mittelmeeres. I. Reise S.M. Schiffes "Pola" im Jahre 1890. Denkschriften der Akademie der Wissenschaften in Wien 59: 83–104.

Natterer, K. 1892b. Chemische Untersuchungen im östlichen Mittelmeer. Berichte der Commission für Erforschung des östlichen Mittelmeeres. II. Reise S.M. Schiffes "Pola" im Jahre 1891. Denkschriften der Akademie der Wissenschaften in Wien 59: 105–118.

Natterer, K. 1893. Chemische Untersuchungen im östlichen Mittelmeer. Berichte der Commission für Erforschung des östlichen Mittelmeers. III. Reise S.M. Schiffes "Pola" im Jahre 1892. Denkschriften der Akademie der Wissenschaften in Wien 60: 49–82.

Natterer, K. 1894. Chemische Untersuchungen im östlichen Mittelmeer. Berichte der Commission für Erforschung des östlichen Mittelmeeres. Reise S.M. Schiffes "Pola" im Jahre 1893. Denkschriften der Akademie der Wissenschaften in Wien 61: 23–63.

Natterer, K. 1895. Tiefsee-Forschungen im Marmara-Meer auf S.M. Schiff "Taurus" im Mai 1894. Denkschriften der Akademie der Wissenschaften in Wien 62: 19–117.

Natterer, K. 1898. Berichte der Commission für oceanographische Forschungen im Rothen Meere (nördliche Hälfte) 1895–96. IX. Chemische Untersuchungen. Denkschriften der Akademie der Wissenschaften in Wien 65: 445–572.

Neatby, L.H. 1973. *Discovery in Russian and Siberian waters*. Athens: Ohio Univ. Press. vii + 226 pp.

Nicolson, M. 1983. The development of plant ecology, 1790–1960. Ph.D. diss., Univ. of Edinburgh. viii + 417 pp.

Nordgaard, O. 1918. *Michael og Ossian Sars*. Kristiania: Steenske Forlag. 96 pp.

Nye, M.J. 1984. Scientific decline. Is quantitative evaluation enough? Isis 75: 697–708.

Olson, R.J. 1981a. ^{15}N tracer studies of the primary nitrite maximum. Journal of Marine Research 39: 203–226.

Olson, R.J. 1981b. Differential photoinhibition of marine nitrifying bacteria: A possible mechanism for the formation of the primary nitrite maximum. Journal of Marine Research 39: 227–238.

Oppenheimer, J.M. 1980. Some historical backgrounds for the establishment of the Stazione Zoologica at Naples. Pp. 179–187 in M. Sears and D. Merriman (eds.). *Oceanography: The past*. New York: Springer-Verlag.

Orton, J.H. 1920. Sea temperature, breeding and distribution in marine animals. Journal of the Marine Biological Association of the United Kingdom 12: 339–366.

Park, T. 1939a. Analytical population studies in relation to general ecology. American Midland, Naturalist 21: 235–255.

Park, T. 1939b. Ecology looks homeward. Quarterly Review of Biology 14: 332–336.

Patten, B.C. 1968. Mathematical models of plankton production. Internationale Revue der gesamten Hydrobiologie 53: 357–408.

Pedersen, O. 1974. Mohn, Henrik. Pp. 442–443 in *Dictionary of Scientific Biography*, vol. 9.

Petersen, C.G.J. 1918. The sea bottom and its production of fish food. Report of the Danish Biological Station 25: 1–62.

Peterson, B.J. 1980. Aquatic primary productivity and the ^{14}C-CO_2 method: A history of the productivity problem. Annual Review of Ecology and Systematics 11: 359–385.

Pettersson, H. 1923. Forteckning över Otto Pettersson's från trycket utgivna skrifter. Pp. 99–114 in *Festchrift tillägnad Otto Pettersson den 12 Februari 1923*. Göteborg: Helsingfors.

Pettersson, O. 1894. A review of Swedish hydrographic research in the Baltic and North seas. Scottish Geographical Magazine 10: 281–302, 352–359, 413–427, 449–462, 525–539, 617–635.

Pettersson, O. 1896. Ueber die Beziehungen zwischen hydrographischen und meteorologischen Phänomenen. Meteorologische Zeitschrift 31: 285–321.

Pettersson, O. 1900. Die Wasserzirkulation im Nordatlantischen Ozean. Petermann's Geographische Mitteilungen 46: 61–65, 81–92.

Pettersson, O. 1905a. On the influence of icemelting upon oceanic circulation. Svenska Hydrografisk-Biologiska Kommissionens Skrifter 2: 1–16.

Pettersson, O. 1905b. On the probable occurrence in the Atlantic Current of variations periodical, and otherwise, and their bearing on meteorological and biological phenomena, with an introduction. Rapports et Procès-Verbaux des Réunions, Conseil International pour l'Exploration de la Mer 3: Appendix A, i–x, 3–26.

Pettersson, O. 1915. Climatic variations in historic and prehistoric time. Svenska Hydrografisk Biologiska Kommissionens Skrifter 5: 1–26.

Pettersson, O. 1926. Hydrography, climate and fisheries in the transition area. Journal du Conseil International pour l'Exploration de la Mer 1: 305–321.

Pettersson, O. 1930. Gustaf Ekman, 1852–1930. Journal du Conseil International pour l'Exploration de la Mer 5: 287–289.

Pettersson, O., and G. Ekman. 1891. Grunddragen af Skageracks och Kattegats hydrografi. Kungliga Svenska Vetenskapsakadamiens Handlingar B 24(11): 1–161, 10 Taf.

Pfeiffer, J.E. 1949. The Office of Naval Research. Scientific American 180(2): 11–15.

Poole, H.H. 1925. On the photoelectric measurement of submarine illumination. Scientific Proceedings of the Royal Dublin Society 18: 99–115.

Poole, H.H. 1936. The photo-electric measurement of submarine illumination in offshore waters. Rapports et Procès-Verbaux des Réunions, Conseil International pour l'Exploration de la Mer 101(II): 1–2.

Poole, H.H. 1960. William Ringrose Gelston Atkins, 1884–1959. Biographical Memoirs of Fellows of the Royal Society 5: 1–22.

Poole, H.H., and W.R.G. Atkins. 1926. On the penetration of light into seawater. Journal of the Marine Biological Association of the United Kingdom 14: 177–198.

Poole, H.H., and W.R.G. Atkins. 1928. Further photo-electric measurements of the penetration of light into sea-water. Journal of the Marine Biological Association of the United Kingdom 15: 455–483.

Poole, H.H., and W.R.G. Atkins. 1929. Photo-electric measurements of submarine illumination throughout the year. Journal of the Marine Biological Association of the United Kingdom 16: 297–324.

Poole, H.H., and W.R.G. Atkins. 1934. The use of a selenium rectifier photo-electric cell for submarine photometry. Journal of the Marine Biological Association of the United Kingdom 19: 727–736.

Poole, H.H., and W.R.G. Atkins. 1935. The standardisation of photo-electric cells for the measurement of visible light. Philosophical Transactions of the Royal Society of London A 235: 1–27.

Porep, R. 1968. Der Anteil der Mediziner an der Frühphase der Planktonforschung. Gesnerus 25: 195–207.

Porep, R. 1970. Der Physiologe und Planktonforscher Victor Hensen (1835–1924). Sein Leben und sein Werk. Kieler Beiträge zur Geschichte der Medizin und Pharmazie 9:1–147.

Porep, R. 1972. Methodenstreit in der Planktologie—Haeckel contra Hensen, Auseinandersetzung um die Anwendung quantitativer Methoden in der Meeresbiologie um 1890. Medizinhistorische Journal 7: 72–83.

Porep, R. 1976. Hensen, Christian Andreas Victor. Pp. 97–99 in *Schleswig-Holsteinisches Biographisches Lexicon*, Band 4.

Provasoli, L., J.J.A. McLaughlin, and M.R. Droop 1957. The development of artificial media for marine algae. Archiv für Mikrobiologie 25: 392–428.

Puff, A. 1890. Das kalte Auftriebwasser an der Ostseite des nordatlantischen und der Westseite des nordindischen Ozeans. Ph.D. diss., Univ. of Marburg.

Raben, E. 1902. *Beiträge zur Kenntnis der Acetalisirung bei den Aldehyden und Ketonen. Inaugural-Dissertation zur Erlangung der Doktorwürde der hohen philosophischen Fakultät der Koniglichen-Christian-Albrechts-Universität zu Kiel.* Kiel: Druck von H. Fiencke. 58 pp.

Raben, E. 1905a. Über quantitative Bestimmung von Stickstoffverbindungen im Meerwasser nebst einem Anhang über die quantitative Bestimmung der im Meerwasser gelösten Kieselsäure. Wissenschaftliche Meeresuntersuchungen, Abteilung Kiel, Neue Folge 8: 81–101.

Raben, E. 1905b. Weitere Mitteilungen über quantitative Bestimmungen von Stickstoffverbindungen und von gelöster Kieselsäure im Meerwasser. Wissenschaftliche Meeresuntersuchungen, Abteilung Kiel, Neue Folge 8: 277–287.

Raben, E. 1910. Dritte Mitteilung über quantitative Bestimmungen von Stickstoffverbindungen und von gelöster Kieselsäure im Meerwasser. Wissenschaftliche Meeresuntersuchungen, Abteilung Kiel, Neue Folge 11: 303–319.

Raben, E. 1914. Vierte Mitteilung über quantitative Bestimmungen von Stickstoffverbindungen im Meerwasser. Wissenschaftliche Meeresuntersuchungen, Abteilung Kiel, Neue Folge 16: 207–229.

Raben, E. 1916. Quantitative Bestimmung der im Meerwasser gelösten Phosphorsäure. Wissenschaftliche Meeresuntersuchungen, Abteilung Kiel, Neue Folge 18: 1–24.

Rakestraw, N. 1933. Studies on the biology and chemistry of the Gulf of Maine. I. Chemistry of the waters of the Gulf of Maine in August, 1932. Biological Bulletin 64: 149–158.

Rakestraw, N. 1956. Herman Wattenberg: A pioneer in a new field of exploration. Journal of Chemical Education 33: 217–222.

Redfield, A.C. 1934. On the proportions of organic derivatives in sea water and their relation to the composition of plankton. Pp. 176–192 in R.J. Daniel (ed.). *James Johnstone memorial volume*. Liverpool: Liverpool Univ. Press.

Redfield, A.C., B.H. Ketchum, and F.A. Richards. 1963. The influence of organisms on the composition of seawater. Pp. 26–77 in M.N. Hill (ed.). *The sea*. Vol. 2: *The composition of sea water. Comparative and descriptive oceanography*. New York: John Wiley and Sons.

Redfield, A.C., H.P. Smith, and B.H. Ketchum. 1937. The cycle of organic phosphorus in the Gulf of Maine. Biological Bulletin 73: 421–443.

Reibisch, J. 1926. Victor Hensen zum Gedächtnis. Archiv für Hydrobiologie 16: i–xiv.

Reibisch, J. 1931. Karl Brandt, gestorben am 7. Januar 1931. Journal du Conseil International pour l'Exploration de la Mer 6: 157–159.

Reibisch, J. 1933. Karl Brandt zum Gedächtnis. Wissenschaftliche Meeresuntersuchungen, Abteilung Kiel, Neue Folge 21: i–vi.

Reif, C.B. 1986. Memories of Raymond Laurel Lindeman. Bulletin of the Ecological Society of America 67(1): 20–25.

Reighard, J. 1898. Methods of plankton investigation in relation to practical problems. Bulletin of the United States Fish Commission 1887, 17: 169–175.

Reincke, J. 1890. Die Preussische Commission zur wissenschaftlichen Untersuchung der deutschen Meere. Berlin, Deutsche Rundschau 65: 64–82.

Remane, A. 1968. Zoologie und Meereskunde. Pp. 161–179 in K. Jordan (ed.). *Geschichte der Christian-Albrechts-Universität Kiel, 1665–1965. Band 6: Geschichte der Mathematik, der Naturwissenschaften und der Landwirtschaftwissenschaften*. Neumünster: Karl Wachholtz Verlag.

Revelle, R. 1969. The age of innocence and war in oceanography. Oceans 1(3): 6–16.

Revelle, R. 1980. The Oceanographic and how it grew. Pp. 10–24 in M. Sears and D. Merriman (eds.). *Oceanography: The past*. New York: Springer-Verlag.

Revelle, R. 1985. How Mary Sears changed the United States Navy. Deep-sea Research 32(7A): 753–754.

Revelle, R. 1987. How I became an oceanographer and other sea stories. Annual Review of Earth and Planetary Sciences 15: 1–23.

Revelle, R., and W. Munk. 1948. Harald Ulrik Sverdrup—an appreciation. Journal of Marine Research 7: 125–138.

Riley, G.A. 1937a. The copper cycle in Connecticut lakes and its biological significance. Ph.D. diss., Yale Univ. 239 pp.

Riley, G.A. 1937b. The significance of the Mississippi River drainage for biological conditions in the northern Gulf of Mexico. Journal of Marine Research 1: 60–74.

Riley, G.A. 1938a. The measurement of phytoplankton. Internationale Revue der gesamten Hydrobiologie 36: 371–373.

Riley, G.A. 1938b. Plankton studies. I. A preliminary investigation of the plankton of the Tortugas region. Journal of Marine Research 1: 335–352.

Riley, G.A. 1938c. Study of the plankton in tropical waters. Carnegie Institution of Washington Yearbook 37: 98.

Riley, G.A. 1939a. Correlations in aquatic ecology. With an example of their application to problems of plankton productivity. Journal of Marine Research 2: 56–73.

Riley, G.A. 1939b. Limnological studies in Connecticut. Ecological Monographs 9: 53–94.

Riley, G.A. 1939c. Plankton studies. II. The western North Atlantic, May–June, 1939. Journal of Marine Research 2: 145–162.

Riley, G.A. 1940. Limnological studies in Connecticut. Part III. The plankton of Linsley Pond. Ecological Monographs 10: 279–306.

Riley, G.A. 1941a. Plankton studies. III. Long Island Sound. Bulletin of the Bingham Oceanographic Collection 7(3): 1–93.

Riley, G.A. 1941b. Plankton studies. IV. Georges Bank. Bulletin of the Bingham Oceanographic Collection 7(4): 1–73.

Riley, G.A. 1941c. Plankton studies. V. Regional summary. Journal of Marine Research 4: 162–171.

Riley, G.A. 1942. The relationship of vertical turbulence and spring diatom flowerings. Journal of Marine Research 5: 67–87.

Riley, G.A. 1943. Physiological aspects of spring diatom flowerings. Bulletin of the Bingham Oceanographic Collection 8: 1–53.

Riley, G.A. 1946. Factors controlling phytoplankton populations on Georges Bank. Journal of Marine Research 6: 54–73.

Riley, G.A. 1947a. Seasonal fluctuations of the phytoplankton population in New England coastal waters. Journal of Marine Research 6: 114–125.

Riley, G.A. 1947b. A theoretical analysis of the zooplankton population of Georges Bank. Journal of Marine Research 6: 104–113.

Riley, G.A. 1951. Oxygen, phosphate, and nitrate in the Atlantic Ocean. Bulletin of the Bingham Oceanographic Collection 13(1): 1–126.

Riley, G.A. 1952. Biological oceanography. Survey of Biological Progress 2: 79–104.

Riley, G.A. 1953a. Letter to the editor. Journal du Conseil International pour l'Exploration de la Mer 19: 85–89.

Riley, G.A. 1953b. Theory of growth and competition in natural populations. Journal of the Fisheries Research Board of Canada 10: 211–223.

Riley, G.A. 1955. Review of the oceanography of Long Island Sound. Deep-sea Research 3(Supplement): 224–238.

Riley, G.A. 1956a. Oceanography of Long Island Sound, 1952–1954. I. Introduction. Bulletin of the Bingham Oceanographic Collection 15: 9–14.

Riley, G.A. 1956b. Oceanography of Long Island Sound, 1952–1954. II. Physical oceanography. Bulletin of the Bingham Oceanographic Collection 15: 15–46.

Riley, G.A. 1956c. Oceanography of Long Island Sound, 1952–1954. IX. Production and utilization of organic matter. Bulletin of the Bingham Oceanographic Collection 15: 324–344.

Riley, G.A. 1957. Phytoplankton of the north central Sargasso Sea, 1950–52. Limnology and Oceanography 2: 252–270.

Riley, G.A. 1959. Eugene Harris 1928–1956. Bulletin of the Bingham Oceanographic Collection 17: 5–7.

Riley, G.A. (ed.). 1963a. *Marine biology*. Vol. 1: *First International Interdisciplinary Conference*. Baltimore: Port City Press. 286 pp.

Riley, G.A. 1963b. Organic aggregates in sea water and the dynamics of their formation and utilization. Limnology and Oceanography 8: 372–381.

Riley, G.A. 1963c. Theory of food chain relations in the ocean. Pp. 438–463 in M.N. Hill (ed.). *The sea*. Vol. 2: *The composition of sea water. Comparative and descriptive oceanography*. New York: John Wiley and Sons.

Riley, G.A. 1965. A mathematical model of regional variations in plankton. Limnology and Oceanography 10(Supplement): 202–215 R.

Riley, G.A. 1967. Aquatic communities and their adaptations to their environment. Pp. 191–202 in *Water resources of Canada. Symposia presented to the Royal Society of Canada in 1966*. Toronto: Univ. of Toronto Press.

Riley, G.A. 1970. Particulate organic matter in sea water. Advances in Marine Biology 8: 1–118.

Riley, G.A. 1972. Patterns of production in marine ecosystems. Pp. 91–112 in J.A. Wiens (ed.). *Ecosystem structure and function*. Corvallis: Oregon State Univ. Press.

Riley, G.A. 1976. A model of plankton patchiness. Limnology and Oceanography 21: 873–880.

Riley, G.A. 1979. Summation. Pp. 74–78 in *Second informal workshop on the oceanography of the Gulf of Maine and adjacent seas*. May 14–17, 1979. Halifax, N.S.: Dalhousie Univ.

Riley, G.A. 1984. Reminiscences of an oceanographer. Unpublished manuscript. Oceanography Department, Dalhousie Univ., Halifax, N.S. 161 pp.

Riley, G.A., and D.F. Bumpus. 1946. Phytoplankton-zooplankton relationships on Georges Bank. Journal of Marine Research 6: 33–47.

Riley, G.A., and S.A.M. Conover. 1956. Oceanography of Long Island Sound, 1952–1954. III. Chemical oceanography. Bulletin of the Bingham Oceanographic Collection 15: 47–61.

Riley, G.A., and S. Gorgy. 1948. Quantitative studies of summer plankton populations of the western North Atlantic. Journal of Marine Research 7: 100–121.

Riley, G.A., H. Stommel, and D.F. Bumpus. 1949. Quantitative ecology of the plankton of the western North Atlantic. Bulletin of the Bingham Oceanographic Collection 12(3): 1–169.

Riley, G.A., and R. Von Arx. 1949. Theoretical analysis of seasonal changes in the phytoplankton of Husan Harbor, Korea. Journal of Marine Research 8: 60–72.

Ringer, F.K. 1969. *The decline of the German mandarins: The German academic community, 1890–1933.* Cambridge, Mass. Harvard Univ. Press. iv + 528 pp.

Ringer, F.K. 1979. The German academic community. Pp. 409–429 in A. Oleson and J. Voss (eds.). *The organization of knowledge in modern America, 1860–1920.* Baltimore: Johns Hopkins Univ. Press.

Rollefsen, G. 1966. Norwegian fisheries research. Fiskeridirektoratets Skrifter, Series Havundersøkelser 14(1): 1–36.

Rosenberg, H. 1976. Political and social consequences of the great depression of 1873–1896 in central Europe. Pp. 36–90 in J.J. Sheehan (ed.). *Imperial Germany.* New York: Franklin Watts.

Rossiter, M.W. 1975. *The emergence of agricultural science. Justus Liebig and the Americans, 1840–1880.* New Haven: Yale Univ. Press. xiv + 275 pp.

Rothschuh, K.E. 1971. Du Bois-Reymond, Emil Heinrich. Pp. 200–205 in *Dictionary of Scientific Biography,* vol. 4.

Rothschuh, K.E. 1972. Hensen, (Christian Andreas) Victor. Pp. 287–288 in *Dictionary of Scientific Biography,* vol. 6.

Rothschuh, K.E. 1973a. *History of physiology.* Trans. Guenther Risse. New York: Krieger. xxi + 379 pp.

Rothschuh, K.E. 1973b. Lotze, Hermann Rudolf. Pp. 513–516 in *Dictionary of Scientific Biography,* vol. 8.

Russell, E.J. 1942. Rothamsted and its experiment station. Agricultural History 16:161–183.

Russell, E.J. 1966. *A history of agricultural science in Great Britain, 1620–1954.* London: George Allen and Unwin. 493 pp.

Russell, F.S. 1935a. On the value of certain plankton animals as indicators of water movements in the English Channel and North Sea. Journal of the Marine Biological Association of the United Kingdom 20: 309–331.

Russell, F.S. 1935b. The seasonal abundance of young fish in the offshore waters of the Plymouth area. Part II. Journal of the Marine Biological Association of the United Kingdom 20: 147–179.

Russell, F.S. 1948a. The Plymouth Laboratory of the Marine Biological Association of the United Kingdom. Journal of the Marine Biological Association of the United Kingdom 27: 761–764.

Russell, F.S. 1948b. Pure and applied science of the sea. Science Progress 36: 423–435.

Russell, F.S. 1955. George Parker Bidder, 1863–1953. Journal of the Marine Biological Association of the United Kingdom 34: 1–13.

Russell, F.S. 1972. Obituary. Dr. Marie V. Lebour. Journal of the Marine Biological Association of the United Kingdom 52: 777–788.

Russell, F.S. 1978. Sheina Macalister Marshall, 20 April 1896–7 April 1977. Biographical Memoirs of Fellows of the Royal Society 24: 369–389.

Russell, F.S., A.J. Southward, G.T. Boalch, and E.I. Butler. 1971. Changes in biological conditions in the English Channel off Plymouth during the last half century. Nature 234: 468–470.

Russell, H.L. 1893. The bacterial flora of the Atlantic Ocean in the vicinity of Woods Hole, Massachusetts. Botanical Gazette 18: 383–395, 411–417, 439–447.

Ruud, B. 1926. Quantitative investigations of the plankton at Lofoten, March–April, 1922–1924. Preliminary report. *Report on Norwegian Fishery and Marine Investigations* 3(7): 1–30.

Ruud, J.T. 1930. Nitrates and phosphates in the Southern Seas. Journal du Conseil International pour l'Exploration de la Mer 5: 347–360.

Sandström, J.W., and B. Helland-Hansen. 1903. Ueber die Berechnung von Meeresströmungen. *Report on Norwegian Fishery and Marine Investigations* 2(4): 1–43.

Sandström, J.W., and B. Helland-Hansen. 1905. On the mathematical investigation of ocean currents. Trans. D'Arcy W. Thompson. Pp. 135–163 in *Report of the North Sea Fisheries Investigation Commission (Northern Area), 1902–1903.* Parliamentary Papers. London: HMSO.

Sars, G.O. 1869. *Indberetninger til Departementet før det Indre fra Cand. G.O. Sars om de af ham i Aarene 1864–69, anstillede praktisk-vidensabelige Undersøgelser angaaende Torskefiskeriet i Lofoten.* Christiania.

Sars, G.O. 1877. Report of practical and scientific investigations of the cod fisheries near the Loffoden Islands, made during the years 1864–1869. Trans. H. Jacobsen. Report of the United States Commissioner of Fish and Fisheries 1877, V: 565–661.

Schlee, S. 1973. *The edge of an unfamiliar world. A history of oceanography.* New York: E.P. Dutton. 398 pp.

Schlee, S. 1978. *On almost any wind. The saga of the oceanographic research vessel Atlantis.* Ithaca, N.Y.: Cornell Univ. Press. 301 pp.

Schlee, S. 1980. The R/V *Atlantis* and her first oceanographic institution. Pp. 49–56 in M. Sears and D. Merriman (eds.). *Oceanography: The past.* New York: Springer-Verlag.

Schloesing, A. 1875. Sur l'ammoniaque de l'atmosphère. Comptes Rendus de l'Académie des Sciences de Paris 80: 175–178.

Schloesing, A., and E. Laurent. 1890. Sur la fixation de l'azote gazeux par les légumineuses. Comptes Rendus de l'Académie des Sciences de Paris 111: 750–753.

Schloesing, J.-J.-T. 1875a. Sur les lois des échanges d'ammoniaque entre les mers, l'atmosphère et les continents. Comptes Rendus de l'Académie des Sciences de Paris 81: 81–84.

Schloesing, J.-J.-T. 1875b. Sur les échanges d'ammoniaque entre les eaux naturelles et l'atmosphère. Comptes Rendus de l'Académie des Sciences de Paris 81: 1252–1254.

Schloesing, J.-J.-T., and A. Muntz. 1877. Sur la nitrification par les ferments organiques. Comptes Rendus de l'Académie des Sciences de Paris 84: 301–303.

Schreiber, E. 1927. Die Reinkultur von marinem Phytoplankton und deren Bedeutung für die Erforschung der Produktionsfähigkeit des Meerwassers. Wissenschaftliche Meeresuntersuchungen, Abteilung Helgoland 16(2): 1–34.

Schreiber, E. 1929. Die Methoden einer physiologischen Meerwasseranalyse. Rapports et Procès-Verbaux des Réunions, Conseil International pour l'Exploration de la Mer 53: 75–79.

Schulze, E. 1888. Die Stickstoffversorgung der Pflanze und der Kreislauf des Stickstoffs in der Natur. Landwirtschaftliches Jahrbuch der Schweiz 2: 74–92.

Schulze, E. 1890. Ueber die Entstehung der salpetersäuren Salze im Boden. Landwirtschaftliches Jahrbuch der Schweiz 4: 109–121.

Schulze, E. 1891. Ueber die Entstehung der salpetersäuren Salze im Boden. Land-wirtschaftliches Jahrbuch der Schweiz 5: 82–86.

Schütt, F. 1892a. Analytische Plankton-Studien. Ziele, Methoden und Anfangs-Resultate der quantitativ-analytischen Planktonforschung. Kiel and Leipzig: Lip-sius und Tischer. viii + 117 pp., 16 tables, chart.

Schütt, F. 1892b. Das Pflanzenleben der Hochsee. Ergebnisse der Plankton-Expedition der Humboldt-Stiftung, Band I.A., pp. 243–314.

Schütt, F. 1904. Kosmologie als Ziel der Meeresforschung. Rede beim Antritt des Rectorats der Universität zu Greifswald am 16. Mai 1904. Jena: Verlag von Gustav Fischer. 25 pp.

Scoresby, W. 1820. An account of the Arctic regions with a history and description of the northern whale-fishery. Vol. 1: The Arctic. xx + 551 pp. + 82 pp. of appen-dixes. Volume II. The whale-fishery. viii + 574 pp. + plates. Edinburgh: Archibald Constable.

Shelford, V., and F.W. Gail. 1922. A study of light penetration into sea-water made with the Kunz photo-electric cell, with particular reference to the distribution of plants. Publications of the Puget Sound Biological Station 3: 141–176.

Shelford, V., F.W. Gail. 1929. Use of photo-electric cells for light measurement in ecological work. Ecology 10: 298–311.

Shelford, V.E., and J. Kunz. 1926. The use of the photoelectric cells of different alkali metals and color screens in the measurement of light penetration into water. Transactions of the Wisconsin Academy of Arts, Sciences and Letters 22: 283–298.

Shor, E.N. 1978. Scripps Institution of Oceanography. Probing the oceans 1936 to 1976. San Diego: Tofua Press. x + 502 pp.

Smetacek, V. 1985. The annual cycle of Kiel Bight plankton: A long-term analysis. Estuaries 8(2A): 145–157.

Solhaug, T., and G. Saetersdal. 1972. The development of fishery research in Norway in the nineteenth and twentieth centuries in the light of the history of the fisheries. Proceedings of the Royal Society of Edinburgh B 73: 399–412.

Southward, A.J. 1980. The western English Channel—an inconstant ecosystem. Nature 285: 361–366.

Southward, A.J., and E.K. Roberts. 1984. The Marine Biological Association, 1884–1984: One hundred years of marine research. Report and Transactions of the Devonshire Association for the Advancement of Science 116: 155–199.

Southward, A.J., and E.K. Roberts. 1987. One hundred years of marine research at Plymouth. Journal of the Marine Biological Association of the United Kingdom 67: 465–506.

Spiess, F. 1935. Henrik Mohn. Zur hundertsten Wiederkehr seines Geburstags. Annalen der Hydrographie, LXIII Jahrgang (1935), Heft V, pp. 181–182.

Spjeldnaes, N. 1976. Sverdrup, Harold Ulrik. Pp. 166–167 in Dictionary of Scientific Biography, vol. 13.

Stahlberg, W. 1940. Zu Ernst Abbes hundertstem Geburtstage. Die Naturwissen-schaften 28(4): 49–54.

Stanbury, F.A. 1931. The effect of light of different intensities, reduced selectively and nonselectively, upon the rate of growth of Nitzschia closterium. Journal of the Marine Biological Association of the United Kingdom 17: 633–653.

Stauffer, R.C. 1957. Haeckel, Darwin and ecology. Quarterly Review of Biology 32: 138–144.

Steele, J.H. 1956. Plant production on the Fladen ground. Journal of the Marine Biological Association of the United Kingdom 35: 1–33.

Steele, J.H. 1958. Plant production in the northern North Sea. Scottish Home Department, Marine Research 1958, no. 7, pp. 1–36.

Steele, J.H. 1959. The quantitative ecology of marine phytoplankton. Biological Reviews 34: 129–158.

Steele, J.H. 1961. Primary production. Pp. 519–538 in M. Sears (ed.). *Oceanography*. AAAS Publ. no. 67. Washington, D.C.

Steele, J.H. 1974. *The structure of marine ecosystems*. Cambridge, Mass.: Harvard Univ. Press. 128 pp.

Steemann Nielsen, E. 1937a. The annual amount of organic matter produced by the phytoplankton in the Sound off Helsingor. Meddelellser fra Danmarks Fiskeri-og Havundersøgelser, Series Plankton, III(3): 1–37.

Steemann Nielsen, E. 1937b. On the relation between the quantities of phytoplankton and zooplankton in the sea. Journal du Conseil International pour l'Exploration de la Mer 12: 147–154.

Steemann Nielsen, E. 1952. The use of radioactive carbon (C^{14}) for measuring organic production in the sea. Journal du Conseil International pour l'Exploration de la Mer 18: 117–140.

Steemann Nielsen, E. 1954. On organic production in the oceans. Journal du Conseil International pour l'Exploration de la Mer 19: 309–328.

Steudel, J. 1974. Müller, Johannes Peter. Pp. 567–574 in *Dictionary of Scientific Biography*, vol. 9.

Steuer, A. 1910. *Planktonkunde*. Leipzig und Berlin: B.G. Teubner. xv + 723 pp.

Stone, N. 1983. *Europe transformed, 1878–1919*. Glasgow: Fontana. 447 pp.

Strickland, J.D.H. 1960. Measuring the production of marine phytoplankton. Bulletin of the Fisheries Research Board of Canada 122: 1–172.

Sulloway, F.J. 1979. *Freud, biologist of the mind. Beyond the psychoanalytic legend*. New York: Basic Books. xxvi + 612 pp.

Sund, O. 1929a. The determination of nitrates in sea water. Rapports et Procès-Verbaux des Réunions, Conseil International pour l'Exploration de la Mer 53: 80–89.

Sund, O. 1929b. Reports of the area committees. North-eastern area 1928. Norway. Rapports et Procès-Verbaux des Réunions, Conseil International pour l'Exploration de la Mer 60: 77–81.

Sverdrup, H.U. 1929. The waters on the north Siberian shelf. Pp. 1–131 + tables in *The Norwegian North Polar Expedition with the "Maud," 1918–1925*. Scientific Results IV, no. 2.

Sverdrup, H.U.1938. On the explanation of the oxygen minima and maxima in the oceans. Journal du Conseil International pour l'Exploration de la Mer 13: 163–172.

Sverdrup, H.U. 1953. On conditions for the vernal blooming of phytoplankton. Journal du Conseil International pour l'Exploration de la Mer 18: 287–295.

Sverdrup, H.U., M.W. Johnson, and R.H. Fleming. 1942. *The oceans*. Englewood Cliffs, N.J.: Prentice-Hall. 1087 pp.

Tanner, Z.L. 1897. Deep-sea exploration: A general description of the steamer *Albatross*, her appliances and methods. Bulletin of the United States Fish Commission for 1896, 16: 257–424, 40 pls., 76 figs.

Taylor, F.J.R. 1980. Phytoplankton ecology before 1900: Supplementary notes to the "Depths of the ocean." Pp. 509–521 in M. Sears and D. Merriman (eds.). *Oceanography: The past*. New York: Springer-Verlag.

Thompson, D.W. 1947. Otto Pettersson, 1848–1941. Journal du Conseil International pour l'Exploration de la Mer 15: 121–125.

Thomsen, P. 1910. Über das Vorkommen von Nitrobakterien im Meere. Wissenschaftliche Meeresuntersuchungen, Abteilung Kiel, Neue Folge 11: 1–27.

Thorade, H. 1935. Henrik Mohn und die Entwicklung der Meereskunde. Annalen der Hydrographie, LXIII Jahrgang (1935), Heft V, pp. 182–186.

Titze, H. 1983. Enrollment expansion and academic overcrowding in Germany. Pp. 57–88 in K.H. Jarausch (ed.). *The transformation of higher learning, 1860–1930*. Stuttgart: Klett-Cotta.

Tizard, T.H., H.N. Moseley, J.Y. Buchanan, and J. Murray. 1885. Narrative of the cruise of H.M.S. *Challenger*, with a general account of the scientific results of the expedition. *Report of the Scientific Results of the Voyage of H.M.S. Challenger, 1873–1876*, Narrative 1 (1,2). 1172 pp.

Tobey, R.C. 1981. *Saving the prairies. The life cycle of the founding school of American plant ecology, 1895–1955*. Berkeley: Univ. of California Press. x + 315 pp.

Turner, F.M. 1978. The Victorian conflict between science and religion: A professional dimension. Isis 69: 356–376.

Turner, F.M. 1980. Public science in Britain, 1880–1919. Isis 71: 589–608.

Turner, R.S. 1971. The growth of professorial research in Prussia, 1818 to 1848—causes and context. Historical Studies in the Physical Sciences 3: 137–182.

Turner, R.S. 1981. The Prussian professoriate and the research imperative, 1790–1840. Pp. 109–121 in H.N. Jahnke and M. Otte (eds.). *Epistemological and social problems of the sciences in the early nineteenth century*. Boston: D. Reidel.

Turner, R.S., E. Kerwin, and D. Woolwine. 1984. Careers and creativity in nineteenth-century physiology. Zloczower Redux. Isis 75: 523–529.

Vanhöffen, E. 1897. Die Fauna and Flora Grönlands. *Grönland-Expedition der Gesellschaft für Erdkunde zu Berlin, 1891–1893, Unter Leitung von Erich von Drygalski*, Band 2(1). Berlin: W.H. Kuhl. 383 pp., 8 Taf.

Van Iterson, G., Jr., L.E. den Dooren de Jong, and A.J. Kluyver. 1983. *Martinus Willem Beijerinck: His life and work*. Foreword by C.B. Van Niel, preface by Thomas D. Brock. Madison, Wis.: Science Tech, Inc. xxix + 181 pp.

Vaughan, T.W. 1930. The oceanographic point of view. Pp. 40–56 in *Contributions to marine biology*. Stanford, Calif.: Stanford Univ. Press.

Vernon, H.M. 1898. The relations between marine animal and vegetable life. Mitteilungen der Zoologischen Station zu Neapel 13(3): 341–425.

Vinogradsky, S.N. 1890a. Sur les organismes de la nitrification. Comptes Rendus de l'Académie des Sciences de Paris 110: 1013–1016.

Vinogradsky, S.N. 1890b. Recherches sur les organismes de la nitrification. Paris, Annales de l'Institut Pasteur 4: 211–231, 257–275, 760–761.

Vinogradsky, S.N. 1891. Sur la formation et l'oxydation des nitrites pendant la nitrification. Comptes Rendus de l'Académie des Sciences de Paris 113: 89–92.

Vinogradsky, S.N. 1893. Sur l'assimilation de l'azote gazeux de l'atmosphère par les microbes. Comptes Rendus de l'Académie des Sciences de Paris 116: 1385–1388.

Vinogradsky, S.N. 1894. Sur l'assimilation de l'azote gazeux de l'atmosphère par les microbes. Comptes Rendus de l'Académie des Sciences de Paris 118: 353–355.

Vinogradsky, S.N. 1895. Recherches sur l'assimilation de l'azote libre de l'atmosphère par les microbes. Archives des Sciences biologiques de l'Institut impériale de Médecine expérimentale à Pétersbourg 3: 297–352.

Volbehr, F.L.C., and R. Weyl. 1956. *Professoren und Dozenten der Christian-Albrechts-Universität zu Kiel, 1665–1965.* Rev. R. Bülcke; comp. H.-J. Neuiger. Kiel: Ferdinand Hirt.

Volterra, V. 1928. Variations and fluctuations of the number of individuals in animal species living together. Trans. M.E. Wells. Journal du Conseil International pour l'Exploration de la Mer 3: 1–51.

Waksman, S.A. 1946. Sergei Nikolaevitch Winogradsky: The story of a great bacteriologist. Soil Science 62: 197–226.

Waksman, S.A. 1952. *Soil microbiology.* New York: John Wiley and Sons. vi + 356 pp.

Waksman, S.A. 1953. *Sergei N. Winogradsky, his life and work.* New Brunswick, N.J.: Rutgers Univ. Press.

Waksman, S.A., M. Hotchkiss, and C.L. Carey. 1933. Marine bacteria and their role in the sea. II. Bacteria concerned in the cycle of nitrogen in the sea. Biological Bulletin 65: 137–167.

Waksman, S.A., H.W. Reuszer, C.L. Carey, M. Hotchkiss, and C.E. Renn. 1933. Studies on the biology and chemistry of the Gulf of Maine. III. Bacteriological investigations of the sea water and marine bottoms. Biological Bulletin 64: 183–205.

Walker, T.A. 1980. A correction to the Poole and Atkins Secchi disc/light-attenuation formula. Journal of the Marine Biological Association of the United Kingdom 60: 769–771.

Walsh, J.J. 1972. Implications of a systems approach to oceanography. Science 176: 969–975.

Ward, B.B., R.J. Olson, and M.J. Perry. 1982. Microbial nitrification rates in the primary nitrite maximum off southern California. Deep-sea Research 29(2A): 247–255.

Watson, W.N.B. 1969. The Scottish Marine Station for Scientific Research, Granton, 1884 to 1903. The Book of the Old Edinburgh Club 33(1): 50–58.

Wattenberg, H. 1926. Bericht über die chemischen Arbeiten. Die Deutsche Atlantische Expedition I. Bericht. Zeitschrift der Gesellschaft für Erdkunde zu Berlin, Jahrgang 1926, no. 1.

Wattenberg, H. 1927. Bericht über die chemischen Arbeiten. Die Deutsche Atlantische Expedition II. Bericht. Zeitschrift der Gesellschaft für Erdkunde zu Berlin, Jahrgang 1927, no. 5–6.

Weber, L. 1889. Eine neue Montirung des Milchglasplattenphotometers. Schriften des Naturwissenschaftlichen Vereins für Schleswig-Holstein 8: 187–198.

Weber, L. 1895. Resultate der Tageslichtmessungen in Kiel in den Jahren 1890 bis 1892. Schriften des Naturwissenschaftlichen Vereins für Schleswig-Holstein 10: 77–94.

Went, A.E.J. 1972a. The history of the International Council for the Exploration of the Sea. Proceedings of the Royal Society of Edinburgh B 73: 351–360.

Went, A.E.J. 1972b. Seventy years agrowing. A history of the International Council for the Exploration of the Sea, 1902–1972. Rapports et Procès-Verbaux des Réunions, Conseil International pour l'Exploration de la Mer 165: 1–252.

Whipple, G.C. 1894. Some observations on the growth of diatoms in surface waters. Technology Quarterly 7: 214–231.

Whipple, G.C. 1895. Some observations on the temperature of surface waters; and the effect of temperature on the growth of microorganisms. Journal of the New England Water Works Association 9: 202–222.

Whipple, G.C. 1896. Some observations on the relation of light to the growth of diatoms. Journal of the New England Water Works Association 11: 1–26.

Whipple, G.C. 1897. Biological studies in Massachusetts. American Naturalist 31: 503–508, 576–581, 1016–1026.

Whipple, G.C. 1899. *The microscopy of drinking water.* New York: John Wiley and Sons. xii + 300 pp., 19 pls.

Whitfield, M. 1974. The hydrolysis of ammonium ions in sea water—a theoretical study. Journal of the Marine Biological Association of the United Kingdom 54: 565–580.

Wieser, W. 1952. Investigations on the microfauna inhabiting seaweeds on rocky coasts. IV. Studies on the vertical distribution of the fauna inhabiting seaweeds below the Plymouth Laboratory. Journal of the Marine Biological Association of the United Kingdom 31: 145–174.

Wille, C. 1882. Historical account. *The Norwegian North-Atlantic Expedition, 1876–1878,* Vol. 1(1). Christiana: Grøndahl & Son. 46 pp., chart.

Worster, D. 1985. *Nature's economy. A history of ecological ideas.* Cambridge: Cambridge Univ. Press. xviii + 404 pp.

Wright, W.R. 1987. Scientific exploration. Pp. 2–9 in R.H. Backus (ed.). *Georges Bank.* Cambridge, Mass.: MIT Press.

Wroblewski, J.S. (ed.). 1982. *Selected works of Gordon A. Riley.* With introductory essays by G.E. Hutchinson, E.L. Mills, R.O. Fournier, P.J. Wangersky, and closing remarks by E.S. Deevey, Jr. Halifax, N.S.: Dalhousie Univ. xxi + 489 pp.

Wüst, G. 1955. Stromgeschwindigkeiten im Tiefen- und Bodenwasser des Atlantischen Ozeans auf Grund dynamischer Berechnung der *Meteor*—Profile der Deutschen Atlantischen Expedition 1925/27. Deep-Sea Research 3(Supplement): 373–397.

Wüst, G. 1958. Über Stromgeschwindigkeiten und Strommengen in der Atlantischen Tiefsee. Geologische Rundschau 47: 187–195.

Wüst, G. 1968. History of investigations of the longitudinal deep-sea circulation (1800–1922). Congrès international d'histoire de l'océanographie 1. Bulletin de l'Institut Océanographique, no. spécial, 21: 109–120.

Yonge, C.M. 1946. Jubilee of the Marine Biological Station, Millport. Nature 158: 506.

Yonge, C.M. 1962. Dr. A. P. Orr. Nature 196: 719.

Young, F.G. 1937. Claude Bernard and the theory of the glycogenic function of the liver. Annals of Science 2:47–83.

Zloczower, A. 1981. *Career opportunities and the growth of scientific discovery in nineteenth-century Germany with special reference to physiology.* New York: Arno Press. 131 pp.

Zo Bell, C. 1946. *Marine microbiology.* Waltham, Mass.: Chronica Botanica. xv + 240 pp.

INDEX

Page numbers in italics indicate illustrations.

tion depth, 154–157, 163, 170; on control of plankton cycle, 130–131, 136, 146n.81, 148, 162–170; on denitrifying bacteria, 68–69, 95–97, 112, 128; opposition to Cleve, 130–131; and origin of light and dark bottle method, 154–157, 224n.52; on phytoplankton experiments, 148, 151, 154–156, 227; on spatial variations of bloom, 116, 146, 148–149, 162–166, 243–244; on spring bloom, 116, 128, 130–131, 149–150, 168–170; statement of his aims, 151–152; use of physical oceanography by, 131, 152; on vertical distribution of phytoplankton, 152–153, 157; and Whipple's hypotheses, 142, 146n.81, 153; work with Nathansohn, 149–151

Grave, Caswell, 259

Grazing: as control of phytoplankton, 106–107, 143, *144*, 151n.11, 231, 238–243, 249, 286–287; estimation of, 242, 286–287, 313, 315; by filtration, 315; seasonal changes in, 136, *241*. See also Zooplankton

Great Barrier Reef Expedition, 239n.12, 255, 256n.62

Gripenberg, Stina, 232

Grisebach, August, 35, 123

Gulf of Maine. *See* Bigelow, H. B.: studies of Gulf of Maine

Günther, Albert, 190

Gunther, E. R., 258n.1

Haeckel, Ernst, 29; criticism of Hensen, 29–31, 66; *Plankton-Studien*, 29; on richness of tropics, 30; suspicion of quantitative approach, 30–31

Hanseatic League, 76

Hardy, A. C., 258n.1, 285–286

Hardy, W. B., 207, 215

Harris, Eugene, 300–301

Harrison, Ross G., 259

Harvey, H. W., 5–6, 136n.45, 190, 207, *211, 228,* 300–301; approach to nature and philosophy, 215; on carbon dioxide system, 232; career, 214–215, 252; comparison with Riley, 305–307; on grazing, 236–243, 285; hydrographic work of, 214–218; on marine production, 229–231, 239–243, 249–250; measurements of plankton abundance by, 233n.76, 234–236, *235;* on nitrate,

217–219; on nitrification, 72, 94; on nutrient techniques, 158, 165–166, 217, 255; on production of English Channel, 250, 300; on relationships in sea, 229–231, 243, 246–250, 253, 256–257; relation with Atkins, 215, 246n.37, 247n.40; on trace substances, 247–248; use of laboratory cultures by, 218, 229, 247–248, 256, 306

Harvey plant pigment unit (HPPU), 234–236, 240

Heape, Walter, 195

Heincke, Friedrich, 83, 87, 318

Helland-Hansen, Bjørn, 80, 86, 110, 152, 220

Hellriegel, J., 61

Helmholtz, Hermann von, 38

Henking, H., 83, 87

Hensen nets. *See* Hensen, Victor: design of nets and equipment by

Hensen, Victor, 1–2, 4, 6, 10, *11,* 27, 47, 113, 257, 310, 322; design of nets and equipment by, 17–21, *18;* dispute with Haeckel, 29–31; early career, 10–13; on embryology and morphology, 12–13; on fish stocks, 4, 16; on food chain in sea, 41, 48; and German physiology, 20, 38–39, 172; on hearing and speech, 13; hexagon theory of, 32–34, *33;* isolation of glycogen by, 11–12; on Kiel Commission, 14–16; on Lohmann's criticism, 37; on marine production, 17, 22, 28, 37, 40–42, 122–123, 127, 172, 310, 317; Physiological Institute of, 12, 16; politics of, 12, 79; publications of, 13; on quantitative techniques, 17–21, 34–35, 123, 152, 172; sampling and statistics of, 32–34, 123, 172; scientific method of, 4, 37–40; on seasonality of plankton, 122–123, 127; on uniformity of distribution, 16, 31–34; use of chemical methods by, 21, 35, 129. *See also* Plankton-Expedition; Production, marine

Hentschel, Ernst, 160

Herdman, G. A., 212

Herdman, W. A., 95; career, 199n.29, 200n.30; on fisheries problems, 209–210; and ICES, 201; on Irish Sea plankton, 32–34, 127, 197, 200; and Kiel Commission, 200; and Lancashire Sea-Fisheries Laboratory, 199, 206; and Liverpool Marine Biology Committee,

374   *Index*

Microzooplankton, 243
Millport laboratory (Scotland), 207, 222.
  See also Scottish Marine Biological As-
  sociation
Ministry of Agriculture and Fisheries (En-
  glish), 214
Mixing: and phytoplankton cultures, 204;
  and plankton production, 131, 141–
  146, 149–150, 153, 157–171, 246–
  247, 277–281, 311–312; and supply of
  nutrients, 219, 313. *See also* Stability;
  Thermocline; Vertical circulation
Möbius, Karl, 14, 43–45
Modeling, mathematical, 3, 262–263,
  311–312, 317n.15. *See also* Riley, G.
  A.; Steele, J. H.
Mohn, Henrik, 77, 110, 114
Montgomery, R. B., 307
Moore, Benjamin, 212
Morley, Earl of, 194
Mörth, Wilhelm, 74n.94
Mulkay, Michael, 173n.2, 183–185,
  186n.34
Müller, Johannes, 11, 29, 31
Müller, Karl, 138, 183
Multiple regression. *See* Riley, G. A.: mul-
  tiple regression models, use of statistics
  by
Muntz, Achille, 60
Murray, John, 83, 197n.22; with Hjort
  on *Michael Sars*, 2, 151
Musée océanographique de Monaco, 109,
  117–118

Nanoplankton, 36, 67; abundance of,
  134, 316; in appendicularians, 28, 35,
  134; naming of, 28, 35, 134n.42; in
  seasonal cycle, 137, 141. *See also*
  Picoplankton
Nansen, Fridtjof, 80, 83–84, 110–114,
  149
Nathansohn, Alexander, 6, 71, 96–99,
  103–106, 108–119, 116, 243, 249,
  310; on equilibrium and adaptation,
  105–106; on geographical patterns of
  production, 113–116; life and career,
  108–110; on mixing and vertical water
  movement, 110–119, 141, 147–149,
  152, 160–162, 170; and physical
  oceanography, 109–110, 120, 173,
  309; work in Monaco, 109, 116–118,
  162; work with Gran, 128n.24, 149–
  151

National (research vessel), 23. *See also*
  Plankton-Expedition
National Academy of Sciences (U.S.),
  Committee on Oceanography, 273,
  320. *See also* Bigelow, H. B.
National Science Foundation (U.S.), 322
Natterer, Konrad, 63, 74, 90, 113
Naturphilosophie, 125
Nelson, E. W., 203–204, 214, 227
Nicholls, A. G., 238
Nitrate, 53–57, 59–65, 88–90, 132;
  analysis of, 88–90, 158, 160n.49, 165,
  217; in deep water, 73, 93–94, 98–
  103, 298, 315; as limiting nutrient,
  126, 157, 160n.49, 204, 217–218; in
  relation to circulation, 93; in relation to
  phosphate, 217–219, 232, 245; in
  southern ocean, 99–102; supply from
  land, 50, 96, 111; and temperature,
  92–93, 103
Nitrification, 184, 243; based on organic
  nitrogen compounds, 73, 94, 104, 174;
  as chemical process, 99; on land, 54,
  59–60; role of *Nitrosomonas* and *Nitrobacter* in, 72; in sea, 54–57, 71–73,
  93, 99–104, 108, 217–218; solution of
  problem of, 57, 73; in southern ocean,
  100–102
Nitrite, 54–57, 60–65, 68, 74, 90–91
Nitrogen cycle, 52; Brandt's knowledge
  of, 50–51, 64, 88–104, 108; Cooper
  on, 244–245; Gazert's model of, 100–
  102; Gebbing's model of, 97–108;
  Harris on, 300–301; on land, 57–62;
  modern knowledge of, 53–55;
  nineteenth-century knowledge of, 57–
  66; role of bacteria in, 59–61; role of
  sediments in, 90, 94; in sea, 55–57, 56,
  58, 108, 217–218, 231–232, 244–245,
  255. *See also* Nitrate; Nitrification; Nitrite; Nitrogen fixation
Nitrogen fixation, 53–57, 59–61, 64, 66,
  70, 95–96, 102; by *Azotobacter* and
  Clostridium, 70, 95–96; by *Oscillatoria*
  (*Trichodesmium*), 55, 57, 70n.81
Nordenskiöld, Otto, 81
Nordsee-Expedition: of 1872, 15; of
  1895, 67
Norwegian North Atlantic Expeditions of
  1876–1878, 77
Nutrients for phytoplankton, 42, 63; in
  Arctic Ocean, 111–112, 159; competi-
  tion for, 145, 315; depletion of, 132,